**Advanced Multilevel Converters and
Applications in Grid Integration**

Advanced Multilevel Converters and Applications in Grid Integration

Edited by

Ali I. Maswood
Nanyang Technological University
Singapore

Hossein Dehghani Tafti
Nanyang Technological University
Singapore

This edition first published 2019
© 2019 John Wiley & Sons Ltd

The right of Ali I. Maswood and Hossein Dehghani Tafti to be identified as the authors of the editorial material in this work has been asserted in accordance with law.

Registered Offices
John Wiley & Sons, Inc., 111 River Street, Hoboken, NJ 07030, USA
John Wiley & Sons Ltd, The Atrium, Southern Gate, Chichester, West Sussex, PO19 8SQ, UK

Editorial Office
The Atrium, Southern Gate, Chichester, West Sussex, PO19 8SQ, UK

For details of our global editorial offices, customer services, and more information about Wiley products, visit us at www.wiley.com.

Wiley also publishes its books in a variety of electronic formats and by print-on-demand. Some content that appears in standard print versions of this book may not be available in other formats.

Limit of Liability/Disclaimer of Warranty
MATLAB® is a trademark of The MathWorks, Inc. and is used with permission. The MathWorks does not warrant the accuracy of the text or exercises in this book. This work's use or discussion of MATLAB® software or related products does not constitute endorsement or sponsorship by The MathWorks of a particular pedagogical approach or particular use of the MATLAB® software.

While the publisher and authors have used their best efforts in preparing this work, they make no representations or warranties with respect to the accuracy or completeness of the contents of this work and specifically disclaim all warranties, including without limitation any implied warranties of merchantability or fitness for a particular purpose. No warranty may be created or extended by sales representatives, written sales materials or promotional statements for this work. The fact that an organization, website, or product is referred to in this work as a citation and/or potential source of further information does not mean that the publisher and authors endorse the information or services the organization, website, or product may provide or recommendations it may make. This work is sold with the understanding that the publisher is not engaged in rendering professional services. The advice and strategies contained herein may not be suitable for your situation. You should consult with a specialist where appropriate. Further, readers should be aware that websites listed in this work may have changed or disappeared between when this work was written and when it is read. Neither the publisher nor authors shall be liable for any loss of profit or any other commercial damages, including but not limited to special, incidental, consequential, or other damages.

Library of Congress Cataloging-in-Publication Data

Names: Maswood, Ali I., 1957- editor. | Tafti, Hossein Dehghani, 1987- editor.
Title: Advanced multilevel converters and applications in grid integration / edited by Ali I. Maswood, Hossein Dehghani Tafti.
Description: Hoboken, NJ : John Wiley & Sons, 2019. | Includes bibliographical references and index. |
Identifiers: LCCN 2018023735 (print) | LCCN 2018033627 (ebook) | ISBN 9781119476016 (Adobe PDF) | ISBN 9781119475897 (ePub) | ISBN 9781119475866 (hardcover)
Subjects: LCSH: Electric current converters. | Electric current rectifiers.
Classification: LCC TK7872.C8 (ebook) | LCC TK7872.C8 A48 2019 (print) | DDC 621.381/3–dc23
LC record available at https://lccn.loc.gov/2018023735

Cover Design: Wiley
Cover Image: © Leonid Katsyka/Shutterstock

Set in 10/12pt WarnockPro by SPi Global, Chennai, India
Printed in Singapore by C.O.S. Printers Pte Ltd

10 9 8 7 6 5 4 3 2 1

This book is dedicated to our families and parents.

Contents

List of Contributors

Ali I. Maswood
Nanyang Technological University
Singapore

Gabriel Heo Peng Ooi
Nanyang Technological University
Singapore

Georgios Konstantinou
University of New South Wales
Australia

Harikrishna Raj Pinkymol
Nanyang Technological University
Singapore

Hossein Dehghani Tafti
Nanyang Technological University
Singapore

Josep Pou
Nanyang Technological University
Singapore

Md Shafquat Ullah Khan
Nanyang Technological University
Singapore

Muhammad M. Roomi
Nanyang Technological University
Singapore

Ziyou Lim
Nanyang Technological University
Singapore

Preface

The ever-increasing demand for energy and the evidence of global warming have forced many nations to divert renewable and clean energy sources into the mainstream for power generation. Statistical results have proven that the total amount of power capacity installed yearly is increasing tremendously in order to meet supply demands. In many countries, government bodies and authorities are also planning future-sustainable cities. Solar power systems and wind power systems are getting more attention in various countries. Clean energy sources, such as fuel cell power systems, are still undergoing testing to serve as the primary energy sources for many housing blocks. Furthermore, most of the available high-power rotating machines in industry require variable speeds and special control algorithms. A power conversion stage, using semiconductor switches, is required in these renewable energy systems and industrial applications. Therefore, a deep understanding of the design of high-power converters is required for researchers and industrial engineers.

Two-level power converters are applied in various industrial applications; however, they have limitations for high-power conversion systems. They also have a low quality of output power, which necessitates the application of large filters with higher losses and lower power conversion efficiencies. On the other hand, semiconductor switches have limitations in terms of high current and high voltages. Therefore, implementing multilevel converters in high-power and high-voltage applications has become a trend recently. With the multilevel approach, better power quality can be achieved by synthesizing a higher number of output voltage levels. Many research works on multilevel topologies have been proposed, but the technology is still not available commercially. However, understanding the principles of advanced multilevel converters with enhanced efficiency, better output voltage quality, and reduced number of switches is necessary for power engineering researchers and industrial engineers, since a comprehensive presentation of these is not available in the books/reports in the market presently.

This book, *Advanced Multilevel Converters and Applications in Grid Integration*, presents the principles of advanced multilevel converters, which require lesser number of components and provide higher power conversion efficiency and output power quality. Their operational principles and control strategies are explained in detail. Furthermore, the mathematical expressions and design procedures of their components are also presented. Their advantages and disadvantages as compared to the classical multilevel and two-level power converters are also provided. Some of the industrial applications of advanced multilevel converters are included in the proposed book, along with a deep explanation of their control strategies. Thereby, this book is able to provide a better understanding of the gap differences between the research conducted and the current industrial needs.

The book is divided into four parts:

Part I includes several modulation algorithms for classical multilevel converters and control strategies for the voltage balancing of capacitors. The reader will be exposed to the basics of multilevel converters in this part, which will help in gaining a better understanding of the principles of advanced multilevel converters.

Part II presents several multilevel rectifiers and their operational principles. Unidirectional and bidirectional multilevel rectifiers are initially discussed in this part, followed by discussions on multilevel diode-clamped rectifiers. Finally, flying capacitor–based multilevel converters are explored. The operational principles of each configuration, along with the relevant mathematical presentations of their operations, are demonstrated in each chapter. A procedure for designing the components of the converter and the related voltage/current stress of the components are also illustrated. The performance of various control and modulation strategies on multilevel rectifiers are evaluated in each chapter using simulation and experimental results.

Part III demonstrates the various topologies of advanced multilevel inverters, including transformerless diode-clamped and flying capacitor–based multilevel inverters. Furthermore, multilevel Z-source inverters are also described in this part. Similar to Part II, each chapter provides the operational principles, mathematical formulations, and design procedures for these converters, as well as the various control strategies and their evaluation results.

Part IV investigates the various industrial applications of the advanced multilevel converters presented in this book. Photovoltaic power plants, wind power plants, fuel cell power generation, and flexible alternating current transmission systems (FACTS) are covered in this chapter. Several control strategies in these systems are presented, followed by their evaluation results during various operation conditions.

Part I

A review on Classical Multilevel Converters

1

Classical Multilevel Converters

Gabriel H. P. Ooi, Ziyou Lim, and Hossein Dehghani Tafti

1.1 Introduction

Power electronic converters are classified mainly based on the current (CS) or voltage source (VS). In the early days around the 1980s, current source inverters (CSI) were popular when thyristor-based semiconductors were first developed [1]. The CSIs are also known as load-commutated inverters (LCI) in the industry, which mainly comprise the gate turn-off thyristor (GTO) or the integrated gate-commutated thyristor (IGCT) in the circuit. Usually, CSIs are operated under short-circuit conditions using current-controlled switching devices; hence, a low switching frequency is required. Besides, the gate driver circuitry design can become too complicated.

In 1964, Ray E. Morgan had proved that the performance of an inverter will be more efficient for a fast switching operation where switching losses are greatly reduced through the experimented bridge chopper inverter. Despite the IGCT allowing a higher switching frequency (up to a few kilohertz) than the GTO, the switching losses are considerably high. Hence, the switching frequency for the IGCT is typically limited to a few hundred hertz (around 500 Hz).

In the mid-1990s, insulated-gate bipolar transistors (IGBTs) designed to be open-circuit voltage-controlled semiconductor devices for fast switching purposes had become available in the market. As a result, IGBTs would have been more favorable in the voltage source inverter (VSI) where a lower switching loss, higher efficiency, and a higher reliability are achieved. Moreover, the VSI has a low current harmonic distortion due to the high switching frequency operation. The VSI is used in industrial applications due to the numerous advantages, and therefore, multilevel VSI is the prime focus of research and will be further discussed in this chapter.

1.2 Classical Two-Level Converters

The configuration of the conventional three-phase two-level VSI as shown in Fig. 1.1 has been commonly used in the early 1960s for motor drives. The 2L-VSIs were constructed using thyristor-based switching devices during those times until the 1990s when IGBTs were used to replace the thyristors for faster switching purposes.

Advanced Multilevel Converters and Applications in Grid Integration, First Edition.
Edited by Ali I. Maswood and Hossein Dehghani Tafti.
© 2019 John Wiley & Sons Ltd. Published 2019 by John Wiley & Sons Ltd.

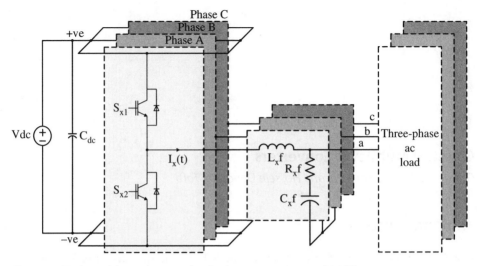

Figure 1.1 Traditional three-phase two-level voltage source inverter (2L-VSI).

The topology of the three-phase 2L-VSI is simple in nature, consisting only of two switching devices in each phase leg, thus allowing the cost of implementation to be relatively low. Although a fast switching operation of up to 20 kHz can be performed with the IGBTs to reduce the switching losses in the 2L-VSI, there is an implication of high stress in motor winding especially when a long cable is connected between the inverter and the motor [2]. The increase in switching frequency operation causes high dV/dt (large voltage spikes) due to the characteristic impedances of the cable; hence, it has a detrimental impact on the motor.

However, electromagnetic interference (EMI) becomes a major issue when the power electronic appliances are operating at a switching frequency range between 10 kHz and 30 MH. Since power electronic converters usually operate below the frequency of 10 MHz (typically around 10 kHz to 150 kHz), EMI is spread by the conduction of the wires. Hence, the conducted EMI is the key concern in the design of power converters instead of the radiated EMI. Thus, the EMI filter is required to suppress the conducted EMI for the 2L-VSI with a 20 kHz switching frequency so that the filter size can also be minimized.

Apart from that, there is a limitation for the 2L-VSI to drive high power ($\geq$100 MW) applications due to the available voltage ratings of the semiconductor devices in the market. Based on the configuration of the 2L-VSI, each of the semiconductor devices has to support the total amount of DC voltage supply. Therefore, the design of 2L-VSI becomes complex when a number of semiconductor components are series-connected together in order to withstand high DC voltage [3, 4]. As a result, the reliability of the 2L-VSI will also be greatly affected due to the increase in the number of switching devices required in the circuit.

1.3 The Need for Multilevel Converters

Multilevel power converters, a new breed of power electronic converters, are being developed and evolved since the 1980s. The ability to overcome the limitations and the challenges faced in the traditional 2L-VSI topology has been favorable in many industrial high-power applications.

True to their name, multilevel converters synthesize the output voltage into a multiple stepped level (staircase-like) waveform. By doing so, the voltage stress across each semiconductor device is greatly reduced. Additionally, by higher number of output voltage stepped levels, the voltage harmonic is decreased. Thus, the size of the output filter can be further reduced or even eliminated. In addition, good output power quality can be achieved even with a much lower switching frequency (less than 10 kHz). Hence, the problems of EMI and high switching losses can be avoided.

Multilevel converters have drawn great research interest and attention over the last decade because of their numerous advantages. Despite having been considered as proven mature technologies in many industries, multilevel converters still encounter many challenges, such as unbalanced capacitor voltages when a higher number of output voltage stepped levels is intended to be achieved. Therefore, an investigation of the existing multilevel topologies used in industries will be discussed in the following sections of this chapter. The advantages along with the drawbacks of each multilevel converter topology will be elaborated as well. These will serve in providing a clearer vision to remain focused on the research area and alternative solutions can be proposed to improve the performance of multilevel converters.

1.4 Classical Multilevel Converters

1.4.1 Multilevel Diode Clamped Converters

The three-level neutral point clamped inverter (3L-NPC), shown in Fig. 1.2, was created by Nabae and colleagues [5] as a multilevel power converter in 1981. The expansion of the 3L-NPCI is discussed in [6] and the first 10 kVA four-level inverter prototypes are investigated in [7].

Based on the fundamental 3L-NPC (Fig. 1.2), the three-level output voltage stepped waveform is synthesized through both the diodes clamped to the middle of both the dc-link capacitors. The middle connection of the dc-link capacitors is referred to as the neutral point and also served as the output phase voltage reference point. Since the output voltage waveform is synthesized from the dc-link capacitor voltage, the number of diode elements required to be clamped will increase when a higher number of output voltage level is desired. Therefore, the topology of 3L-NPC evolves into a multilevel diode clamped inverter (MDCI).

The number of switches i and clamping diodes D required for the m-level (mL) output phase voltage of the MDCIs can be determined using equations (1.1) and (1.2), respectively:

$$i_{\text{switches(MDCI)}} = 2[(mL) - 1] \tag{1.1}$$

$$D_{\text{diodes(MDCI)}} = 2[(mL) - 2] \tag{1.2}$$

The number of dc-link capacitors Q needed for mL-MDCIs can be known based on the following equation (1.3):

$$Q_{\text{capacitors(MDCI)}} = (mL) - 1 \tag{1.3}$$

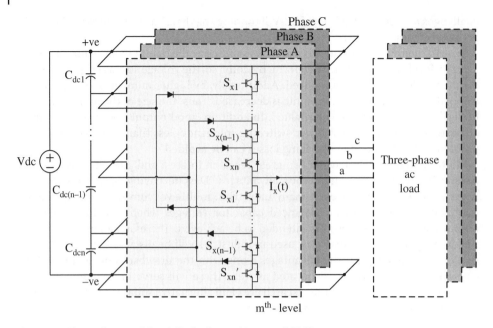

Figure 1.2 Three-phase multilevel diode clamped inverter (MDCI).

Since there are no passive elements required for the MDCIs (excluding the dc-link capacitors), only conduction losses and switching losses are accounted for the clamped diodes and switches in the converter.

When a higher number of output voltage stepped levels is desired, a higher number of capacitors will be needed in the dc-link. Theoretically, the dc VS should be distributed equally among the number of dc-link capacitors. However, in practical cases, the parameters of both active and passive elements can never be identically the same, due to which, an unbalanced condition occurs in the dc-link capacitor voltages. Besides, the amount of charging and discharging current flowing through the diodes clamped between each pair of dc-link capacitors is unequal because of the switching strategy used [6].

The unbalanced condition of the dc-link capacitor voltages makes the MDCI topology become disadvantageous and challenging to achieve a higher number of output phase voltage levels. Several methods have been proposed to overcome the dc-link unbalanced condition based on an active control [8], an active balancing circuit [9–11], or a passive RLC technique [12, 13].

Even though many research studies have contributed to resolve the unbalanced condition for three- to five-level MDCI, [8] makes the control a little more complex or more losses are incurred from the active balancing circuit [9–11] and the passive RLC technique [12, 13]. Additional costs are also incurred to implement these proposed methods for a higher number of output phase voltage levels.

These factors could be the reason that MDCI with more than a three-level output phase voltage is still not available in any of the industrial markets yet. Currently, 3L-NPC is the only MDCI topology popularly available in the market. This is due to its simplicity in design, low costs, and high reliability as compared with other multilevel topologies.

The 3L-NPC is widely used in industrial medium power motor drives as well as power inverters for wind and solar systems.

1.4.2 Multilevel Flying Capacitor Converter

A multilevel flying capacitor inverter (MFCI) as shown in Fig. 1.3 was named "the versatile multilevel commutation cell" originally in 1992 [14]. MFCI offers an alternative approach for multilevel conversion. Based on its original name, the topology of the MFCI is constructed with the multiple cells concept. By increasing the number of cells in the MFCI, a higher number of output phase voltage levels can be obtained as well. The design does not only allow the topology to be versatile but also achieves the objective of a multilevel approach with the property of voltage sharing among the switches.

The MFCI has a similar structure as the MDCI, where capacitors are clamped between the switches instead of the diode elements. Hence, each cell in the MFCI consists of two semiconductor devices with a capacitor clamped. The total number of switches i and capacitors p required for mL-MFCI can be determined based on the number of cells (n_{cells}) implemented as shown in equations (1.4) and (1.5), respectively:

$$i_{\text{switches(MFCI)}} = n_{cells} \times 2 \tag{1.4}$$

$$p_{\text{capacitors(MFCI)}} = n_{cells} - 1 \tag{1.5}$$

Based on the topology of the MFCI as shown in Fig. 1.3, the synthesized output phase voltage levels are not dependent on the dc-link capacitors. Hence, the MFCI, unlike the MDCI, will not face the same challenge of the unbalanced dc-link capacitor voltages when a higher number of output phase voltage levels is desired.

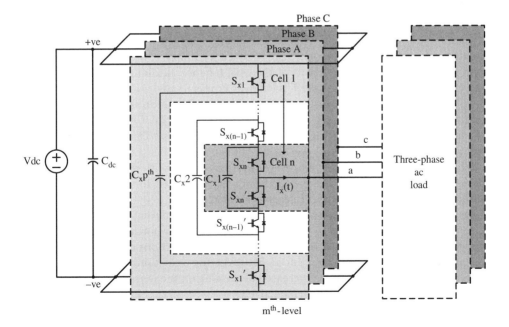

Figure 1.3 Three-phase *n*-cell multilevel flying capacitor inverter (MFCI).

However, more capacitors are needed for the mL-MFCI with a higher voltage level. Capacitors usually lead to a large inrush of current during the start-up period. Therefore, pre-charge circuits are needed to avoid high current during the transient period which may damage the power rating of the semiconductor devices [15]. In order to minimize the current from overstressing the switches during the initial operation, the common practice in any converter is to place a resistor in between the DC source and the dc-link capacitors. High losses may be experienced during this process, thus a switch is also connected in parallel to the pre-charge resistor and will be turned on after pre-charging. So excess losses can now be avoided during normal operations by shorting the pre-charge resistor through the switch.

Besides that, capacitors are passive elements that also serve as energy storage devices. The amount of energy stored in a capacitor will affect the voltage quality across it. The analytical and experimental results obtained in [16] have proven that a higher carrier frequency can reduce or minimize the capacity of the flying capacitors, and low voltage ripples can be achieved. However, a higher carrier frequency will limit the MFCI's high power conversion and also lead to issues such as EMI.

Besides that, a large amount of energy stored may affect the losses of the MFCI converter in addition to the switching and conduction losses of the switches. The reliability of capacitors, especially electrolytic capacitors required for a higher power rating is another major concern in the MFCI converter. Additionally, the space taken up by a number of flying capacitors will also increase as compared with the size of the diodes clamped in the MDCI.

As a result, these mentioned factors are a great disadvantage in designing higher mL-MFCI for industrial use. Therefore, the MFCI is currently attractive for high speed drives and those applications that require low current ripples. The MFCI is only available in ALSTOM.

1.4.3 Multilevel Cascaded H-Bridge Converter

An alternative multilevel approach was proposed by Zheng *et al.* [17] in 1996 which used a similar concept of modularity as the MFCI; hence, flexibility for this multilevel converter can be allowed. The multilevel converter was named the multilevel cascaded H-bridge inverter (MCHBI) based on the topology configuration shown in Fig. 1.4.

Each cell comprises four switches to form a conventional full-bridge (H-bridge) VSI and an isolated dc capacitor. A higher output phase voltage level can be achieved by cascading more cells together. The staircase-like voltage waveforms can only be synthesized using square pulses instead of chopped waveforms (pulse-width modulation – PWM) that are commonly used in other multilevel converters. Hence, this results in the experience of higher conduction losses in each switching device.

The MCHBI is advantageous when a higher number of output phase voltage level is attained (more than 11-level), and the output voltage waveform appears closer to a pure sinusoidal-like waveform. Thus, lower harmonic distortions can be achieved and the output filter size can also be reduced. However, the total number of switches i and capacitors p required will also be increased depending on the number of cells required based on equations (1.6) and (1.7), respectively:

$$i_{switches(MCHB)} = n_{cells} \times 4 \tag{1.6}$$

$$p_{capacitors(MCHB)} = n_{cells} \tag{1.7}$$

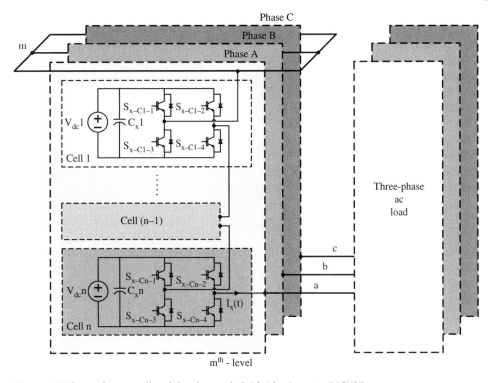

Figure 1.4 Three-phase *n*-cell multilevel cascaded H-bridge inverter (MCHBI).

The MCHBI does not have the problem of unbalanced capacitor voltages like the MDCI and also does not require excessive flying capacitors like the MFCIs to achieve multilevel conversion. Nevertheless, the MCHBI experiences a large circulating current in each cell due to its configuration. In addition, the MCHBI also requires isolated dc sources to supply across the dc capacitors in each cell. This makes MCHB very disadvantageous; hence, an alternative solution like a phase-shifted transformer is required to provide isolated dc voltages for each cell.

Phase-shifted transformers are usually hard to manufacture, which necessitates that a costlier solution be implemented. Besides that, high losses and heat dissipated are incurred from this bulky transformer when the load is not operated at its rated power. Even though a phase-shifted transformer takes up a large volume and space of the converter, many industries use the MCHBI to achieve a higher output phase voltage level for high power motor drives.

1.4.4 Modular Multilevel Converter

The modular multilevel inverter (M^2I) as shown in Fig. 1.5 was introduced [18, 19] and the prototype of the 2MW power rating with 17L output voltage was experimented and verified [20]. The configuration of M^2I is similar to that of the MCHBI; the only difference between them is the placement of the DC source. Based on Fig. 1.5, M^2I only requires a single DC source unlike the MCHB with many isolated DC sources. Hence, M^2I is highly suitable for HVDC systems.

Each cell consists of two semiconductor devices and a capacitor which forms a single phase half-bridge VSI (chopper cell). The configuration of each cell can also be done with full-bridge VSI (bridge cell) same as those used in the MCHB modular cell. Hence, the number of switches i for both chopper cell and bridge cell configuration can be determined based on the number of cells with equations (1.8) and (1.9), respectively:

$$i_{\text{switches(M}^2\text{I–Chopper)}} = n_{cells} \times 2 \tag{1.8}$$

$$i_{\text{switches(M}^2\text{I–Bridge)}} = n_{cells} \times 4 \tag{1.9}$$

While the capacitors p required for both chopper cell and bridge cell configurations can be known based on equation (1.10):

$$p_{\text{capacitors(M}^2\text{I)}} = n_{cells} \tag{1.10}$$

A bulky phase-shifted transformer is not required, but M^2I experiences a greater circulating current than the MCHBI. Therefore, a single coupled or center-tapped inductor is required for each phase leg, connected between the positive and negative dc rail connected modular cells [21–23] as shown in Fig. 1.5.

The M^2I would offer a more interesting and an attractive larger power application as compared with the MCHBI, but the implementation and design of either a single coupled or center-tapped inductor would make the M^2I a bit complex and tedious. The 9L-M^2I was first commissioned by SIEMENS for the HVDC PLUS transmission system. The M2I converters are also commercialized for the HVDC systems by other companies like ABB (HVDC Light) and ALSTOM (HVDC MaxSine).

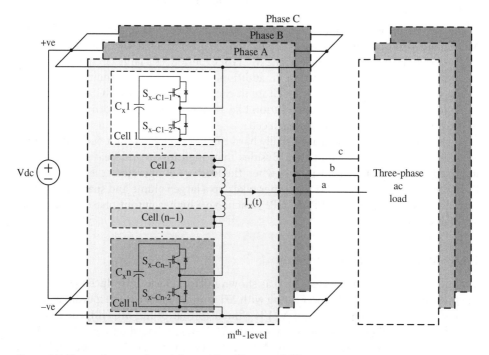

Figure 1.5 Three-phase n-cell modular multilevel inverter (M^2I).

1.4.5 Multilevel Active Neutral Point Clamped Inverter

An alternative multilevel inverter as shown in Fig. 1.6 was designed in 2005 by Bernet *et al.* [23, 24] and was named "the active neutral point clamped converter (ANPC)" initially. The original design was proposed to overcome the unequal loss distribution among the switches in 3L-NPC.

Later, the origin was expanded with multiple cells and multilevel ANPC was developed thereafter [25]. A higher number of output phase voltage levels can be achieved with the addition of commutation cells like those in the MFCI. Therefore, each cell comprises two switches with a capacitor clamped.

The total number of switches *i* and capacitors *p* in each phase leg can be known based on the number of cells implemented using equations (1.11) and (1.12), respectively:

$$i_{\text{switches(ANPC)}} = 6 + (n_{cells} \times 2) \tag{1.11}$$

$$p_{\text{capacitors(ANPC)}} = n_{cells} \tag{1.12}$$

The configuration of the ANPC as shown in Fig. 1.6 is basically a combination of both MDCI and MFCI. As a result, the multilevel ANPC has kept the advantageous characteristics of both MDCI and MFCI. This shows that the ANPC will not be affected by the dc-link capacitors when the number of output phase voltage level is increased. The modularity of commutation cells allows the flexibility of the ANPC in extending to a higher number of voltage levels without encountering the unbalanced dc-link condition.

The ANPC in this case does not require any bulky phase-shifted transformers, multiple isolated dc sources or a balancing circuit for dc-link capacitor voltages. Hence, the ANPC offers an attractive method for medium-to-high industrial motor drives as compared with the other previously mentioned multilevel converters. A back-to-back configuration of 5L-ANPC is available in the current industrial market, ABB, for regenerative motor drives.

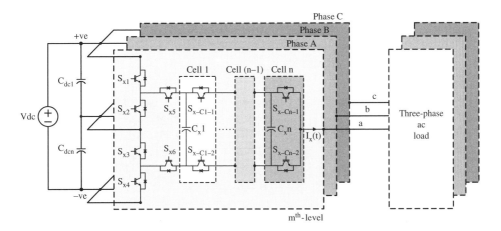

Figure 1.6 Three-phase *n*-cell multilevel active neutral point clamped inverter.

1.5 Multilevel Applications and Future Trends

Multilevel converter topologies are well-known for its applicability in medium-to-high power drives and are popularly used for the industrial motor drives application. Moreover, many academia research studies have also explored the application potential of these multilevel converters in other areas such as transportation, renewable energy, utilities grid, and even low power Class D amplifiers as shown in Table 1.1 [26].

Multilevel converter applications play an important role in our daily lives and power electronic converters are a likely trend. Since the cost of semiconductor devices have been depreciating, more industries have also been investing a lot in developing new multilevel converters for other possible areas that are not listed in Table 1.1 such as more electric aircraft (MEA), heat ventilation and air conditioning (HVAC), and others.

With the involvement of the markets, the cost of implementation is always the prime concern, which explains why some of the existing multilevel converters like MDCI and MFCI are not developed for higher multilevel output voltage due to the total number of components required as shown in Fig. 1.7. Therefore, only multilevel topologies like MCHB, MMI, and ANPC are developed with more than 6-level output phase voltages. Even though many academia research works have contributed alternative solutions to overcome the limitations of these existing multilevel converters to achieve higher multilevel output voltages, but these proposed solutions are still not very welcome in the industry. Thus, recently many research and development works are motivated to focus on designing new multilevel converters with a reduced number of component counts to achieve better power quality more efficiently.

Table 1.1 High-performance multilevel converter applications based on topology [26].

Application		Multilevel topology		
		MDCI/ANPC	MCHBI	MFCI
Transportation	Marine propulsion	[78, 79, 82, 83]	[78, 83]	[83]
	Train traction	[84–86]	[80, 81, 84, 87–89]	[209, 210]
	Automotive	[90]	[91–214]	[215–218]
Renewable energy	Solar photovoltaic	[121, 181–185]	[181, 186–193]	[181]
	Wind power	[165, 194–198]	[199]	—
	Hydro power	[224]	—	—
Utilities grid	FACTS	[225]	[117, 225–227]	[225]
	STATCOM	[228–230]	[55, 62, 192, 228, 231–234]	—
	UPFC/UPQC/DVR	[245–247]	[245, 248, 249]	[250]
	HVDC	[175]	[55, 56, 180]	—
	Active filters	[235–239]	[240–242]	[243, 244]
Industrial/ commercial use	Pumps/motors	[219, 220]	—	—
Small signals	Class D amplifier	—	[221–223]	—

Note: The information in this Table 1.1 is extracted from [26]; hence, the reference numbering in the table refers to the references in [26].

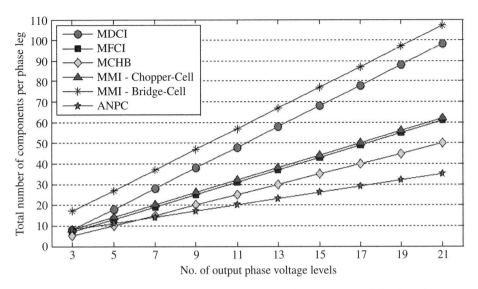

Figure 1.7 Comparison of total component counts based on the respective multilevel topologies. (Note: The total number of components per phase leg presented in this figure includes dc-link capacitors for MDCI, MFCI, and ANPC and coupled inductors for MMI).

The advantages of reducing the amount of components needed do not only benefit the industrial interests, but also help increase the global reliability of the converters. Low maintainability is required since lesser component counts are used in multilevel converters. Furthermore, lower losses are incurred during the power conversion process; hence, a higher energy efficiency can be achieved. In addition, reducing component counts enable the converters to be of lighter weight and beneficial for many applications with space constraints such as MEAs, wind turbines, solar PV grid-connected systems, and HVAC systems.

As a result, the developing trends in new multilevel converters in the near future will be involving cost, size, weight, losses, and power quality [27]. It is factually known that the advancement of multilevel converter designs is closely co-related with the development of power electronic devices as shown in Fig. 1.8.

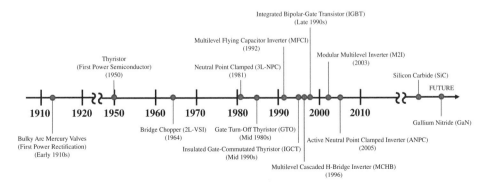

Figure 1.8 The history of power electronics.

As such, new electronic switching devices based on silicon carbide (SiC) and gallium nitride (GaN) shall be the upcoming technologies to support a higher switching frequency operation as well as to sustain a higher current handling capability. With that, a higher power density converter can be achieved. These will be served as guidelines in optimizing the design of power electronic converters for both research and development studies.

References

1 R.E. Morgan, "Bridge-Chopper inverter for 400 CPS sine wave power," *Aerospace, IEEE Transactions on*, vol. 2, 993–997, 1964.

2 A.H. Bonnett, "Analysis of the impact of pulse-width modulated inverter voltage waveforms on AC induction motors," *Industry Applications, IEEE Transactions on*, vol. 32, pp. 386–392, 1996.

3 T. A. Meynard and H. Foch, "Multilevel converters and derived topologies for high power conversion," in *Industrial Electronics, Control, and Instrumentation, 1995, Proceedings of the 1995 IEEE IECON 21st International Conference on*, pp. 21–26, vol. 1, 1995.

4 B. Backlund, M. Rahimo, S. Klaka, and J. Siefken, "Topologies, voltage ratings and state of the art high power semiconductor devices for medium voltage wind energy conversion," in *Power Electronics and Machines in Wind Applications, 2009. PEMWA 2009. IEEE*, pp. 1–6, 2009.

5 A. Nabae, I. Takahashi, and H. Akagi, "A new neutral-point-clamped PWM inverter," *Industry Applications, IEEE Transactions on*, vol. IA-17, pp. 518–523, 1981.

6 N. S. Choi, J. G. Cho, and G. H. Cho, "A general circuit topology of multilevel inverter," in *Power Electronics Specialists Conference, 1991. PESC '91 Record., 22nd Annual IEEE*, pp. 96–103, 1991.

7 O. Apeldoorn and L. Schulting, "10 kVA four level inverter with symmetrical input voltage distribution," in *Power Electronics and Applications, 1993, Fifth European Conference on*, pp. 196–201, vol. 3, 1993.

8 S. Ogasawara and H. Akagi, "Analysis of variation of neutral point potential in neutral-point-clamped voltage source PWM inverters," in *Industry Applications Society Annual Meeting, 1993., Conference Record of the 1993 IEEE*, pp. 965–970, vol. 2, 1993.

9 K.A. Corzine, J. Yuen, and J.R. Baker, "Analysis of a four-level DC/DC buck converter," *Industrial Electronics, IEEE Transactions on*, vol. 49, pp. 746–751, 2002.

10 N. Hatti, Y. Kondo, and H. Akagi, "Five-level diode-clamped PWM converters connected back-to-back for motor drives, "*Industry Applications, IEEE Transactions on*, vol. 44, pp. 1268–1276, 2008.

11 N. Hatti, K. Hasegawa, and H. Akagi, "A 6.6-kV Transformerless motor drive using a five-level diode-clamped PWM inverter for energy savings of pumps and blowers," *Power Electronics, IEEE Transactions on*, vol. 24, pp. 796–803, 2009.

12 H. Du Toit Mouton, "Natural balancing of three-level neutral-point-clamped PWM inverters," *Industrial Electronics, IEEE Transactions on*, vol. 49, pp. 1017–1025, 2002.

13 Z. Mohzani, B.P. McGrath, and D.G. Holmes, "Natural balancing of the Neutral Point voltage for a three-phase NPC multilevel converter," in *IECON 2011 – 37th Annual Conference on IEEE Industrial Electronics Society*, pp. 4445–4450, 2011.

14 T. A. Meynard and H. Foch, "Multi-level conversion: high voltage choppers and voltage-source inverters," in *Power Electronics Specialists Conference, 1992. PESC '92 Record, 23rd Annual IEEE*, pp. 397–403, vol. 1, 1992.

15 S. Thielemans, A. Ruderman, and J. Melkebeek," Self-precharge in single-leg flying capacitor converters," in *Industrial Electronics, 2009. IECON '09. 35th Annual Conference of IEEE*, pp. 812–817, 2009.

16 S.S. Fazel, S. Bernet, D. Krug, and K. Jalili, "Design and comparison of 4-kV neutral-point-clamped, flying-capacitor, and series-connected H-bridge multilevel converters," *Industry Applications, IEEE Transactions on*, vol. 43, pp. 1032–1040, 2007.

17 P. Fang Zheng, L. Jih-Sheng, J. W. McKeever, and J. VanCoevering, "A multilevel voltage-source inverter with separate DC sources for static VAr generation," *Industry Applications, IEEE Transactions on*, vol. 32, pp. 1130–1138, 1996.

18 A. Lesnicar and R. Marquardt, "An innovative modular multilevel converter topology suitable for a wide power range," in *Power Tech Conference Proceedings, 2003 IEEE Bologna*, pp. 1–6, vol. 3, 2003.

19 M. Glinka and R. Marquardt, "A new AC/AC-multilevel converter family applied to a single-phase converter," in *Power Electronics and Drive Systems, 2003. PEDS 2003. The Fifth International Conference on*, pp. 16–23, vol. 1, 2003.

20 M. Glinka, "Prototype of multiphase modular-multilevel-converter with 2 MW power rating and 17-level-output-voltage," in *Power Electronics Specialists Conference, 2004. PESC 04. 2004 IEEE 35th Annual*, pp. 2572–2576, vol. 4, 2004.

21 M. Hagiwara, R. Maeda, and H. Akagi, "Control and analysis of the modular multilevel cascade converter based on double-star chopper-cells (MMCC-DSCC)," *Power Electronics, IEEE Transactions on*, vol. 26, pp. 1649–1658, 2011.

22 H. Akagi, "Classification, terminology, and application of the modular multilevel cascade converter (MMCC)," *Power Electronics, IEEE Transactions on*, vol. 26, pp. 3119–3130, 2011.

23 T. Bruckner, S. Bernet, and H. Guldner, "The active NPC converter and its loss-balancing control," *Industrial Electronics, IEEE Transactions on*, vol. 52, pp. 855–868, 2005.

24 B. Thomas and B. Steffen, "The active NPC converter for medium-voltage applications," in *Industry Applications Conference, 2005. Fourtieth IAS Annual Meeting. Conference Record of the 2005*, pp. 84–91, vol. 1, 2005.

25 P., Barbosa, P. Steimer, J. Steinke, M. Winkelnkemper, and N. Celanovic, "Active-neutral-point-clamped (ANPC) multilevel converter technology," in *Power Electronics and Applications, 2005 European Conference on*, pp. 1–10, 2005.

26 S. Kouro, M. Malinowski, K. Gopakumar, J. Pou, L. G. Franquelo, W. Bin *et al.*, "Recent advances and industrial applications of multilevel converters," *Industrial Electronics, IEEE Transactions on*, vol. 57, pp. 2553–2580, 2010.

27 J. W. Kolar, U. Drofenik, J. Biela, M. L. Heldwein, H. Ertl, T. Friedli, *et al.*, "PWM converter power density barriers," in *Power Conversion Conference – Nagoya, 2007. PCC '07*, pp. 1–9, 2007.

2

Multilevel Modulation Methods

Ziyou Lim, Hossein Dehghani Tafti, and Harikrishna R. Pinkymol

2.1 Introduction

The design and implementation of modulation techniques are relevant and important with the growth of multilevel converter topologies. Modulation schemes are applied to synthesize a good output voltage quality while allowing better control over both the magnitude and frequency of the output voltage.

Multilevel modulation methods presented in Fig. 2.1 are basically classified based on (i) carrier-based modulation (CBM) and (ii) space vector modulation (SVM). Both of these modulations belong to the same family of pulse-width modulation (PWM). Both methods seem to be very different from each other, but there are no clear-cut superior performances proven to distinguish one from the other. Both methods do have their respective pros and cons.

SVM can generate additional switching states whereby CBM cannot, hence a higher degree of freedom is provided. Many SVM 2D and 3D algorithms have been proposed for m-level and up to four-leg multilevel converters. However, any of the SVM methods presented would still require to go through the three-step process as mentioned in [1]: (i) vector selection (ii) duty cycle computation, and finally, (iii) the sequence of the vectors generated. A simplified method has been proposed in [2] using the single-phase modulator (1DM) technique that can be extended for any number of output phase voltage levels and phase legs. Equivalent performances have been realized by 1DM as compared with other conventional SVM methods [3]. But a lower computational cost is achieved for 1DM only when compared with multi-dimensional SVM (M-SVM).

SVM is heavily dependent on the computational calculations; hence, there is no need to increase the number of carriers for higher m-level multilevel converters unlike CBM. Nevertheless, SVM, except the nearest vector modulation in this case, is not applicable for high power applications due to the high switching frequency operations required.

Despite the many research studies that have explored SVM methods, CBM is the industry's favorite. This could partly be attributed to the computational methods required in SVM that must be programmed into the low-cost digital processors to generate the gating signals. Thus, this process makes SVM complex. The gating signals of CBM could be generated by either programmed digital processors (results are more complicated than that for SVM) or the simplest way is through logic gates and analog comparators.

Advanced Multilevel Converters and Applications in Grid Integration, First Edition.
Edited by Ali I. Maswood and Hossein Dehghani Tafti.
© 2019 John Wiley & Sons Ltd. Published 2019 by John Wiley & Sons Ltd.

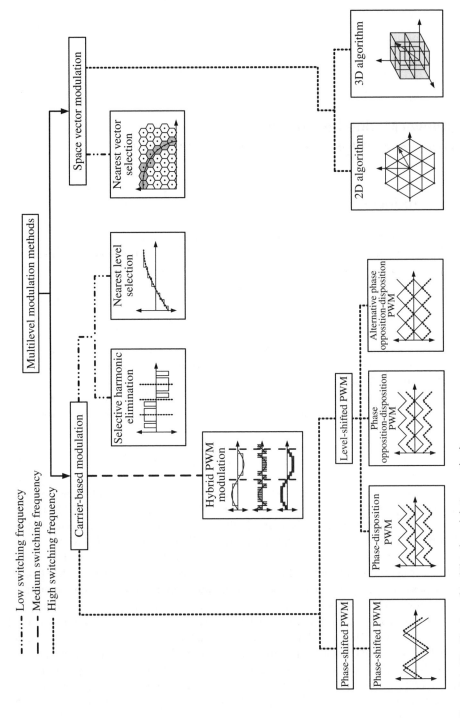

Figure 2.1 Classification of multilevel modulation methods.

Comparative studies between both SVM and CBM methods have been evaluated in [4] and [5] for the neutral point clamped (NPC) converter and the multilevel cascaded H-bridge (MCHB) inverter, respectively; the results show that the switching losses are comparable among the two methods. This results in a clearer choice between CBM and SVM for industrial applications.

Even though the carrier-based sinusoidal PWM (SPWM) usually operates at high switching frequencies (above 1 kHz), SPWM could go for a much lower switching frequency (about 500 Hz) to meet the high power application requirements. The area between SPWM and the hybrid PWM are not fully explored yet. Hence, the problems and advantages of each traditional SPWM schemes will be elaborated in this chapter. New hybrid carrier-based SPWMs are proposed and analyzed in the following sections. Better performances and power quality are achieved and verified experimentally.

2.2 Carrier-Based Sinusoidal Pulse-Width Modulation Methods

Sinusoidal pulse-width modulation (SPWM) is proven to be the simplest method that can be achieved through comparator circuits directly without any tedious effort needed for programming the digital processors. Hence, SPWM has been not only of research interest and but also of use in industrial applications.

Comparative studies between the existing SPWM methods will be carried out in this section. In addition to that, the limitation and problems of SPWM for multilevel converter topologies will also be discussed. These have further motivated interest in exploring and analyzing the new proposed SPWM methods that will be explained in the following sections with the respective multilevel converters.

2.2.1 Operation Principles

The gating signals of SPWM are generally achieved by comparing the reference sinusoidal wave with the triangular carriers. When the reference signal is greater than the gradient of the triangular carrier, the switching pulse will be at the '"turn on" state, and vice versa: when the reference signal is smaller than the gradient of the carrier, the switching pulses will be at the "turn off" state.

Hence, the duty ratio of the switching pulse is varied according to the reference sinusoidal signal $\sin(\omega t)$, which can be observed in the Fig. 2.2. Equation (2.1) expresses the state of the switching pulse during the modulation period as follows:

$$S_x(t) = \begin{cases} 1 & \text{if } \sin(\omega t) > \text{tri}(t) \\ 0 & \text{if } \sin(\omega t) < \text{tri}(t) \end{cases} \tag{2.1}$$

For any SPWM modulation techniques used, the number of carriers ($n_{carrier}$) required can be determined using equation (2.2) based on the m-level (mL) output phase voltage levels desired to synthesize:

$$n_{carrier} = (mL) - 1 \tag{2.2}$$

According to equation (2.2), the $n_{carrier}$ needed for SPWM will be increased linearly with the number of output phase voltage levels produced. However, an increase in

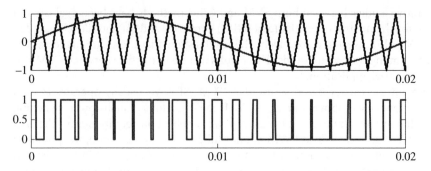

Figure 2.2 Gating signal of SPWM.

the value of the $n_{carrier}$ complicates the analog circuit designs for higher mL multilevel converters especially when more than seven-level (7L) output phase voltages are synthesized. Besides that, the overall switching frequency of the multilevel converters will also be affected by the $n_{carrier}$ implemented, depending on the inherent characteristics of the SPWM methods used.

2.2.2 Limitations of Sinusoidal Pulse-Width Modulation in Multilevel Converters

Regardless of any SPWM techniques used in the multilevel converters, it should be noted that the modulation range is being restricted when higher mL output phase voltage levels are produced. It is relatively important as it affects the degree of freedom in controlling the amplitude of output voltage. Modulation index is usually referred to as the modulation amplitude ratio (Mx) as defined in equation (2.3):

$$\text{Mx} = \frac{\hat{V}_{control}\,(\text{Peak amplitude of reference sinusoidal signal})}{\hat{V}_{tri}\,(\text{Peak amplitude of carrier signal})} \qquad (2.3)$$

Fig. 2.3 illustrates the limitation of Mx and the degree of freedom in control with respect to the output voltage levels. The controllable range of Mx ratio becomes smaller between 0.9 Mx and 1.0 Mx when more than a 21-level output phase voltage is being synthesized. This constraint controllability is undesirable for many applications. When Mx falls below the minimum required index for the respective voltage levels, SPWM is considered as under-modulation. As a consequence, the output phase voltage of the converter will become [(mL)-2] voltage levels with higher harmonic distortion and may cause undesirable large dV/dt and dI/dt during the operation.

2.2.3 Performances of Level-Shifted PWM and Phase-Shifted PWM

The two common modulation strategies under SPWM are the level-shifted carrier PWM (LSC-PWM) and phase-shifted carrier PWM (PSC-PWM) as shown in Fig. 2.1. There are three types of carrier disposition PWM strategies that are further classified under LSC-PWM, namely, the phase-disposition PWM (PD-PWM)

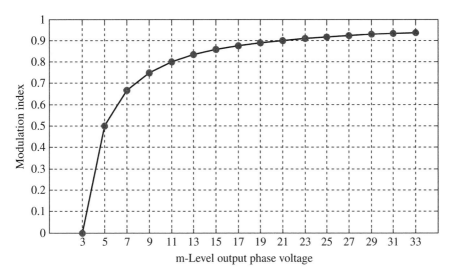

Figure 2.3 Limitation of the modulation index (Mx) vs *mL*-output phase voltage.

(Fig. 2.4(a)), phase-opposition-disposition (POD-PWM) (Fig. 2.4(b)), and alternative phase-opposition-disposition (APOD-PWM) (Fig. 2.4(c)).

A comparative study of PD-PWM, POD-PWM and APOD-PWM has been based on the number of output phase voltage levels in [6], and the detailed analytical studies of the harmonic performances of LSC-PWM and PSC-PWM for MDCI and MCHB, respectively, are also evaluated in [7]. Both have proven that PD-PWM has superior performance in terms of lower harmonic components. Because of that, the LSC-PWM is sometimes directly referred to as the popularly implemented PD-PWM.

However, one can observe that the carrier harmonics of LSC-PWM will always be at the same harmonic order despite any $n_{carrier}$ used for comparing. Meanwhile, the carrier harmonics of PSC-PWM will be shifted accordingly depending on the $n_{carrier}$ and the carrier frequency. The harmonic characteristics of PSC-PWM will be proven again subsequently in the following section.

The amplitude of carrier harmonics is often an important factor to be taken note of, since the conducted EMI is concerned with the switching frequency operation between 10 kHz and 20 kHz, which is typically used in power electronic converters. In this case, the conducted EMI for LSC-PWM methods applied to high power applications (switching frequency range between 500 Hz and 700 Hz) will not be a major concern.

Despite its superior performances as compared with the other SPWM schemes, there is a concerning issue when LSC-PWM is performed during the under-modulation condition. This problem can be highlighted using the gating signals obtained from the LSC-PWM for the five-level output phase voltage (four triangular carriers required for modulation) as shown in Fig. 2.5. Two conditions (i) 0.9 Mx (normal operation) and (ii) 0.3 Mx (under-modulation) are considered for a clearer illustration.

It can be seen from Fig. 2.3 that the normal operation range of Mx to synthesize the five-level output phase voltage should fall between 0.5 Mx and 1.0 Mx.

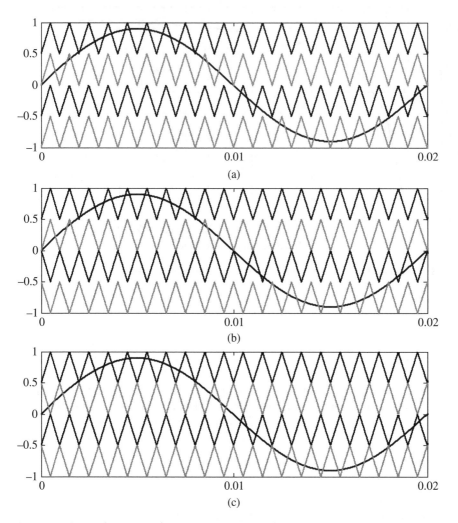

Figure 2.4 Carrier disposition of LSC-PWM. (a) PD-PWM. (b) POD-PWM. (c) APOD-PWM.

Therefore, Fig. 2.5(a) shows the gating signals generated for normal operations. The under-modulation condition in Fig. 2.5(b) shows that the top gating signal (GS-1) will be at the "turn off" state all the way while the bottom gating signal (GS-4) will be at the "turn on" state all the way.

The constant "switch on" state of GS-4 in the LSC-PWM under-modulation condition will bring undesirable harm to the semiconductor devices (especially the switches). As a result, the electronic devices will breakdown easily due to excess voltage and current stresses. Thus, this affects the overall reliability of the multilevel converters.

On the other hand, PSC-PWM will not encounter the same problem as LSC-PWM during the under-modulation condition as proven in Fig. 2.6. It can be observed that all of the gating signals are still generated even in the event of the under-modulation

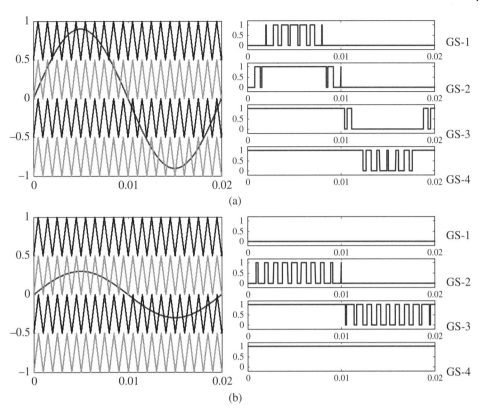

Figure 2.5 Performances of LSC-PWM – Five-level modulation. (a) Modulation index (Mx) = 0.9. (b) Modulation index (Mx) = 0.3.

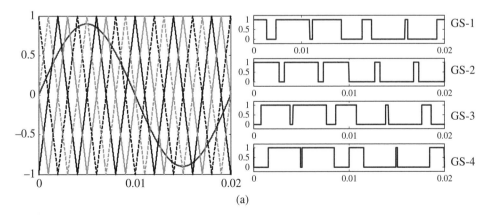

Figure 2.6 Performances of PSC-PWM – five-level modulation. (a) Modulation index (Mx) = 0.9. (b) Modulation index (Mx) = 0.3.

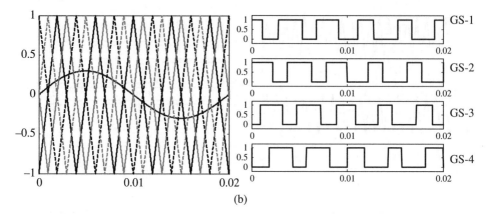

(b)

Figure 2.6 (*Continued*)

condition. Thus, a higher reliability is achievable with PSC-PWM as compared with LSC-PWM.

Besides that, the gating signals generated from PSC-PWM are always symmetrical to each other be it during normal operations or under the under-modulation condition. Hence, stress and loss are equally distributed among the semiconductor devices [5]. These evidences have strongly driven industries to choose PSC-PWM over LSC-PWM for many multilevel converter topologies except 3L-NPC or the family of mL-MDCI. Therefore, LSC-PWM is commonly used for those diode-clamped related multilevel topologies.

2.3 Space Vector Modulation (SVM)

In this section, the SVM for the three-level NPC converter is discussed. The circuit configuration of the converter is shown in Fig. 2.7.

The output phase voltages V_{An}, V_{Bn}, and V_{Cn} in Fig. 2.7 are given by:

$$V_{Xn} = V_{XO} - V_{nO} \ for \ X = A, B, C. \tag{2.4}$$

The equivalent space vector generated by the phase voltages is:

$$V_s = \frac{2}{3} \left(V_{An} + V_{Bn} e^{j\frac{2\pi}{3}} + V_{Cn} e^{j\frac{4\pi}{3}} \right) = V_r e^{j\theta} \tag{2.5}$$

Substituting (2.4) in (2.5):

$$V_s = \frac{2}{3} \left(V_{AO} + V_{BO} e^{j\frac{2\pi}{3}} + V_{CO} e^{j\frac{4\pi}{3}} \right) = V_r e^{j\theta} \tag{2.6}$$

Pole voltages V_{AO}, V_{BO}, V_{CO} in (2.6) can take three distinct levels based on the switching states. The resulting 19 voltage space vectors and 27 inverter switching states can be arranged in a three-dimensional representation as shown in Fig. 2.8(a). The distribution of voltage vectors forms two hexagons, and for analysis, they are divided into six triangular sections of 60° each. In Fig. 2.8(a), (A-F) are the sectors and each sector is divided into four sub-sectors S-1, S-2, S-3, and S-4. The voltage vectors are divided into four groups: large vector (LV), medium vector (MV), small vector (SV), and zero vector (ZV) based

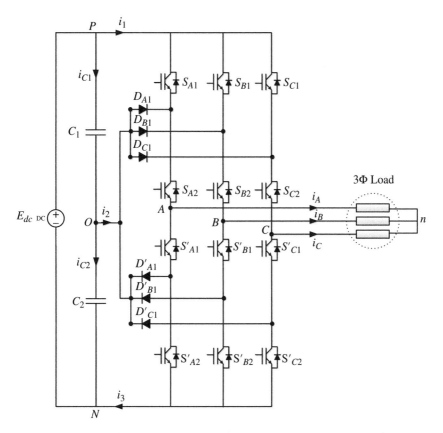

Figure 2.7 Three-level neutral-point-clamped (NPC) converter.

on their magnitudes. The magnitudes of voltage vectors are:

$$\text{LV}: (V_1, V_2, V_3, V_4, V_5, V_6) = \frac{2}{3}E_{dc}; \; \text{MV}: (V_{12}, V_{23}, V_{34}, V_{45}, V_{56}) = \frac{E_{dc}}{\sqrt{3}};$$

$$\text{SV}: (V_{01}, V_{02}, V_{03}, V_{04}, V_{05}, V_{06}) = \frac{E_{dc}}{3} \text{ and } \text{ZV}: (V_0) = \text{zero}.$$

Three phase reference voltage output at the inverter can be synthesized by generating the equivalent space vector by switching a three voltage vector where the tip of the vectors becomes the vertex of a triangle in which the reference space vector resides. For example, in Fig. 2.8(b), the reference space vector V_S lying in the sub-sector S-4 is synthesized by selecting V_2, V_{02}, and V_{12} over one sampling time. The duration of each selected vectors is obtained by the time-averaging of V_S over one sampling time T_s. Let $V_i, V_j,$ and V_k represent the left side vector, right side vector, and the middle vector in each sub-sectors, respectively, and $d_i, d_j,$ and d_k are that the respective duty ratios. Then, The volt-second balance equation is given as:

$$V_s = (d_i.V_i + d_j.V_j + d_k.V_k) = V_r e^{j\theta} \tag{2.7}$$

Then, applying (2.7) in Fig. 2.6,

$$V_s = (d_i.V_{02} + d_j.V_{12} + d_k.V_2) \tag{2.8}$$

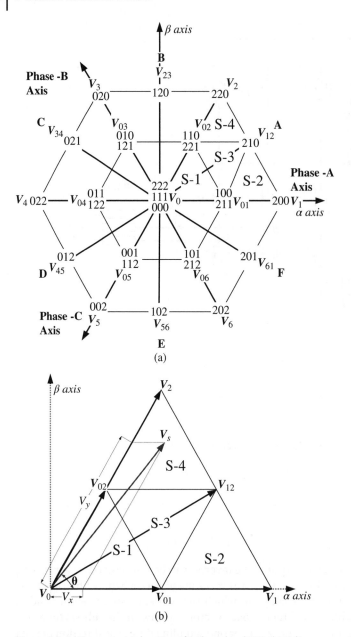

Figure 2.8 (a) Space vector locations for three-level DCI. (b) Voltage space vector lying in sector A.

Substituting the values of the voltage vectors and separating the real and imaginary parts,

$$V_s = \left(d_i \cdot \frac{E_{dc}}{3} e^{j60} + d_j \cdot \frac{E_{dc}}{\sqrt{3}} e^{j30} + d_k \cdot \frac{2E_{dc}}{3} e^{j60} \right) = V_r e^{j\theta}$$

$$V_r \cos\theta = \frac{E_{dc}}{3} \left(\frac{d_i}{2} + \frac{3}{2} d_j + d_k \right) \quad and \quad V_r \sin\theta = \frac{E_{dc}}{2\sqrt{3}} (1 + d_k) \tag{2.9}$$

Table 2.1 Relativity to time of voltage vectors in Sector A.

Duty ratio	Subsector			
	S-1	S-2	S-3	S-4
d_i	$1-d_j-d_k$			
d_j	$2d_X$	$2d_X-1$	$2d_X+2d_Y-1$	0
d_k	$2d_Y$	$2d_Y$	$1-2d_Y$	$2d_Y-1$

Also, from Fig. 2(b),

$$V_X = V_r\left(\cos\theta - \frac{1}{\sqrt{3}}\sin\theta\right)$$

$$V_Y = V_r\frac{2}{\sqrt{3}}\sin\theta \tag{2.10}$$

Let

$$d_X = \frac{V_X}{\frac{2}{3}E_{dc}} \quad and \quad d_Y = \frac{V_Y}{\frac{2}{3}E_{dc}} \tag{2.11}$$

From (2.9) and (2.11)

$$d_i = (1 - d_j - d_k); \quad d_j = 2d_X \quad and \quad d_k = 2d_Y - 1 \tag{2.12}$$

The switching vectors and their respective duty ratios are calculated in the same manner as described above for synthesizing the reference space vector lying in any of the subsectors in sector A and it is listed in Table 2.1.

The subsector identification is related to d_X and d_Y as follows:

$$\text{if } (d_X + d_Y) \leq 0.5 \text{ then S-1}$$
$$\text{if } d_X > 0.5 \text{ then S-2}$$
$$\text{if } d_Y > 0.5 \text{ then S-4}$$
$$\text{else S-3.} \tag{2.13}$$

The reference vector V_S residing in any of the sectors can be synthesized by switching the nearest three vectors over one sampling time where the duty ratios of the respective vectors are obtained by virtually transferring V_S to sector A and applying Table 2.1. While transferring V_S lying in any other sector than sector A to calculate the duty ratios, the angle θ should vary between 0° and 60°.

2.4 Summary

Two types of modulation schemes for generating the switching signals of multilevel converters have been addressed in this chapter. Sinusoidal PWM compares triangular waveforms with voltage reference to generate the switching signals. In the SVMs scheme, the nearest voltage vectors to the reference voltage vector are identified and accordingly the switching signals are generated. The details of the implementation of these algorithms have been discussed in this chapter.

References

1 S. Kouro, M., Malinowski, K. Gopakumar, J. Pou, L. G. Franquelo, W. Bin *et al.*, "Recent advances and industrial applications of multilevel converters," *Industrial Electronics, IEEE Transactions on*, vol. 57, pp. 2553–2580, 2010.

2 J.I. Leon, R. Portillo, S. Vazquez, J. J. Padilla, L. G. Franquelo, and J. M. Carrasco, "Simple unified approach to develop a time-domain modulation strategy for single-phase multilevel converters," *Industrial Electronics, IEEE Transactions on*, vol. 55, pp. 3239–3248, 2008.

3 J.I. Leon, S. Vazquez, J.A. Sanchez, R. Portillo, L. G. Franquelo, J. M. Carrasco *et al.*, "Conventional space-vector modulation techniques versus the single-phase modulator for multilevel converters," *Industrial Electronics, IEEE Transactions on*, vol. 57, pp. 2473–2482, 2010.

4 D. Andler, S. Kouro, M. Perez, J. Rodriguez, and W. Bin, "Switching loss analysis of modulation methods used in neutral point clamped converters," in *Energy Conversion Congress and Exposition, 2009. ECCE 2009. IEEE*, pp. 2565–2571, 2009.

5 S. Kouro, M. Perez, H. Robles, and J. Rodriguez, "Switching loss analysis of modulation methods used in cascaded H-bridge multilevel converters," in *Power Electronics Specialists Conference, 2008. PESC 2008. IEEE*, pp. 4662–4668, 2008.

6 G. Carrara, S. Gardella, M. Marchesoni, R. Salutari, and G. Sciutto, "A new multilevel PWM method: a theoretical analysis," *Power Electronics, IEEE Transactions on*, vol. 7, pp. 497–505, 1992.

7 B.P. McGrath and D.G. Holmes, "Multicarrier PWM strategies for multilevel inverters," *Industrial Electronics, IEEE Transactions on*, vol. 49, pp. 858–867, 2002.

3

Mathematical Modeling of Classical Three-Level Converters
Gabriel H. P. Ooi

3.1 Introduction

This chapter demonstrates the mathematical analysis of the operating modes of the two basic multilevel power converter topologies: the three-level diode-clamped inverter and the flying capacitor inverter. Mathematical analysis in this chapter is examined on balanced load and balanced dc bus voltage under a steady-state condition, whereas the analysis of a three-level diode-clamped inverter in Section 3.2 is adopted based on [1]. Moreover, the theoretical analysis in this section is verified with the simulation and experimental results. To obtain a better performance, the voltage across each dc capacitor must be equally distributed. Achieving a balanced dc-link and a balanced load has several perceived advantages such as good output voltage/current quality and the distribution of the voltage stress equally across the semiconductor devices [2].

3.2 Three-Level Diode-Clamped Inverter Topology

A three-level diode-clamped inverter circuit is shown in Fig. 3.1. This topology requires six identical fast recovery diodes (i.e. D_{s+} and D_{s-}) connected to the virtual ground node "m," two dc capacitors, and 12 identical insulated gate bipolar transistors (IGBT) to generate a three-level output voltages waveform. The switching pattern of a three-level diode-clamped inverter requires two unipolar triangular carriers and a modulated control signal ($M_a(t)$, $M_b(t)$ and $M_c(t)$) as shown in Fig. 3.2. The modulating signal is compared with level-shifted triangular waves to generate gate signals for switches in each leg. Thus, the switching requirement for the per-phase leg "a" is determined as:

$$\begin{cases} S_{a1} + S_{a3} = 1 \\ S_{a2} + S_{a4} = 1 \end{cases} \tag{3.1}$$

S_{a1}, S_{a2}, S_{a3}, and S_{a4} represent the switching logic in 1 or 0 (turn on = 1 and turn off = 0).

Fig. 3.2 shows the switching pattern for a positive modulation signal with a level-shifted pulse width modulation (LSPWM) technique for both the three-level diode-clamped inverter and the flying capacitor inverter. The modulation signals of phase "a" are separated into two sub-modulations ($M_{a1}(t)$ and $M_{a2}(t)$) to operate in

Advanced Multilevel Converters and Applications in Grid Integration, First Edition.
Edited by Ali I. Maswood and Hossein Dehghani Tafti.

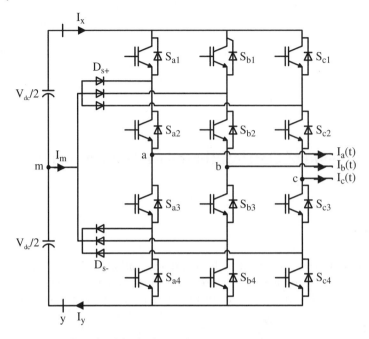

Figure 3.1 Three-level diode-clamped inverter circuit.

a linear region within the carrier frequency range. The positive modulation equation is expressed as:

$$M_{a1}(t) = \frac{-2}{T_s}t_{a1} + 1$$

$$M_{a2}(t) = \frac{-2}{T_s}t_{a2} \tag{3.2}$$

T_s is the time period of the carrier wave. Assume the duty ratio D is at any value between 0 and 1, then at a specific amplitude of the modulation signal, the time interval of $t = T_s(1-D)/2$ is shown in Fig. 3.2. Hence, equation (3.2) is simplified as:

$$m_{a1} = \frac{-2}{T_s}\frac{T_s(1 - D_{a1})}{2} + 1 = D_{a1}$$

$$m_{a2} = \frac{-2}{T_s}\frac{T_s(1 - D_{a2})}{2} = D_{a2} - 1 \tag{3.3}$$

Positive amplitude of modulation for both topologies (the three-level diode-clamped inverter and the three-level flying capacitor inverter) is expressed as:

$$m = m_{a1} + m_{a2} = D_{a1} + D_{a2} - 1 = S_{a1} + S_{a2} - 1 \tag{3.4}$$

Fig. 3.3 shows the switching pattern on a negative modulation signal with a LSPWM technique for the three-level diode-clamped inverter. The switching patterns on a negative modulation signal for a flying capacitor inverter are contrasted with those of the three-level diode-clamped inverter, equations (3.5) to (3.7) are depicted for the diode-clamped inverter. The derivation of negative modulation signals applied on

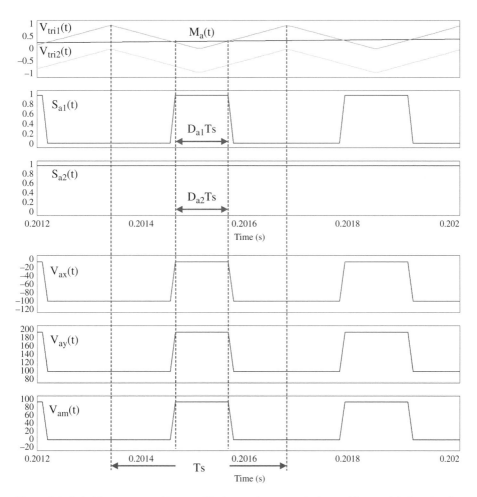

Figure 3.2 Switching pattern with a specific duty ratio at a particular positive modulation signal.

phase "a" is similar to the method presented in equations (3.2) to (3.4). The negative sub-modulation equation is expressed as:

$$M_{a3}(t) = \frac{-2}{T_s}t_{a3} + 1$$

$$M_{a4}(t) = \frac{-2}{T_s}t_{a4} \tag{3.5}$$

Let $t = DT_s/2$, and substitute "t" into equation (3.5):

$$m_{a3} = \frac{-2}{T_s}\frac{D_{a3}T_s}{2} + 1 = 1 - D_{a3}$$

$$m_{a4} = \frac{-2}{T_s}\frac{D_{a4}T_s}{2} = -D_{a4} \tag{3.6}$$

Hence, the negative amplitude modulation signal for the three-level diode-clamped inverter is:

$$m = m_{a3} + m_{a4} = 1 - D_{a3} - D_{a4} = 1 - S_{a3} - S_{a4} \tag{3.7}$$

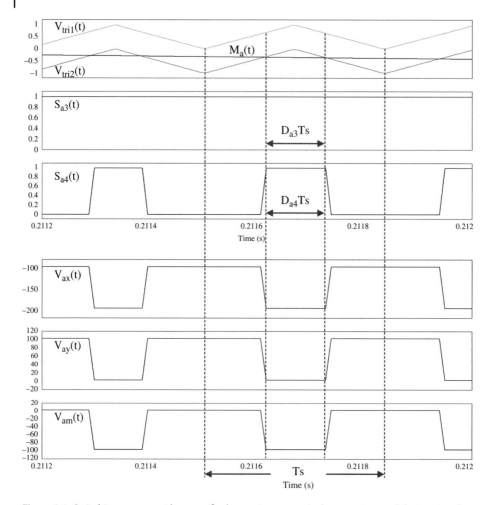

Figure 3.3 Switching pattern with a specific duty ratio at a particular negative modulation signal.

For the three-level flying capacitor inverter, the negative amplitude modulation signal is:

$$m = m_{a3} + m_{a4} = 1 - D_{a3} - D_{a4} = 1 - S_{a3} - S_{a4} \tag{3.8}$$

From the above details, the output pole voltage is derived from the two consecutive equations as:

$$
\begin{cases}
V_{sx}(t) = V_{xm}[S_{s1}(t) + S_{s2}(t)] - V_{dc}(t) = \dfrac{V_{dc}(t)}{2}[S_{s1}(t) + S_{s2}(t)] - V_{dc}(t) \\[2mm]
V_{sy}(t) = V_{ym}[S_{s3}(t) + S_{s4}(t)] + V_{dc}(t) = -\dfrac{V_{dc}(t)}{2}[S_{s3}(t) + S_{s4}(t)] + V_{dc}(t) \\[2mm]
V_{sm}(t) = \dfrac{V_{sx}(t) + V_{sy}(t)}{2} = \dfrac{V_{xm}}{2}[S_{s1}(t) + S_{s2}(t)] + \dfrac{V_{ym}}{2}[S_{s3}(t) + S_{s4}(t)] \\[2mm]
\qquad\quad = \dfrac{V_{dc}(t)}{4}\{[S_{s1}(t) + S_{s2}(t)] - [S_{s3}(t) + S_{s4}(t)]\}
\end{cases} \tag{3.9}
$$

$V_{sx}(t)$ and $V_{sy}(t)$ are the voltages occurring at the lower and upper terminal of the three-level diode-clamped inverter with respect to node "x" and "y," respectively, whereas $V_{sm}(t)$ is the phase voltage for the individual leg of the inverter with respect to the neutral point "m." From equations (3.1) and (3.4), the average phase voltage $\bar{v}_{sm}(t)$ is regulated between the duty ratio and the amplitude of the modulating signals. If m_s changes from -1 to 1, then the average phase voltage $\bar{v}_{sm}(t)$ changes linearly from $-V_{dc}/2$ to $+V_{dc}/2$. Thus, the average phase voltage $\bar{v}_{sm}(t)$ is known as:

$$
\begin{aligned}
\bar{v}_{sm}(t) &= \frac{V_{dc}}{4}\{[S_{s1}(t) + S_{s2}(t)] - [1 - S_{s1}(t) + 1 - S_{s2}(t)]\} \\
&= \frac{V_{dc}}{2}[S_{s1}(t) + S_{s2}(t) - 1] \\
&= \frac{V_{dc}}{2}m_s
\end{aligned}
\tag{3.10}
$$

where $\bar{v}_{sm}(t)$ is the individual average phase voltage ($\bar{v}_{am}(t)$, $\bar{v}_{bm}(t)$, and $\bar{v}_{cm}(t)$) and m_s is the modulation signal (m_a, m_b, and m_c) as stated in equation (3.4). For the simplicity of control, a sinusoidal voltage/current is given at the ac terminal of the three-level diode-clamped inverter. The control loop must be able to change the amplitude of a modulation signal with sinusoidal time function with the angular operating frequency ωt and the initial phase angle θ_0. Thus, the phase voltages of the three-phase terminal are expressed as:

$$
\begin{cases}
V_{am}(t) = \dfrac{V_{dc}}{2}M_a(t) = \dfrac{V_{dc}m\cos(\omega t + \theta_0)}{2} \\[2mm]
V_{bm}(t) = \dfrac{V_{dc}}{2}M_b(t) = \dfrac{V_{dc}m\cos\left(\omega t + \theta_0 - \frac{2\pi}{3}\right)}{2} \\[2mm]
V_{cm}(t) = \dfrac{V_{dc}}{2}M_c(t) = \dfrac{V_{dc}m\cos\left(\omega t + \theta_0 - \frac{4\pi}{3}\right)}{2}
\end{cases}
\tag{3.11}
$$

In Table 3.1, the DC terminal current is determined by the switching states and both node "x" and node "y" currents. If the switching pattern occurs in state 1, $I_y(t) = 0A$, and

Table 3.1 Three-level diode-clamped inverter corresponding switching states.

State	Switching states				Per-phase leg voltage		
	S_{s1}	S_{s2}	S_{s3}	S_{s4}	V_{sx}	V_{sy}	V_{sm}
1	1	1	0	0	0	V_{dc}	$V_{dc}/2$
2	0	1	1	0	$-V_{dc}/2$	$V_{dc}/2$	0
3	0	0	1	1	$-V_{dc}$	0	$-V_{dc}/2$

"s" represents phases a, b, and c and S_{s1} to S_{s4} are presented as the IGBT switching devices for an individual leg. Logic 1 represents "turn on" and logic 0 represents "turn off" for S_{s1}, S_{s2}, S_{s3}, and S_{s4}.

$I_x(t)$ is expressed as:

$$I_x(t) = \begin{bmatrix} S_{a1}I_a(t)\text{sgn}(M_a(t)) \\ +S_{b1}I_b(t)\text{sgn}(M_b(t)) \\ +S_{c1}I_c(t)\text{sgn}(M_c(t)) \end{bmatrix} \tag{3.12}$$

If the switching pattern occurs in state 3, Ix(t) = 0A, and Iy(t) is formulated as:

$$I_y(t) = -\begin{bmatrix} S_{a4}I_a(t)\text{sgn}(-M_a(t)) \\ +S_{b4}I_b(t)\text{sgn}(-M_b(t)) \\ +S_{c4}I_c(t)\text{sgn}(-M_c(t)) \end{bmatrix} \tag{3.13}$$

where the sgn(.) function is defined as:

$$\text{sgn}(M_s(t)) = \begin{cases} 1, & \text{if } M_s(t) \geq 0 \\ 0, & \text{otherwise} \end{cases} \tag{3.14}$$

Applying the same modulation analysis as stated in equations (3.2) to (3.7) into equations (3.12) and (3.13), one can conclude that the dc current is changed by the modulation signal, which is shown in the following equation:

$$I_x(t) = \begin{bmatrix} I_a(t)M_a(t)\text{sgn}(M_a(t)) \\ +I_b(t)M_b(t)\text{sgn}(M_b(t)) \\ +I_c(t)M_c(t)\text{sgn}(M_c(t)) \end{bmatrix}$$

$$I_y(t) = \begin{bmatrix} I_a(t)M_a(t)\text{sgn}(-M_a(t)) \\ +I_b(t)M_b(t)\text{sgn}(-M_b(t)) \\ +I_c(t)M_c(t)\text{sgn}(-M_c(t)) \end{bmatrix} \tag{3.15}$$

From equation (3.15), the neutral point clamped current is formulated by KCL and it is shown in equations (3.16) and (3.17). The neutral point clamped current consists of two current components, one is the three-phase current at the ac terminal $I_a(t) + I_b(t) + I_c(t)$, and the other is the harmonic dc injection current $fcn_a(t) + fcn_b(t) + fcn_c(t)$. The derivation of this neutral point current will be further elaborated in the following equations:

$$I_m(t) = I_a(t) + I_b(t) + I_c(t) - I_x(t) + I_y(t) \tag{3.16}$$

$$I_m(t) = [I_a(t) + I_b(t) + I_c(t)] - [fcn_a(t) + fcn_b(t) + fcn_c(t)] \tag{3.17}$$

where $I_a(t)$, $I_b(t)$, and $I_c(t)$ are the output current of a three-phase three-level diode-clamped inverter with the power factor angle of δ and $fcn_a(t)$, $fcn_b(t)$, and $fcn_c(t)$ are the dc current functions with respect to the three-phase ac terminal current and modulation index, respectively. Both current equations are written as:

$$I_a(t) = i\cos(\omega t + \theta_0 - \delta)$$

$$I_b(t) = i\cos\left(\omega t + \theta_0 - \delta - \frac{2\pi}{3}\right)$$

$$I_c(t) = i\cos\left(\omega t + \theta_0 - \delta - \frac{4\pi}{3}\right) \tag{3.18}$$

$$fcn_a(t) = I_a(t)M_a(t)[sgn(M_a(t)) - sgn(-M_a(t))]$$
$$fcn_b(t) = I_b(t)M_b(t)[sgn(M_b(t)) - sgn(-M_b(t))]$$
$$fcn_c(t) = I_c(t)M_c(t)[sgn(M_c(t)) - sgn(-M_c(t))] \qquad (3.19)$$

where $sgn(.) - sgn(-.)$ in one cycle is defined as:

$$sgn(.) - sgn(-.) = \begin{cases} -1, & -\pi \le t < -\dfrac{\pi}{2} \\[2mm] +1, & -\dfrac{\pi}{2} \le t \le \dfrac{\pi}{2} \\[2mm] -1, & \dfrac{\pi}{2} < t \le \pi \end{cases} \qquad (3.20)$$

In equation (3.19), $sgn(.) - sgn(-.)$ is extended into a simple Fourier series as follows:

$$\begin{cases} sgn(M_a(t)) - sgn(-M_a(t)) = \dfrac{1}{\pi} \displaystyle\sum_{n=1,2,3,...}^{\infty} \left\{ \int_{-\pi}^{\pi} \begin{bmatrix} sgn(M_a(t)) \\ -sgn(-M_a(t)) \end{bmatrix} cos(nt)dt \right\} cos[n(\omega t + \theta_0)] \\[4mm] sgn(M_b(t)) - sgn(-M_b(t)) = \dfrac{1}{\pi} \displaystyle\sum_{n=1,2,3,...}^{\infty} \left\{ \int_{-\pi}^{\pi} \begin{bmatrix} sgn(M_b(t)) \\ -sgn(-M_b(t)) \end{bmatrix} cos(nt)dt \right\} cos\left[n\left(\omega t + \theta_0 - \dfrac{2\pi}{3}\right)\right] \\[4mm] sgn(M_c(t)) - sgn(-M_c(t)) = \dfrac{1}{\pi} \displaystyle\sum_{n=1,2,3,...}^{\infty} \left\{ \int_{-\pi}^{\pi} \begin{bmatrix} sgn(M_c(t)) \\ -sgn(-M_c(t)) \end{bmatrix} cos(nt)dt \right\} cos\left[n\left(\omega t + \theta_0 - \dfrac{4\pi}{3}\right)\right] \end{cases}$$

$$\begin{cases} sgn(M_a(t)) - sgn(-M_a(t)) = \dfrac{4}{\pi} \displaystyle\sum_{n=1,3,5,...}^{\infty} \dfrac{1}{n} cos(n(\omega t + \theta_0)) sin\left(\dfrac{n\pi}{2}\right) \\[4mm] sgn(M_b(t)) - sgn(-M_b(t)) = \dfrac{4}{\pi} \displaystyle\sum_{n=1,3,5,...}^{\infty} \dfrac{1}{n} cos\left(n\left(\omega t + \theta_0 - \dfrac{2\pi}{3}\right)\right) sin\left(\dfrac{n\pi}{2}\right) \\[4mm] sgn(M_c(t)) - sgn(-M_c(t)) = \dfrac{4}{\pi} \displaystyle\sum_{n=1,3,5,...}^{\infty} \dfrac{1}{n} cos\left(n\left(\omega t + \theta_0 - \dfrac{4\pi}{3}\right)\right) sin\left(\dfrac{n\pi}{2}\right) \end{cases}$$

$$(3.21)$$

Substituting equations (3.14) and (3.17) into equation (3.15), one can obtain

$$fcn_a(t) = \dfrac{mi}{\pi} \begin{bmatrix} 2\cos\delta \displaystyle\sum_{n=1,3,5,...}^{\infty} \dfrac{sin\left(\frac{n\pi}{2}\right) cos(n(\omega t + \theta_0))}{n} \\[4mm] + \displaystyle\sum_{n=1,3,5,...}^{\infty} \dfrac{sin\left(\frac{n\pi}{2}\right) cos((n-2)(\omega t + \theta_0) + \delta)}{n} \\[4mm] + \displaystyle\sum_{n=1,3,5,...}^{\infty} \dfrac{sin\left(\frac{n\pi}{2}\right) cos((n+2)(\omega t + \theta_0) - \delta)}{n} \end{bmatrix}$$

$$fcn_b(t) = \frac{mi}{\pi} \begin{bmatrix} 2\cos\delta \sum_{n=1,3,5,\ldots}^{\infty} \frac{\sin\left(\frac{n\pi}{2}\right)\cos\left(n\left(\omega t + \theta_0 - \frac{2\pi}{3}\right)\right)}{n} \\ + \sum_{n=1,3,5,\ldots}^{\infty} \frac{\sin\left(\frac{n\pi}{2}\right)\cos\left((n-2)\left(\omega t + \theta_0 - \frac{2\pi}{3}\right) + \delta\right)}{n} \\ + \sum_{n=1,3,5,\ldots}^{\infty} \frac{\sin\left(\frac{n\pi}{2}\right)\cos\left((n+2)\left(\omega t + \theta_0 - \frac{2\pi}{3}\right) - \delta\right)}{n} \end{bmatrix} \quad (3.22)$$

$$fcn_c(t) = \frac{mi}{\pi} \begin{bmatrix} 2\cos\delta \sum_{n=1,3,5,\ldots}^{\infty} \frac{\sin\left(\frac{n\pi}{2}\right)\cos\left(n\left(\omega t + \theta_0 - \frac{4\pi}{3}\right)\right)}{n} \\ + \sum_{n=1,3,5,\ldots}^{\infty} \frac{\sin\left(\frac{n\pi}{2}\right)\cos\left((n-2)\left(\omega t + \theta_0 - \frac{4\pi}{3}\right) + \delta\right)}{n} \\ + \sum_{n=1,3,5,\ldots}^{\infty} \frac{\sin\left(\frac{n\pi}{2}\right)\cos\left((n+2)\left(\omega t + \theta_0 - \frac{4\pi}{3}\right) - \delta\right)}{n} \end{bmatrix}$$

If the ac terminal of the inverter is an underbalanced condition, then the first term of the neutral point current equation as stated in equation (3.17) is neglected and the neutral point clamped current is equal to the second component current as stated in the following equation:

$$I_m(t) = -[fcn_a(t) + fcn_b(t) + fcn_c(t)] \quad (3.23)$$

$$I_m(t) = -\frac{3mi}{\pi} \begin{bmatrix} 2\cos\delta \sum_{n=3,9,15,\ldots}^{\infty} \frac{\sin\left(\frac{n\pi}{2}\right)\cos(n(\omega t + \theta_0))}{n} \\ - \sum_{n=3,9,15,\ldots}^{\infty} \frac{\sin\left(\frac{n\pi}{2}\right)\cos(n(\omega t + \theta_0) + \delta)}{n+2} \\ - \sum_{n=3,9,15,\ldots}^{\infty} \frac{\sin\left(\frac{n\pi}{2}\right)\cos(n(\omega t + \theta_0) - \delta)}{n-2} \end{bmatrix} \quad (3.24)$$

Substituting equation (3.21) into (3.23), the neutral point current is obtained in equation (3.24). As a result, the three-level diode-clamped inverter will experience triplen harmonic currents and zero dc components in the neutral point clamped as shown in Figs. 3.4 and 3.5. Hence, to obtain a balanced dc voltage across the dc capacitors, the dc components at the node of two series connected dc capacitor must be zero.

Based on equation (3.24), the high harmonic current is dominated by third order harmonics. This is expressed in a simplified equation based on trigonometric properties expressed in the following equation:

$$I_m(t) \approx \frac{12mi}{5\pi} \left[-\frac{2}{3}\cos(\delta)\cos(3(\omega t + \theta_0)) - \sin(\delta)\sin(3(\omega t + \theta_0)) \right] \quad (3.25)$$

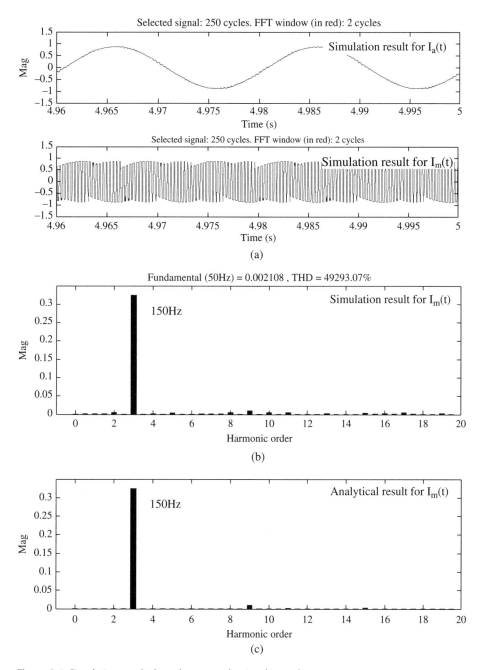

Figure 3.4 Simulation results based on neutral point clamped current.

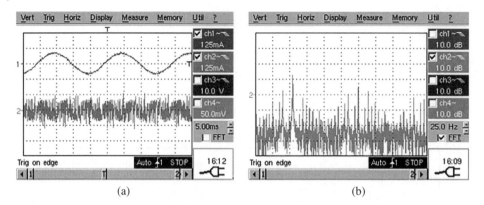

Figure 3.5 Experimental results based on neutral point clamped current.

From the above equation (3.25), the neutral point current depends on the load angle which is also known as the power factor angle (δ). Hence, the neutral point clamped current is regulated within $0 \leq \delta \leq 90°$ and this is presented as:

$$[0.5093mi]\cos(3\omega t + 3\theta_0 - \pi) \leq I_m(t) \leq [0.7639mi]\sin(3\omega t + 3\theta_0 - \pi) \quad (3.26)$$

The mathematical expression of I_m in Equation (3.26) is defined to determine the maximum current required for designing the PCB trace current capacity.

3.3 Three-Level Flying-Capacitor Inverter Topology

A three-phase three-level flying-capacitor inverter is shown in Fig. 3.6 and this type of topology usually requires 12 identical IGBT devices and a minimum of four dc capacitors to generate a three-level output voltage. The switching strategy for this capacitor-clamped inverter is similar to that of the three-level diode-clamped inverter as presented in Fig. 3.1 and the switching states for the upper leg and lower leg per phase is depicted in Table 3.2. To generate all gating signals for the safety operation, the

Table 3.2 Three-level flying capacitor inverter corresponding switching states.

State	Switching states				Per phase leg voltage		
	Ss1	Ss2	Ss3	Ss4	Vsx	Vsy	Vsm
1	1	1	0	0	0	Vdc	Vdc/2
2	0	1	0	1	-Vdc/2	Vdc/2	0
3	0	0	1	1	-Vdc	0	-Vdc/2

"s" represents phases a, b, and c and S_{s1} to S_{s4} represent the IGBT switching devices for an individual leg. Logic 1 represents "turn on" and logic 0 represents "turn off" for S_{s1}, S_{s2}, S_{s3}, and S_{s4}.

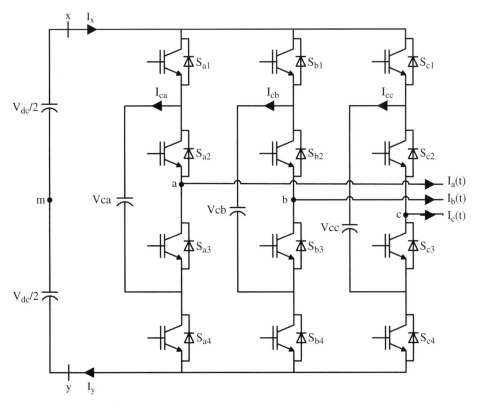

Figure 3.6 Three-level flying-capacitor inverter circuit.

switching device per phase leg is expressed in the following equation:

$$\begin{cases} S_{a1} + S_{a4} = 1 \\ S_{a2} + S_{a3} = 1 \end{cases} \tag{3.27}$$

The output phase voltage of a three-level flying capacitor inverter is expressed in terms of the switching functions, similar to that in Section 3.2. Since both topologies employ the LSPWM technique, the output voltage expression of the flying capacitor is the same as that of the diode-clamped inverter, and it is:

$$\begin{cases} V_{sx}(t) = V_{xm}[S_{s1}(t) + S_{s2}(t)] - V_{dc}(t) = \dfrac{V_{dc}(t)}{2}[S_{s1}(t) + S_{s2}(t)] - V_{dc}(t) \\[3mm] V_{sy}(t) = V_{ym}[S_{s3}(t) + S_{s4}(t)] + V_{dc}(t) = -\dfrac{V_{dc}(t)}{2}[S_{s3}(t) + S_{s4}(t)] + V_{dc}(t) \\[3mm] V_{sm}(t) = \dfrac{V_{sx}(t) + V_{sy}(t)}{2} = \dfrac{V_{xm}}{2}[S_{s1}(t) + S_{s2}(t)] + \dfrac{V_{ym}}{2}[S_{s3}(t) + S_{s4}(t)] \\[3mm] \qquad\quad = \dfrac{V_{dc}(t)}{4}\{[S_{s1}(t) + S_{s2}(t)] - [S_{s3}(t) + S_{s4}(t)]\} \end{cases}$$

$$\tag{3.28}$$

Based on the switching states given in Table 3.2, the instantaneous upper and lower dc current are solely dependent on the switching functions of that particular switch. For a stable and balanced floating capacitor voltage, the instantaneous dc current for the upper and lower terminal ($I_x(t)$ and $I_y(t)$) must flow complementing each other (i.e. $I_x(t)$ = const, $I_y(t) = 0$ and $I_x(t) = 0$, $I_y(t)$ = const). Thus, the dc link current is known as:

$$\begin{cases} I_x(t) = \begin{bmatrix} S_{a1}I_a(t)\mathrm{sgn}(M_a(t)) \\ +S_{b1}I_b(t)\mathrm{sgn}(M_b(t)) \\ +S_{c1}I_c(t)\mathrm{sgn}(M_c(t)) \end{bmatrix} \\[3em] I_y(t) = \begin{bmatrix} S_{a3}I_a(t)\mathrm{sgn}(-M_a(t)) \\ +S_{b3}I_b(t)\mathrm{sgn}(-M_b(t)) \\ +S_{c3}I_c(t)\mathrm{sgn}(-M_c(t)) \end{bmatrix} \end{cases} \tag{3.29}$$

The inner capacitor current of phases a, b, and c ($I_{ca}(t)$, $I_{cb}(t)$ and $I_{cc}(t)$) of a three-level capacitor clamped inverter is defined by the current division law and the superposition theorem. Fig. 3.7 shows the equivalent circuit for the derivation of the inner capacitor current of phase a.

From Fig. 3.7, the inner capacitor current is found to be

$$\begin{cases} I_{ca}(t) = I_{ca1}(t) + I_{ca2}(t) \\ I_{cb}(t) = I_{cb1}(t) + I_{cb2}(t) \\ I_{cc}(t) = I_{cc1}(t) + I_{cc2}(t) \end{cases} \tag{3.30}$$

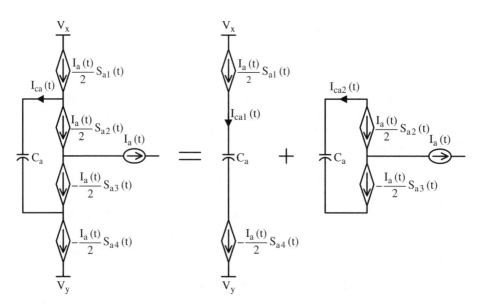

Figure 3.7 Equivalent circuit of phase "a" three-level flying capacitor inverter with the derivation of an inner capacitor current of phase "a".

On rearranging equation (3.30) in terms of the switching function, we get:

$$
\begin{cases}
I_{ca}(t) = \dfrac{I_a(t)}{2}[S_{a1}(t) - S_{a2}(t) + S_{a3}(t) - S_{a4}(t)] \\[4pt]
\qquad = I_a(t)[-1 + S_{a1}(t)\mathrm{sgn}(M_a(t)) + S_{a3}(t)\mathrm{sgn}(-M_a(t))] \\[6pt]
I_{cb}(t) = \dfrac{I_b(t)}{2}[S_{b1}(t) - S_{b2}(t) + S_{b3}(t) - S_{b4}(t)] \\[4pt]
\qquad = I_b(t)[-1 + S_{b1}(t)\mathrm{sgn}(M_b(t)) + S_{b3}(t)\mathrm{sgn}(-M_b(t))] \\[6pt]
I_{cc}(t) = \dfrac{I_c(t)}{2}[S_{c1}(t) - S_{c2}(t) + S_{c3}(t) - S_{c4}(t)] \\[4pt]
\qquad = I_c(t)[-1 + S_{c1}(t)\mathrm{sgn}(M_c(t)) + S_{c3}(t)\mathrm{sgn}(-M_c(t))]
\end{cases}
\tag{3.31}
$$

On simplifying equation (3.31),

$$
\begin{cases}
I_{ca}(t) = I_a(t)\{-1 + M_a(t)[\mathrm{sgn}(M_a(t)) - \mathrm{sgn}(-M_a(t))]\} \\
I_{cb}(t) = I_b(t)\{-1 + M_b(t)[\mathrm{sgn}(M_b(t)) - \mathrm{sgn}(-M_b(t))]\} \\
I_{cc}(t) = I_c(t)\{-1 + M_c(t)[\mathrm{sgn}(M_c(t)) - \mathrm{sgn}(-M_c(t))]\}
\end{cases}
$$

$$
\begin{cases}
I_{ca}(t) = -I_a(t) + \mathrm{fcn}_a(t) \\
I_{cb}(t) = -I_b(t) + \mathrm{fcn}_b(t) \\
I_{cc}(t) = -I_c(t) + \mathrm{fcn}_c(t)
\end{cases}
\tag{3.32}
$$

The functions $\mathrm{fcn}_a(t)$, $\mathrm{fcn}_b(t)$, and $\mathrm{fcn}_c(t)$ are the same as discussed in equation (3.18). Hence, equation (3.32) can be extended to a compact Fourier series form. This is expressed as the following equation:

$$
I_{ca}(t) = -i\cos(\omega t + \theta_0 - \delta) + \frac{mi}{\pi}
\begin{bmatrix}
2\cos\delta \displaystyle\sum_{n=1,3,5,\ldots}^{\infty} \dfrac{\sin\left(\frac{n\pi}{2}\right)\cos(n(\omega t + \theta_0))}{n} \\[12pt]
+ \displaystyle\sum_{n=1,3,5,\ldots}^{\infty} \dfrac{\sin\left(\frac{n\pi}{2}\right)\cos((n-2)(\omega t + \theta_0) + \delta)}{n} \\[12pt]
+ \displaystyle\sum_{n=1,3,5,\ldots}^{\infty} \dfrac{\sin\left(\frac{n\pi}{2}\right)\cos((n+2)(\omega t + \theta_0) - \delta)}{n}
\end{bmatrix}
$$

$$
I_{cb}(t) = -i\cos\left(\omega t + \theta_0 - \frac{2\pi}{3} - \delta\right) + \frac{mi}{\pi}
\begin{bmatrix}
2\cos\delta \displaystyle\sum_{n=1,3,5,\ldots}^{\infty} \dfrac{\sin\left(\frac{n\pi}{2}\right)\cos\left(n\left(\omega t + \theta_0 - \frac{2\pi}{3}\right)\right)}{n} \\[12pt]
+ \displaystyle\sum_{n=1,3,5,\ldots}^{\infty} \dfrac{\sin\left(\frac{n\pi}{2}\right)\cos\left((n-2)\left(\omega t + \theta_0 - \frac{2\pi}{3}\right) + \delta\right)}{n} \\[12pt]
+ \displaystyle\sum_{n=1,3,5,\ldots}^{\infty} \dfrac{\sin\left(\frac{n\pi}{2}\right)\cos\left((n+2)\left(\omega t + \theta_0 - \frac{2\pi}{3}\right) - \delta\right)}{n}
\end{bmatrix}
$$

$$I_{cc}(t) = -i\cos\left(\omega t + \theta_0 - \frac{4\pi}{3} - \delta\right) + \frac{mi}{\pi}\left[\begin{array}{c} 2\cos\delta \displaystyle\sum_{n=1,3,5,\ldots}^{\infty} \dfrac{\sin\left(\frac{n\pi}{2}\right)\cos\left(n\left(\omega t + \theta_0 - \frac{4\pi}{3}\right)\right)}{n} \\[3mm] + \displaystyle\sum_{n=1,3,5,\ldots}^{\infty} \dfrac{\sin\left(\frac{n\pi}{2}\right)\cos\left((n-2)\left(\omega t + \theta_0 - \frac{4\pi}{3}\right) + \delta\right)}{n} \\[3mm] + \displaystyle\sum_{n=1,3,5,\ldots}^{\infty} \dfrac{\sin\left(\frac{n\pi}{2}\right)\cos\left((n+2)\left(\omega t + \theta_0 - \frac{4\pi}{3}\right) - \delta\right)}{n} \end{array}\right]$$

$$(3.33)$$

Equation (3.33) shows a general expression for the inner capacitor current pertaining to each phase of the three-level flying capacitor inverter. Based on equation (3.33), the inner capacitor current consists of the fundamental and odd harmonic components. To simplify the equation, the inner capacitor current is determined by the dominant current component that is shown in Figs. 3.8 and 3.9. Therefore, the inner current is written as:

$$I_{ca}(t) \approx -i\cos(\omega t + \theta_0 - \delta) + \frac{mi}{\pi}\left[\begin{array}{l} 2\cos(\omega t + \theta_0 - \delta) \\[2mm] +\dfrac{2}{3}\cos(\omega t + \theta_0 + \delta) \\[2mm] +\dfrac{2}{3}\cos[3(\omega t + \theta_0) - \delta] \\[2mm] -\dfrac{1}{3}\cos[3(\omega t + \theta_0) + \delta] \end{array}\right]$$

$$I_{cb}(t) \approx -i\cos\left(\omega t + \theta_0 - \frac{2\pi}{3} - \delta\right) + \frac{mi}{\pi}\left[\begin{array}{l} 2\cos\left(\omega t + \theta_0 - \dfrac{2\pi}{3} - \delta\right) \\[2mm] +\dfrac{2}{3}\cos\left(\omega t + \theta_0 - \dfrac{2\pi}{3} + \delta\right) \\[2mm] +\dfrac{2}{3}\cos\left[3\left(\omega t + \theta_0 - \dfrac{2\pi}{3}\right) - \delta\right] \\[2mm] -\dfrac{1}{3}\cos\left[3\left(\omega t + \theta_0 - \dfrac{2\pi}{3}\right) + \delta\right] \end{array}\right]$$

$$(3.34)$$

$$I_{cc}(t) \approx -i\cos\left(\omega t + \theta_0 - \frac{4\pi}{3} - \delta\right) + \frac{mi}{\pi}\left[\begin{array}{l} 2\cos\left(\omega t + \theta_0 - \dfrac{4\pi}{3} - \delta\right) \\[2mm] +\dfrac{2}{3}\cos\left(\omega t + \theta_0 - \dfrac{4\pi}{3} + \delta\right) \\[2mm] +\dfrac{2}{3}\cos\left[3\left(\omega t + \theta_0 - \dfrac{4\pi}{3}\right) - \delta\right] \\[2mm] -\dfrac{1}{3}\cos\left[3\left(\omega t + \theta_0 - \dfrac{4\pi}{3}\right) + \delta\right] \end{array}\right]$$

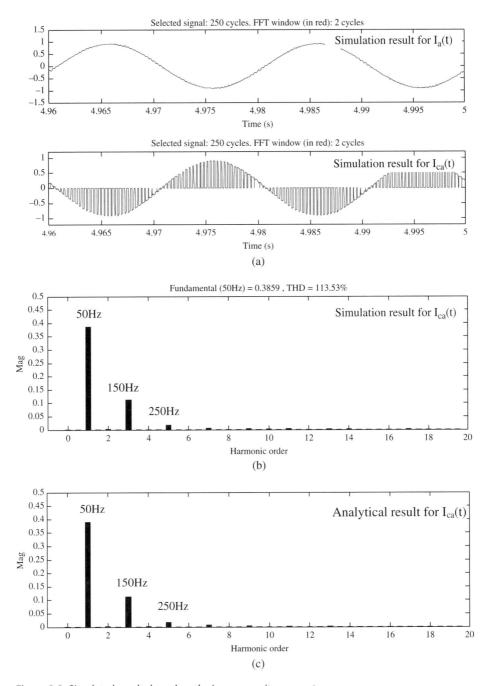

Figure 3.8 Simulated results based on the inner capacitor current.

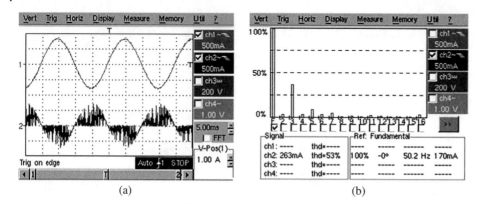

(a)   (b)

Figure 3.9 Experimental results for load current and inner capacitor current.

3.4 Summary

The current expressions derived based on the switching function for the three-level inverters are presented in this chapter. The derived switching function of the LS-PWM is verified with the relationship between the pole voltage and the modulation. After expanding the pole voltage expressions, the final mathematical expression of the modulation is expressed as $2V_{am}/V_{dc}$. In conclusion, the modulation index for any voltage level power converters is similar to the two-level inverter as long as the inverter is operated under the voltage-source inverter (VSI) operation.

Besides that, the amplitude of the low frequency harmonic order of the diode-clamped and flying capacitor is also expressed by expanding the switching function. Analytical results based on the calculation method are verified through the experimental and simulation results. According to the obtained results, the Fourier series analysis only provides the correct amplitude for a low harmonic order. To achieve the exact solution for the amplitude of the remaining high frequency components, a double Fourier series analysis can be adopted in a future study. However, the method of achieving the switching function expression is sufficient for other mathematical analyses such as power loss analysis and stress analysis of the semiconductor devices.

References

1 A. Yazdani and R. Iravani, *Voltage-Sourced Converters In Power Systems: Modeling, Control, and Applications.* IEEE Press/John Wiley, 2010.
2 R. Stala, "Application of balancing circuit for DC-Link voltages balance in a single-phase diode-clamped inverter with two three-level legs," *Industrial Electronics, IEEE Transactions on,* vol. 58: 4185–4195, 2011

4

Voltage Balancing Methods for Classical Multilevel Converters

Gabriel H. P. Ooi, Hossein Dehghani Tafti, and Harikrishna R. Pinkymol

4.1   Introduction

This chapter investigates the different voltage balancing schemes for dc-link capacitors used in three-level inverters. The major unbalanced voltage condition across dc capacitors is caused as a series of connected diodes are clamped to the neutral point of both the series of connected capacitors. This gives rise to an unequal flow of charging and discharging currents in dc capacitors. Besides, in a practical environment, the active/passive elements also contribute to the unbalanced condition due to the existence of unequal parameters. This introduces a dc component in the output voltage spectrum. This high amplitude dc component is carried through the load terminal and results in excessive heating [1].

Voltage balancing with resistance–inductance–capacitance (RLC) elements for a single-phase full bridge three-level diode-clamped inverter (DCI) has been proposed by Stala [2] and an analytical study based on the LC filter for the three-phase three-level DCI has been discussed by Mouton [3]. However, the balancing method in [3] is required to connect the common virtual ground between the LC filter and the resistive load, which does not practically solve any resistive-inductive load (RL load) application issues and does not have any common connected virtual ground. In most of the cases, practical RL loads usually occur in the ac bus, which is known as "electrical drives." Furthermore, the analysis of the RLC elements for the natural balancing properties proposed by Stala [2] has not been discussed in detail. Hence, the next section will introduce the dc voltage balancing method in terms of both the control and the passive filter (RC) balancing method.

4.2   Active Balancing by Adding dc Offset Voltage to Modulating Signals

The circuit configuration of the three-level DCI is shown in Fig. 4.1. The pole voltage across phases "a–c" and the center point "m" will experience an additional dc component $V_o(t)$, and the high frequency components are created by the switching devices. Thus,

Advanced Multilevel Converters and Applications in Grid Integration, First Edition.
Edited by Ali I. Maswood and Hossein Dehghani Tafti.
© 2019 John Wiley & Sons Ltd. Published 2019 by John Wiley & Sons Ltd.

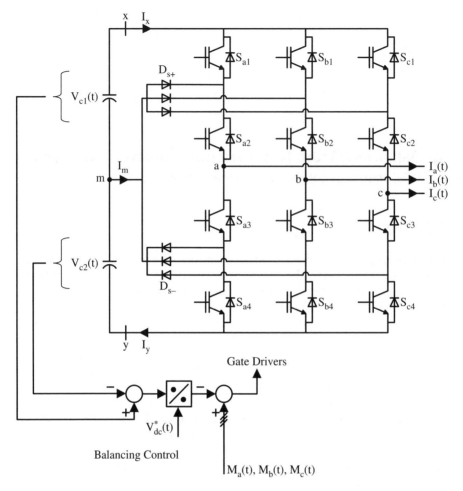

Figure 4.1 Three-phase three-level diode-clamped inverter with dc voltage balancing control without the RC filter.

the pole voltages of the three-phase three-level DCI are expressed as follows:

$$\begin{cases} V_a(t) = V_{am}(t) + V_o(t) \\ V_b(t) = V_{bm}(t) + V_o(t) \\ V_c(t) = V_{cm}(t) + V_o(t) \end{cases} \tag{4.1}$$

If the dc link capacitor voltages are not equal ($V_{c1}(t) \neq V_{c2}(t)$) due to the tolerance value of capacitance, the three-phase pole voltages are shown as:

$$\begin{cases} V_{am}(t) = \dfrac{V_{c1}(t)}{2}[S_{a1}(t) + S_{a2}(t)] - \dfrac{V_{c2}(t)}{2}[2 - S_{a1}(t) - S_{a2}(t)] \\[3mm] V_{bm}(t) = \dfrac{V_{c1}(t)}{2}[S_{b1}(t) + S_{b2}(t)] - \dfrac{V_{c2}(t)}{2}[2 - S_{b1}(t) - S_{b2}(t)] \\[3mm] V_{cm}(t) = \dfrac{V_{c1}(t)}{2}[S_{c1}(t) + S_{c2}(t)] - \dfrac{V_{c2}(t)}{2}[2 - S_{c1}(t) - S_{c2}(t)] \end{cases} \tag{4.2}$$

By replacing the switching function with the modulation function, the phase voltages are illustrated as a linear modulation function as follows:

$$
\begin{cases}
V_{am}(t) \approx \dfrac{V_{c1}(t)}{2}[M_a(t) + 1] - \dfrac{V_{c2}(t)}{2}[1 - M_a(t)] \\[2mm]
V_{bm}(t) \approx \dfrac{V_{c1}(t)}{2}[M_b(t) + 1] - \dfrac{V_{c2}(t)}{2}[1 - M_b(t)] \\[2mm]
V_{cm}(t) \approx \dfrac{V_{c1}(t)}{2}[M_c(t) + 1] - \dfrac{V_{c2}(t)}{2}[1 - M_c(t)]
\end{cases}
\tag{4.3}
$$

Equation (4.3) is derived from the following properties with the linear modulation function under a steady-state condition.

$$
S_{x1}(t) + S_{x2}(t) = S_{x1}(t)\mathrm{sgn}(M_x(t)) - S_{x4}(t)\mathrm{sgn}(-M_x(t)) + 1
\tag{4.4}
$$

Then,

$$
\begin{aligned}
& S_{x1}(t)\mathrm{sgn}(M_x(t)) - S_{x4}(t)\mathrm{sgn}(-M_x(t)) + 1 \\
& \approx M_x(t)\mathrm{sgn}(M_x(t)) + M_x(t)\mathrm{sgn}(-M_x(t)) + 1
\end{aligned}
\tag{4.5}
$$

using $\mathrm{sgn}(M_x(t)) + \mathrm{sgn}(-M_x(t)) = 1$, the updated three-phase modulation signal is shown in the following:

$$
\begin{cases}
M_a^*(t) = 2\dfrac{V_{am}(t)}{V_{dc}(t)} - \dfrac{V_{c1}(t) - V_{c2}(t)}{V_{dc}(t)} = M_a(t) - M_0(t) \\[2mm]
M_b^*(t) = 2\dfrac{V_{bm}(t)}{V_{dc}(t)} - \dfrac{V_{c1}(t) - V_{c2}(t)}{V_{dc}(t)} = M_b(t) - M_0(t) \\[2mm]
M_c^*(t) = 2\dfrac{V_{cm}(t)}{V_{dc}(t)} - \dfrac{V_{c1}(t) - V_{c2}(t)}{V_{dc}(t)} = M_c(t) - M_0(t)
\end{cases}
\tag{4.6}
$$

4.3 Measurement Results for dc Offset Modulation Control

The simulated results shown in Figs. 4.2 and 4.3 are used to evaluate the dc voltage balancing with and without adding the dc offset voltage value while considering various dc capacitor parameters. Fig. 4.2 shows the simulated results of the dc voltage across each of the capacitors, C1 and C2, and their capacitor voltages are balanced at 0.5 s with the proposed balancing control as given in Fig. 4.1. The unbalanced state of the dc voltage without the balancing control is shown in Fig. 4.2. All the results are based on parameters C1 = 1050 μF, R_{ESR1} = 120 mΩ and C2 = 950 μF, and R_{ESR2} = 105 mΩ.

The nominal capacitance value of 1000 μF with a tolerance value of ±20% in Fig. 4.3 shows an extreme unbalanced dc capacitor voltage. Such an unbalanced scheme will tend to inject dc components of a higher value into the ac load as compared with the case shown in Fig. 4.2.

To evaluate the performance of the balancing control, experimental results are shown in Fig. 4.4. Both dc capacitors of 1000 μF value are selected for the laboratory prototype.

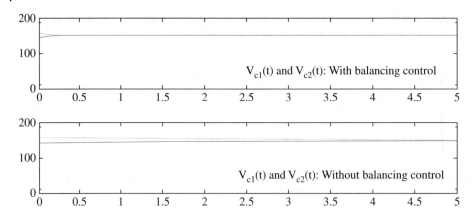

Figure 4.2 Simulation results with the parameters C1= 1050 µF, C2 = 950 µF, R_{ESR} = 120 mΩ, and R_{ESR} = 105 mΩ of the dc voltage across each capacitor $V_{c1}(t)$ and $V_{c2}(t)$.

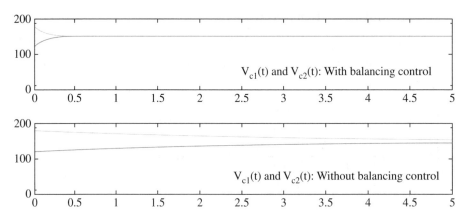

Figure 4.3 Simulation results with the parameters C1= 1200 µF, C2 = 800 µF, R_{ESR} = 110 mΩ, and R_{ESR} = 88 mΩ of the dc voltage across each capacitor $V_{c1}(t)$ and $V_{c2}(t)$.

Fig. 4.4(c) shows the voltage across each capacitor in the dc bus is unbalanced without any balancing properties. With the proposed balancing control, the dc voltage across each capacitor is balanced as shown in Figs. 4.4(a) and (b). To achieve a balanced dc voltage bus, the controller is required to track the voltage error precisely. This requires an accurate measurement from the voltage sensors. Such accuracy voltage sensor is tuned by the gain of the analog amplifier circuit.

The neutral-point-clamped current waveform shown in Figs. 4.5 (simulated results) and 4.6 (experimental results) are almost same as in Fig. 4.4. It is worth noting that the dc voltage for each of the capacitors is not affected by the high frequency harmonic component and third-order harmonic component. With the proposed balancing control, a third harmonic current component will pass through the neutral-point-clamped without any dc current component into the ac load during the switching state 2.

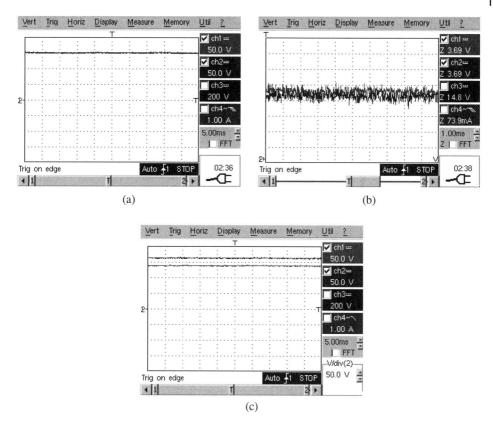

Figure 4.4 Experimental results with the specification of C1 ≈ C2 ≈ ±1.2 × 1000µF and R_{ESR1} ≈ R_{ESR2} ≈ 110 mΩ of the dc capacitor voltages of Vc1(t) and Vc2(t). (a) With balancing modulation control, (b) Zoom in of Fig. 4.4(a), and (c) Without balancing modulation control.

The simulated and experimental results of the pole voltage (phase "a") are illustrated in Figs. 4.7(a) and 4.8(a), respectively. The harmonic spectrum of phase "a" voltage in Fig. 4.7(a) shows that no dc component has been injected into the ac load terminal. The simulated and experimental results of the output load current waveform and its harmonic order are shown in Figs. 4.7(b) and 4.8(b), respectively.

4.4 Natural Balancing by using Star Connected RC Filter

Fig. 4.10 shows the natural balancing technique for dc capacitor voltage with a star-connected passive element for the three-phase three-level DCI. Fig. 4.11 shows the current through the neutral point is based on a zero crossing voltage occurring in either of the phase terminals of a three-level DCI as shown in Fig. 4.9. The variation of both dc-link capacitor voltages is observed by the neutral point current $I_m(t)$ flowing in and out of the node. It can be inferred that the neutral point "m" is a floating point

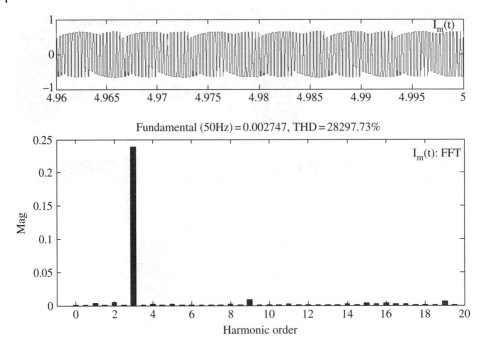

Figure 4.5 Simulation results with the neutral-point-clamped current waveform and FFT analyzer of the two series connected dc capacitor.

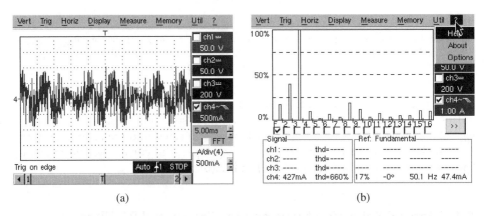

(a)   (b)

Figure 4.6 Experimental results with the neutral-point-clamped current waveform and FFT analyzer of the two series connected dc capacitor. (a) Neutral-point-clamped current waveform, and (b) FFT of the neutral-point-clamped current.

of dc voltage across each dc capacitor and generates a zero sequence output voltage in a three-level DCI. With the natural balancing star-connected RC filter circuit, the zero sequence output voltage can be diverted into the RC filter circuit, whereby the high switching harmonic current will flow through the RC filter. The effect of a few high frequency harmonic currents entering the load terminal can be neglected as the magnitude of such currents is very low. The expression of the output per-phase RC

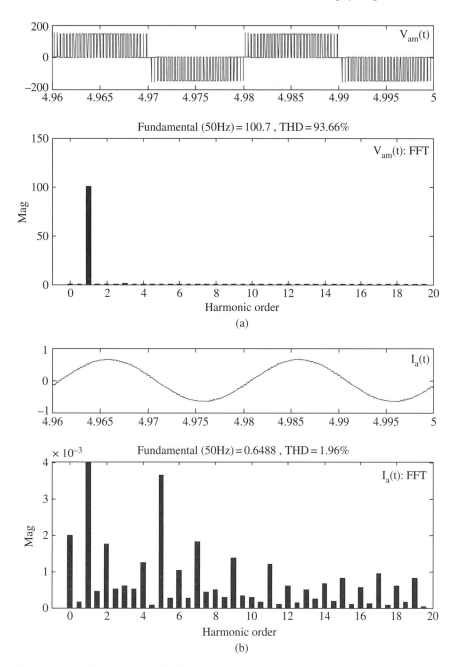

Figure 4.7 Simulation results with the output voltage and current of a three-phase three-level diode-clamped inverter with the voltage balancing controller. (a) Voltage waveform (top figure) and the respective FFT (bottom figure), and (b) Current waveform (top figure) and the respective FFT (bottom figure).

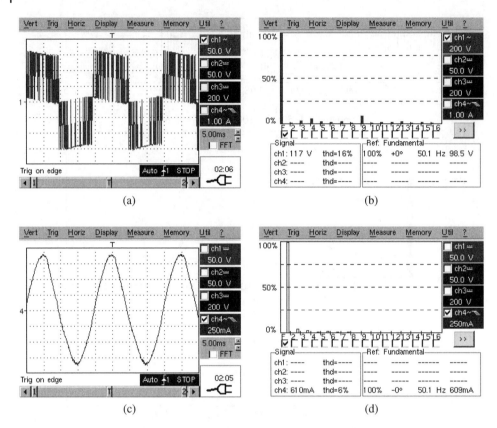

(a)   (b)

(c)   (d)

Figure 4.8 Experimental results with the output voltage and current of a three-phase three-level diode-clamped inverter with the voltage balancing controller. (a) Phase or pole voltage waveform, (b) FFT of the Fig. 4.8(a), (c) Output load current waveform (R = 150 Ω and L = 122 mH), and (d) FFT of Fig. 4.8(c).

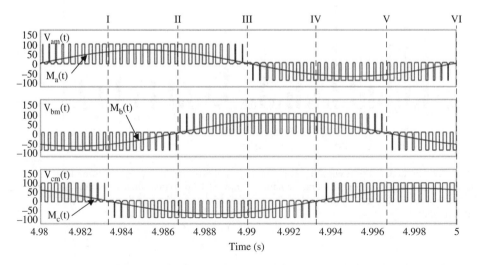

Figure 4.9 Waveform of the output phase voltage and modulation signals [Vam(t), Vbm(t), and Vcm(t)] and [Ma(t), Mb(t), and Mc(t)], respectively, with a zero crossing per-phase leg output voltage analysis.

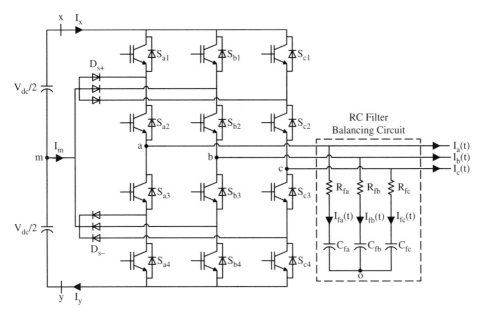

Figure 4.10 Three-phase three-level diode-clamped inverter with a star-connected RC filter balancing circuit.

filter voltages are given in the following equation:

$$
\begin{cases}
V_{ao}(t) = V_{am}(t) - V_{om}(t) \\
V_{bo}(t) = V_{bm}(t) - V_{om}(t) \\
V_{co}(t) = V_{cm}(t) - V_{om}(t)
\end{cases}
\tag{4.7}
$$

where, $V_{om}(t)$ is the zero sequence output voltage between "o" and "m." The zero sequence output voltage with the natural balancing RC filter circuit is expressed as:

$$
V_{om}(t) = \frac{V_{am}(t) + V_{bm}(t) + V_{cm}(t)}{3}
\tag{4.8}
$$

As discussed in the previous sub-section on the balancing control method, the voltage across the RC filter is expressed in terms of the given switching function equation. By substituting equation (4.8) into equation (4.7), the output three phase voltages of the inverter with the star connected with the RC filter is expressed in the following form:

$$
\begin{cases}
V_{ao}(t) = \dfrac{2V_{am}(t) - V_{bm}(t) - V_{cm}(t)}{3} \\[2mm]
V_{bo}(t) = \dfrac{-V_{am}(t) + 2V_{bm}(t) - V_{cm}(t)}{3} \\[2mm]
V_{co}(t) = \dfrac{-V_{am}(t) - V_{bm}(t) + 2V_{cm}(t)}{3}
\end{cases}
\tag{4.9}
$$

On extending equation (4.9) and by substituting equation (4.3), we get

$$
\begin{cases}
V_{ao}(t) \approx \dfrac{1}{3}\left\{ \begin{aligned} &[V_{c1}(t) + V_{c2}(t)]\left[M_a(t) - \frac{1}{2}M_b(t) - \frac{1}{2}M_c(t)\right] \\ &+ [V_{c1}(t) - V_{c2}(t)]\left[1 - \frac{1}{2} - \frac{1}{2}\right] \end{aligned} \right\} \\[4ex]
V_{bo}(t) \approx \dfrac{1}{3}\left\{ \begin{aligned} &[V_{c1}(t) + V_{c2}(t)]\left[M_b(t) - \frac{1}{2}M_a(t) - \frac{1}{2}M_c(t)\right] \\ &+ [V_{c1}(t) - V_{c2}(t)]\left[1 - \frac{1}{2} - \frac{1}{2}\right] \end{aligned} \right\}
\end{cases}
$$

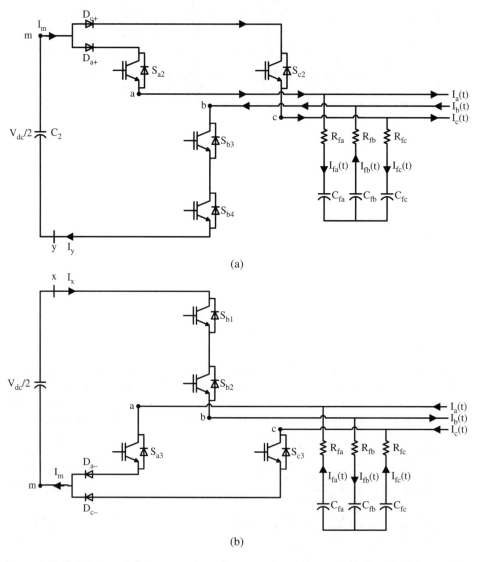

Figure 4.11 Switching state for zero crossing voltage per-phase leg as stated in Fig. 4.9. (a) Intersection points I and VI, (b) intersection point IV, (c) intersection points II and III, and (d) intersection point V.

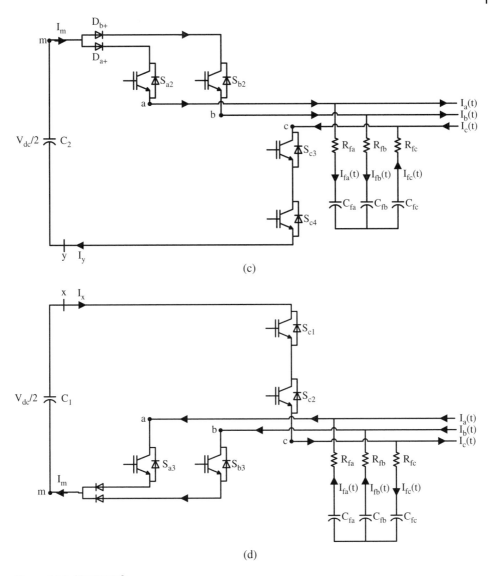

Figure 4.11 (*Continued*)

$$
\left\{ V_{co}(t) \approx \frac{1}{3} \left\{ \begin{array}{l} [V_{c1}(t) + V_{c2}(t)] \left[M_c(t) - \frac{1}{2}M_a(t) - \frac{1}{2}M_b(t) \right] \\ + [V_{c1}(t) - V_{c2}(t)] \left[1 - \frac{1}{2} - \frac{1}{2} \right] \end{array} \right\} \right. \tag{4.10}
$$

Based on equation (4.10), the dc component term on the right hand side of $(V_{C1}(t) - V_{C2}(t))$ is zero. This type of star-connected RC filter is able to perform the self-cancellation of the dc component during the initial period of the dc capacitor charge and discharge. Hence, the output phase voltage of a three-level DCI can operate at the balanced voltage waveform under the steady-state condition.

Since the dc component of the output phase voltage is cancelled by the RC filter property, the voltage across each capacitor of the dc-link terminal is approximately equal. Then, the expression of the output voltage with the RC filter is written as:

$$
\begin{cases}
\begin{aligned}
V_{ao}(t) &\approx \frac{V_{dc}}{3}\left[M_a(t) - \frac{1}{2}M_b(t) - \frac{1}{2}M_c(t)\right] \\
&= \frac{V_{dc}}{3}\left[\begin{matrix} m\cos(\omega t + \theta_0) - \frac{1}{2}m\cos\left(\omega t + \theta_0 - \frac{2\pi}{3}\right) \\ -\frac{1}{2}m\cos\left(\omega t + \theta_0 - \frac{4\pi}{3}\right) \end{matrix}\right] \\[6pt]
V_{bo}(t) &\approx \frac{V_{dc}}{3}\left[M_b(t) - \frac{1}{2}M_a(t) - \frac{1}{2}M_c(t)\right] \\
&= \frac{V_{dc}}{3}\left[\begin{matrix} m\cos\left(\omega t + \theta_0 - \frac{2\pi}{3}\right) - \frac{1}{2}m\cos(\omega t + \theta_0) \\ -\frac{1}{2}m\cos\left(\omega t + \theta_0 - \frac{4\pi}{3}\right) \end{matrix}\right] \\[6pt]
V_{co}(t) &\approx \frac{V_{dc}}{3}\left[M_c(t) - \frac{1}{2}M_a(t) - \frac{1}{2}M_b(t)\right] \\
&= \frac{V_{dc}}{3}\left[\begin{matrix} m\cos\left(\omega t + \theta_0 - \frac{4\pi}{3}\right) - \frac{1}{2}m\cos(\omega t + \theta_0) \\ -\frac{1}{2}m\cos\left(\omega t + \theta_0 - \frac{2\pi}{3}\right) \end{matrix}\right]
\end{aligned}
\end{cases} \tag{4.11}
$$

Based on the above approximation, the dominant harmonic present in the output voltage across the RC filter is the fundamental component without any dc component occurring in the system. The simulated and analytical results are illustrated in Fig. 4.12, and both results show that the fundamental components of the voltage are equal.

The RC filter current is expressed in terms of the rate of change of capacitor voltages and by using Laplace transformation. The final expression of the capacitor voltages of the RC filter network is derived as:

$$
\begin{cases}
V_{Cfa}(t) = \dfrac{V_{ao}(t)}{R_{fa}C_{fa}}e^{\frac{-(t+\theta_0)}{\tau}} \\[10pt]
V_{Cfb}(t) = \dfrac{V_{bo}(t)}{R_{fb}C_{fb}}e^{\frac{-\left(t+\theta_0 - \frac{2\pi}{3}\right)}{\tau}} \\[10pt]
V_{Cfc}(t) = \dfrac{V_{co}(t)}{R_{fc}C_{fc}}e^{\frac{-\left(t+\theta_0 - \frac{4\pi}{3}\right)}{\tau}}
\end{cases} \tag{4.12}
$$

Assume the three-phase RC filter parameters are of equal value, then $R_{fa} = R_{fb} = R_{fc} = R$ and $C_{fa} = C_{fb} = C_{fc} = C$. Thus, (4.12) is rewritten as:

$$
\begin{cases}
V_{Cfa}(t) = \dfrac{V_{ao}(t)}{RC}e^{\frac{-1}{RC}(t+\theta_0)} \\[10pt]
V_{Cfb}(t) = \dfrac{V_{bo}(t)}{RC}e^{\frac{-1}{RC}\left(t+\theta_0 - \frac{2\pi}{3}\right)} \\[10pt]
V_{Cfc}(t) = \dfrac{V_{co}(t)}{RC}e^{\frac{-1}{RC}\left(t+\theta_0 - \frac{4\pi}{3}\right)}
\end{cases} \tag{4.13}
$$

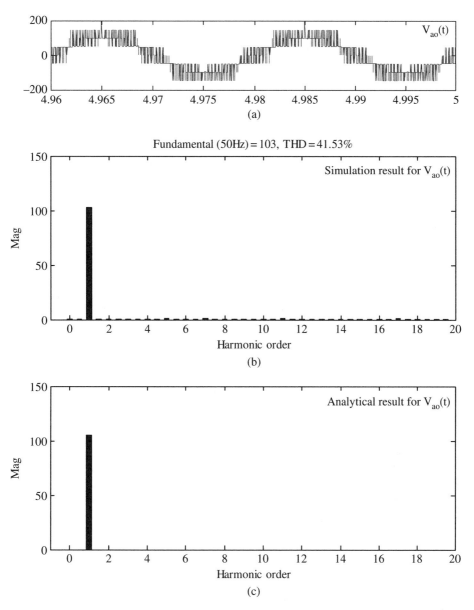

Figure 4.12 RC filter voltage across node "a" to node "o". (a) Simulation result for the instantaneous pole voltage, (b) simulated result for the FFT analysis of the pole voltage, and (c) the analytical result of equation (4.11) in the FFT analysis.

Based on the above assumption, the filter current expression with the periodic time interval is shown in the following:

$$\left\{ I_{fa}(t) = -\frac{V_{dc}m}{2R}e^{-\frac{t+\theta_0}{RC}}\left[\omega\sin(\omega t + \theta_0) + \frac{1}{RC}\cos(\omega t + \theta_0)\right]\right.$$

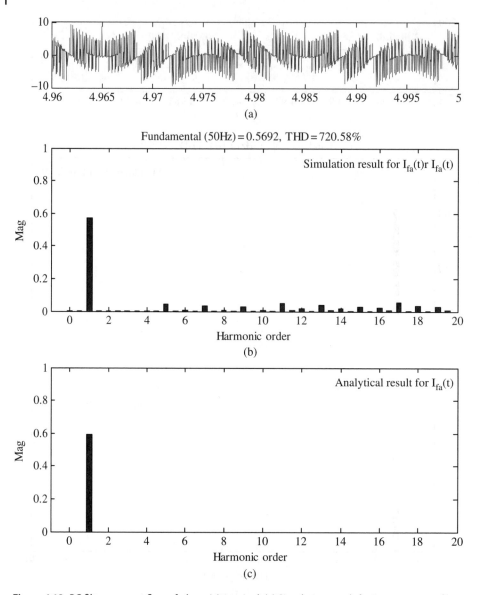

Figure 4.13 RC filters current flow of phase 'a' terminal. (a) Simulation result for instantaneous filter current, (b) simulated result for FFT analysis of the filter current, and (c) the analytical result of equation (4.14) following the FFT analysis.

$$
\begin{cases}
I_{fb}(t) = -\dfrac{V_{dc}m}{2R}e^{-\frac{t+\theta_0-\frac{2\pi}{3}}{RC}}\left[\omega \sin\left(\omega t + \theta_0 - \frac{2\pi}{3}\right) + \frac{1}{RC}\cos\left(\omega t + \theta_0 - \frac{2\pi}{3}\right)\right] \\
I_{fc}(t) = -\dfrac{V_{dc}m}{2R}e^{-\frac{t+\theta_0-\frac{4\pi}{3}}{RC}}\left[\omega \sin\left(\omega t + \theta_0 - \frac{4\pi}{3}\right) + \frac{1}{RC}\cos\left(\omega t + \theta_0 - \frac{4\pi}{3}\right)\right]
\end{cases}
$$

$$(4.14)$$

4.5   Measurement Results for the Natural Balancing Method

The simulated results are used to evaluate the analytical solution of the three-level DCI topology with a mismatch in dc capacitance (C1 = 1200 µF and C2 = 800 µF). The analytical solution of equation (4.14) is shown in Fig. 4.13(c) and the simulated harmonic order of the filter current is shown in Fig. 4.13(b). By comparing both results, the analytical solution shows the order of the dominant harmonic component of the filter current, where high switching frequency components are not included in the Fourier analysis.

Fig. 4.14(a) shows the output pole voltage is stable and balanced under the steady-state condition. Ripple current of the load is low due to the highly inductive load and RC filter as shown in Fig. 4.14 (b). This method can eliminate the additional sensors for hardware implementation and the production cost is also reduced.

To verify the balancing circuit with the RC filter, the experimental results are used to evaluate the output performance of the three-phase/three-level DCI. Fig. 4.15(a) presents the output pole voltage waveform and the respective harmonic order of a three-level DCI. From the obtained results, one can observe that the RC filter circuit has the balancing feature and does not comprise any dc component in the output voltage. Thus, both the capacitor voltages in the dc bus are equally distributed as shown in Figs. 4.14(c) and 4.15(c).

The limitations of the RC filter balancing technique in multilevel inverters are the losses caused by the resistors and the value of the RC parameter is to be selected according to the load power requirement. However, the RC filter balancing technique is a simple and cost-effective solution for obtaining a balanced dc-link voltage.

4.6   Space Vector Modulation with the Self-Balancing Technique

In this section, the space vector modulation, which was introduced in Chapter 2 in Section 2.3, is investigated for the self-balancing of the capacitor voltages in the three-level neutral-point-clamped(NPC) converter. The effect of the switching states on the dc-link capacitor voltages are evaluated and summarized in Table 4.1. The equivalent circuit representation of three-level DCI while switching various vector groups are shown in Fig. 4.9. It can be seen from Fig. 4.9(b) that, when SV with switching state 211 is selected, the load is connected between the positive dc-link terminal and the neutral point which will discharge the top capacitor and charge the bottom capacitor. When SV with a switching state 100 (Fig. 4.9(b)) is selected for switching, the load is distributed between the neutral point and the negative dc-link terminal and the neutral point current i_2 flows in the opposite direction when compared with the switching state 211. Both states generate the same voltage vector; however, they have a contrary effect on the dc-link capacitors. This redundant property of SVs can be used for NPP control. The SV are classified into +ve SV (PSV) and −ve SV (NSV) based on the direction of the NP current generated during switching.

It can be seen from Fig. 4.15 that the SV and MV affect the neutral point current and the other two vector groups do not control the neutral point current. Hence, to control the NPP, the voltage difference between the top and bottom capacitor is obtained and the duty ratio calculated for ZV is distributed among PSV and NSV over

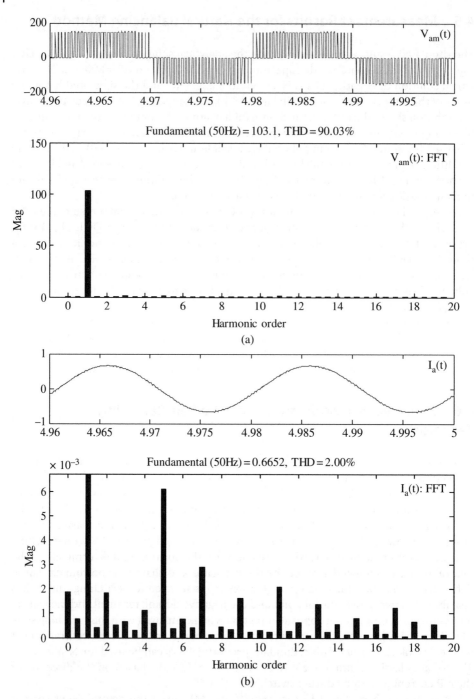

Figure 4.14 Simulation results with the output voltage and current of a three-phase three-level diode-clamped inverter with the balancing RC filter circuit. (a) Voltage waveform (top figure) and the respective FFT (bottom figure), (b) current waveform (top figure) and the respective FFT (bottom figure), and (c) the distribution voltage level of each capacitor in the dc bus.

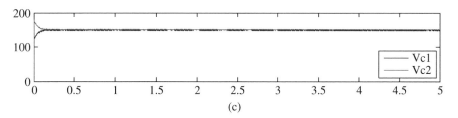

Figure 4.14 (Continued)

Table 4.1 Effect of switching states on dc-link capacitors.

Vector group		Voltage vector	Switching states	Capacitor C_1	Capacitor C_2
ZV		V_0	222,111,000	No effect	
SSV	+veSV (PSV)	V^+_{01}, V^+_{02}, V^+_{03}, V^+_{04}, V^+_{05}, V^+_{06}	211,221,121,122, 112,212	Discharging	Charging
	−ve SV (NSV)	V^-_{01}, V^-_{02}, V^-_{03}, V^-_{04}, V^-_{05}, V^-_{06}	100,110,010,011, 001,101	Charging	Discharging
MV		V_{12}, V_{23}, V_{34}, V_{45}, V_{56}	210,120,021,012, 102,201	Less charging or discharging (depends on load current)	
LV		V_1, V_2, V_3, V_4, V_5, V_6	200,220,020,022, 002,202	No effect	

one sampling time to minimize the voltage error. The selection of redundant vectors and the switching sequence in each of the sub-sectors of sectors A and B over one sampling interval are shown in Table 4.2. In other sectors, the sequence is replaced with equivalent switching vectors.

Let d_{SV} be the relative turn-on time of the SV over one sampling interval T_S. Then, to control the neutral point potential to half of the de-link voltage, d_{SV} is distributed among PSV and NSV as follows through a weighting coefficient k.

$$d_{PSV} = \frac{1+k}{2}d_{SV}$$
$$d_{PSV} = \frac{1-k}{2}d_{SV} \tag{4.14}$$

where $-1 < k < 1$; d_{PSV} and d_{PSV} are the duty ratios of PSV and NSV, respectively.

4.7 Summary

A short description of voltage balancing for a three-level DCI topology is introduced. The unbalanced voltage in the dc link is critical for the load, especially the high voltage stress across each IGBT. This unbalanced phenomenon can cause an even order

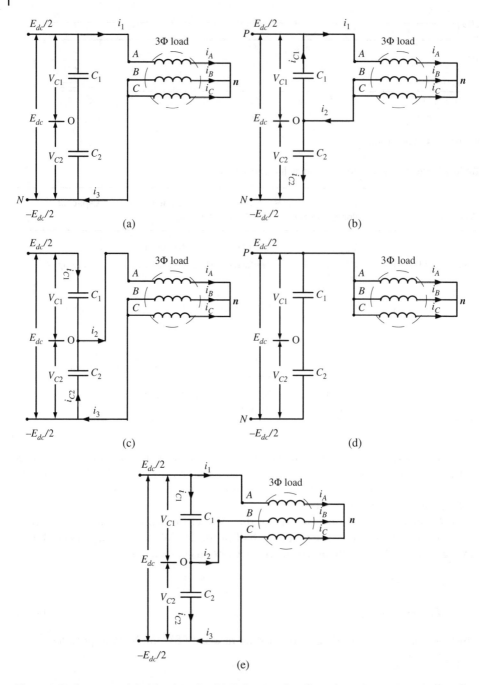

Figure 4.15 Current model of the three-level DCI showing the effect of switching vectors on dc-link capacitors. (a) LV [V_1:200] (b) PSV [V^+_{01}:211] (c) NSV [V^-_{01}:100] (d) ZV [V_0:222] (e) MV [V_{12}:210].

Table 4.2 Switching sequence in Sectors A and B.

Sector	Sub-sector		Switching sequence
A	S-1	$\theta < \dfrac{\pi}{6}$	$V^-_{01}\ V^-_{02}\ V_0\ V^+_{01}\ V_0\ V^-_{02}\ V^-_{01}$
		$\theta > \dfrac{\pi}{6}$	$V^-_{02}\ V_0\ V^+_{01}\ V^+_{02}\ V^+_{01}\ V_0\ V^-_{02}$
	S-2		$V^-_{01}\ V_1\ V_{12}\ V^+_{01}\ V_{12}\ V_1\ V^-_{01}$
	S-3	$\theta < \dfrac{\pi}{6}$	$V^-_{01}\ V^-_{02}\ V_{12}\ V^+_{01}\ V_{12}\ V^-_{02}\ V^-_{01}$
		$\theta > \dfrac{\pi}{6}$	$V^-_{02}\ V_{12}\ V^+_{01}\ V^+_{02}\ V^+_{01}\ V_{12}\ V^-_{02}$
	S-4		$V^-_{02}\ V_{12}\ V_2\ V^+_{02}\ V_2\ V_{12}\ V^-_{02}$
B	S-1	$\theta < \dfrac{\pi}{6}$	$V^-_{02}\ V_0\ V^+_{03}\ V^+_{02}\ V^+_{03}\ V_0\ V^-_{02}$
		$\theta > \dfrac{\pi}{6}$	$V^-_{03}\ V^-_{02}\ V_0\ V^+_{03}\ V_0\ V^-_{02}\ V^-_{03}$
	S-2		$V^-_{02}\ V_{23}\ V_2\ V^+_{02}\ V_2\ V_{23}\ V^-_{02}$
	S-3	$\theta < \dfrac{\pi}{6}$	$V^-_{02}\ V_{23}\ V^+_{03}\ V^+_{02}\ V^+_{03}\ V_{23}\ V^-_{02}$
		$\theta > \dfrac{\pi}{6}$	$V^-_{03}\ V^-_{02}\ V_{23}\ V^+_{03}\ V_{23}\ V^-_{02}\ V^-_{03}$
	S-4		$V^-_{03}\ V_3\ V_{23}\ V^+_{03}\ V_{23}\ V_3\ V^-_{03}$

harmonic distortion in the neutral-point clamp, as well as the output ac terminal, which is severe on the motor winding. Therefore, a voltage balancing in the dc link with the active and passive balancing methods is proposed. By comparing both methods, the active balancing method does not contain any additional loss in the system. But this method requires high resolution for the voltage measurement especially in implementing the DSP. When the dc link voltage level is high, a good measurement of both capacitors must be precisely tuned. However, this active-balancing method does not require any additional cost for the RC filter and it can also reduce the space requirement.

References

1 S. Ogasawara and H. Akagi, "Analysis of variation of neutral point potential in neutral-point-clamped voltage source PWM inverters," in *Industry Applications Society Annual Meeting, 1993, Conference Record of the 1993 IEEE*, pp. 965–970, vol. 2, 1993.
2 R. Stala, "Application of balancing circuit for DC-link voltages balance in a single-phase diode-clamped inverter with two three-level legs," *Industrial Electronics, IEEE Transactions on*, vol. 58, pp. 4185–4195, 2011.
3 H. Du Toit Mouton, "Natural balancing of three-level neutral-point-clamped PWM inverters," *Industrial Electronics, IEEE Transactions on*, vol. 49, pp. 1017–1025, 2002.

Part II

Advanced Multilevel Rectifiers and their Control Strategies

5

Unidirectional Three-Phase Three-Level Unity-Power Factor Rectifier

Gabriel H. P. Ooi and Hossein Dehghani Tafti

5.1 Introduction

This chapter introduces a three-phase three-level unity power factor control rectifier as shown in Fig. 5.1 with the power-balanced control strategy based on decoupling hysteresis current control. Such decoupling current control is the new reference current generated by the grid current control and zero sequence current control. The proposed controller is able to minimize the input current total harmonic distortion (THD) as well as achieving a near unity power factor under an unbalanced and distorted grid voltage condition with a wide load variation. Such a control scheme is designed to maintain a power balance between the ac and the dc bus, a stable input current with low THD, and a fast dc-link voltage tracking. To obtain the least possible THD, the voltage across the capacitors C_1 and C_2 must be balanced. In order to achieve equal dc voltages across both the capacitors for a wide load range, a decoupling hysteresis current control is implemented as shown in Fig. 5.2. The control algorithm is divided into two control loops, an inner loop with a decoupling current control for a balanced capacitor voltage and a dc-link voltage regulator as an outer loop. The detailed theoretical analysis of the controller design is presented in the following sections and is verified by extensive simulation and experimental results.

5.2 Circuit Configuration

The rectifier topology is shown in Fig. 5.1(a): this topology consists of two series of connected capacitors and three bi-directional switches S_b, S_b, and S_c. The bi-directional switch in each phase is constructed with four diodes and an insulated-gate bipolar transistor (IGBT) to accommodate for a bi-directional current flow as can be seen in Fig. 5.1(b) [1, 2]. With reference to Fig. 5.1, the rectifier circuit is examined under a

Advanced Multilevel Converters and Applications in Grid Integration, First Edition.
Edited by Ali I. Maswood and Hossein Dehghani Tafti.

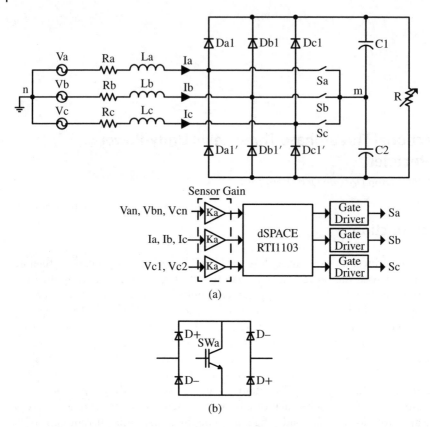

Figure 5.1 A three-phase three-level unity-PF control rectifier with unidirectional power flow. (a) Rectifier configuration with bi-directional switches and (b) implementation of bi-directional switches Sa, Sb, and Sc.

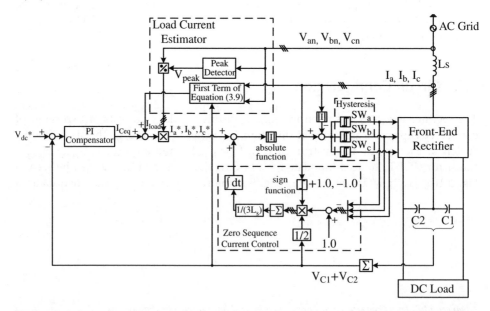

Figure 5.2 A decouple current control with instantaneous power balanced control strategy based on fixed hysteresis band algorithm.

balanced or an unbalanced supply condition without considering the switching losses. The analysis of a three-phase instantaneous voltage supply is formulated as:

$$
\begin{bmatrix} V_{an}(t) \\ V_{bn}(t) \\ V_{cn}(t) \end{bmatrix} = \begin{bmatrix} V_{am}(t) + V_{mn}(t) + L_a \dfrac{dI_a(t)}{dt} + I_a(t)R_a \\ V_{bm}(t) + V_{mn}(t) + L_b \dfrac{dI_b(t)}{dt} + I_b(t)R_b \\ V_{cm}(t) + V_{mn}(t) + L_c \dfrac{dI_c(t)}{dt} + I_c(t)R_c \end{bmatrix}
\tag{5.1}
$$

where $V_{am}(t)$, $V_{bm}(t)$, and $V_{cm}(t)$ are the rectifier input voltage of nodes a, b, and c referring to node m. $V_{mn}(t)$ is the voltage of the virtual ground measured across the dc link midpoint (node m) and the supply ground (node n). The rectifier phase voltage is expressed as a function of the switching states as follows:

$$
\begin{bmatrix} V_{am}(t) \\ V_{bm}(t) \\ V_{cm}(t) \end{bmatrix} = \begin{bmatrix} (1 - S_a(t))\dfrac{V_{dc}(t)}{2}\mathrm{sign}(I_a(t)) \\ (1 - S_b(t))\dfrac{V_{dc}(t)}{2}\mathrm{sign}(I_b(t)) \\ (1 - S_c(t))\dfrac{V_{dc}(t)}{2}\mathrm{sign}(I_c(t)) \end{bmatrix}
\tag{5.2}
$$

$$
V_{mn}(t) = \frac{-V_{dc}(t)}{6} \left\{ \begin{array}{l} [1 - S_a(t)]\mathrm{sign}(I_a(t)) \\ +[1 - S_b(t)]\mathrm{sign}(I_b(t)) \\ +[1 - S_c(t)]\mathrm{sign}(I_c(t)) \end{array} \right\}
\tag{5.3}
$$

$S_a(t)$, $S_b(t)$, and $S_c(t)$ denote the switching function of $SW_a(t)$, $SW_b(t)$, and $SW_c(t)$ of the bi-directional switch, respectively; whereas the function of sign(.) defines the direction of the filter inductor current flow, which is represented as:

$$
\mathrm{sign}(I_a(t)) = \begin{cases} 1 & \text{if } I_a(t) \geq 0 \\ -1 & \text{if } I_a(t) < 0 \end{cases}
\tag{5.4}
$$

The total per-phase input current is the sum of currents flowing through the diodes and the bidirectional switch of that particular phase. This is expressed in equation (5.5).

$$
\begin{cases} I_{SWa}(t) = S_a(t)I_a(t) \\ I_{SWb}(t) = S_b(t)I_b(t) \\ I_{SWc}(t) = S_c(t)I_c(t) \end{cases}
$$

$$
\begin{cases} I_{Da1}(t) = [1 - S_a(t)]I_a(t)\mathrm{fcn}(V_{am}(t)) \\ I_{Db1}(t) = [1 - S_b(t)]I_b(t)\mathrm{fcn}(V_{bm}(t)) \\ I_{Dc1}(t) = [1 - S_c(t)]I_c(t)\mathrm{fcn}(V_{cm}(t)) \end{cases}
$$

$$
\begin{cases} I_{Da1'}(t) = [1 - S_a(t)]I_a(t)\mathrm{fcn}(-V_{am}(t)) \\ I_{Db1'}(t) = [1 - S_b(t)]I_b(t)\mathrm{fcn}(-V_{bm}(t)) \\ I_{Dc1'}(t) = [1 - S_c(t)]I_c(t)\mathrm{fcn}(-V_{cm}(t)) \end{cases}
\tag{5.5}
$$

where $I_{SWa}(t)$, $I_{SWb}(t)$, and $I_{SWc}(t)$ are the bi-directional switch currents flowing through the star-connected point. $I_{Da1}(t)$, $I_{Db1}(t)$, and $I_{Dc1}(t)$ are the current through the upper diodes and $I_{Da1'}(t)$, $I_{Db1'}(t)$, and $I_{Dc1'}(t)$ are the currents flowing through their respective complementary diodes. Where the function of fcn(.) is written as:

$$fcn(V_{am}(t)) = \begin{cases} 1 & \text{if } V_{am}(t) > 0 \\ 0 & \text{if } V_{am}(t) \leq 0 \end{cases} \tag{5.6}$$

The injected current at the midpoint of the dc link is written as:

$$I_m(t) = S_a(t)I_a(t) + S_b(t)I_b(t) + S_c(t)I_c(t) \tag{5.7}$$

Based on the above analysis, the voltage $V_{mn}(t)$ is assumed to be balanced. However, an unbalanced and distorted grid supply can create an unbalanced neutral point; hence the expression of voltage $V_{mn}(t)$ is different from equation (5.3). Such a unbalanced grid condition can complicate the control algorithm if direct voltage control is applied for this system and yields to unbalanced capacitor voltages. Therefore, a simple instantaneous power balanced control strategy is used, based on the decoupled current control for dc voltage balancing.

5.3 Proposed Controller Scheme

Fig. 5.2 shows the power balanced algorithm with the unity power factor control. From the proposed control algorithm, the decoupled current control based on equation (5.7) is derived from an ac equivalent circuit of the star-connected bi-directional switches. In the case of a balance grid condition and identical capacitor parameters, the injected current through the virtual ground is zero. However, in practical cases, the supply is unbalanced due to the non-identical parameters of the passive elements. Hence, a zero sequence current component is injected into a virtual ground of two series connected capacitors. This additional zero sequence current will cause the virtual ground to float at a certain dc offset value, which results in an unbalanced dc-link. Moreover, the instantaneous peak current of the grid does not shape uniformly as no switching occurs during the transition of the fundamental peak current [3].

To achieve a unity power factor under a balanced or unbalanced grid condition, the net input reactive power of the rectifier must be zero with the power angle, $\theta = 0$. With this scheme, the net input active power is the sum of the output dc power and the rate of change of the energy stored in the passive elements irrespective of the grid condition. The active power balance expression is obtained in the following equation (5.8).

$$V_{dc}(t)I_{dc}(t) = \begin{bmatrix} V_{an}(t) & V_{bn}(t) & V_{cn}(t) \end{bmatrix} \begin{bmatrix} I_a(t) \\ I_b(t) \\ I_c(t) \end{bmatrix} - L_s \begin{bmatrix} I_a(t)\dfrac{dI_a(t)}{dt} \\ I_b(t)\dfrac{dI_b(t)}{dt} \\ I_c(t)\dfrac{dI_c(t)}{dt} \end{bmatrix}$$

$$- R_s \begin{bmatrix} I_a^2(t) \\ I_b^2(t) \\ I_c^2(t) \end{bmatrix} - C_{eq}V_{dc}(t)\dfrac{dV_{dc}(t)}{dt} \tag{5.8}$$

L_s and R_s denote three phase filter inductance and filter resistance $[L_a \ L_b \ L_c]$ and $[R_a \ R_b \ R_c]$, respectively; whereas, C_{eq} is the equivalent dc capacitance value of the output rectifier.

Assume that the switching loss is zero and the active power after filtering is fully transferred to the dc-link. Based on this assumption, the dc load current is directly estimated from the net active power of the ac side. The generated reference dc current is formulated with equation (5.8) in summation with the rate of the change in voltage across the equivalent capacitors. According to the power-balancing principle, the dc reference current of the inner control loop is obtained from equation (5.9).

$$I_{dc}^*(t) = I_{load}(t) + I_{Ceq}(t) = \underbrace{\frac{\sum \left(V_{sn}(t) - L_s \frac{dI_s(t)}{dt} - R_s I_s(t) \right) I_s(t)}{V_{dc}(t)}}_{\text{First term}} + \underbrace{C_{eq} \frac{dV_{dc}(t)}{dt}}_{\text{Second term}}$$

(5.9)

$I_s(t)$ is the input grid current. $I_a(t)$, $I_b(t)$, and $I_c(t)$ are the phase current. $V_{sn}(t)$ is the grid voltage of phases a, b, and c when referred to node n. The desired instantaneous reference grid current in Fig 5.2 is formulated as shown in equation (5.10):

$$\begin{cases} I_a^*(t) = [I_{load}(t) + I_{Ceq}(t)]\sin_a \text{ reference} \\ I_b^*(t) = [I_{load}(t) + I_{Ceq}(t)]\sin_b \text{ reference} \\ I_c^*(t) = [I_{load}(t) + I_{Ceq}(t)]\sin_c \text{ reference} \end{cases}$$

(5.10)

where $I_{Ceq}(t)$ is the dc current flow in the energy stored in the passive elements of the rectifier as stated in the second term of equation (5.9). The dc current $I_{Ceq}(t)$ is obtained by a simple first order PI(s) controller as mentioned in Section 5.2.1.

Under an asymmetrical grid condition, the phase angles between any two phases of a three-phase supply system are unequal. The measurement of unequal phase angles under such conditions is difficult, and using the phase lock loop as the design of PLL gets complicated. Therefore, a sine template generator for the proposed converter is calculated from the reference peak voltage and the phase voltage of the grid. While the phase voltage of the grid is measured using the hall-effect voltage transducers, and the reference peak voltage is obtained by the peak detector based on the following equation:

$$V_{peak} = \left\{ \left[\frac{2}{3}V_{an}(t) - \frac{1}{3}V_{bn}(t) - \frac{1}{3}V_{cn}(t) \right]^2 + \left[\frac{1}{\sqrt{3}}V_{bn}(t) - \frac{1}{\sqrt{3}}V_{cn}(t) \right]^2 \right\}^{1/2}$$

(5.11)

where $V_{an}(t)$, $V_{bn}(t)$, and $V_{cn}(t)$ are the instantaneous phase voltage of the input grid supply.

Using equation (5.11), a simple reference sine template for an unbalanced grid supply is obtained in equation (5.12).

$$\sin_s \text{ reference} = \frac{V_{sn}(t)}{V_{peak}}$$

(5.12)

sin$_s$ reference is known as the reference sine wave template for the instantaneous reference current in equation (5.10). The small letter "s" denotes phases a, b, and c of the three-phase grid voltage measurement.

5.3.1 dc-Link Voltage Control

Fig. 5.5 shows the closed loop model of the outer loop voltage control. The PI regulator tracks the dc-link voltage with respect to the reference control value. The disturbance shown in Fig. 5.5 is the injected current during the load change. To design the parameters of the PI regulator, the rate of change of the stored energy in the dc-link is formulated as the dc current of the equivalent capacitors is given by the following equation:

$$I_{Ceq}(s) = [V_{dc}(s) - V_{C1}(s) - V_{C2}(s)] \left[K_p + \frac{K_p}{T_i s} \right] \tag{5.13}$$

where K_p is the proportional gain constant of the voltage control loop and T_i is the settling time of the dc-link voltage. The measured and estimated load current waveforms are presented in Fig. 5.3. There is a deviation between the estimated and measured values. This indicates that the dc capacitor is charged by the three-phase instantaneous grid power during the time interval $(t_0 - t_1)$ as shown in Fig. 5.4. This results in a dc-link voltage variation due to the rate of change of the energy stored, ΔE_{dc}. The proportional gain is obtained by the following equation:

$$\Delta E_{dc} = \frac{1}{2} C_{eq} [V_{dc}^2(t_1) - V_{dc}^2(t_0)] \tag{5.14}$$

Based on the three-phase power-balancing principle, the energy stored in the inductor is assumed to be equal to that of the energy stored in the dc capacitors during the power transfer. Hence, the peak of the fundamental input current is equivalent to the peak of the capacitor current. With this assumption, the amplitude of the instantaneous dc

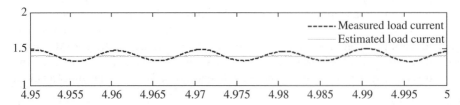

Figure 5.3 DC load current referred to measured and estimated value under ±10% unbalanced grid condition.

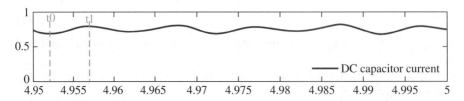

Figure 5.4 Output control signal of the outer voltage controller under ±10% unbalanced grid condition.

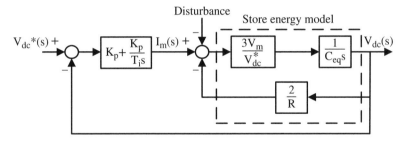

Figure 5.5 Outer loop voltage controller of Fig. 5.2.

current $I_{dc}(t)$ is obtained from the instantaneous grid power and expressed as follows:

$$V_{an}(t)I_a(t) + V_{bn}(t)I_b(t) + V_{cn}(t)I_c(t) = \frac{3}{2}V_m I_m \tag{5.15}$$

where V_m and I_m are the amplitude of the instantaneous ac voltage and current parameters, respectively. Based on equations (5.14) and (5.15), the rate of change of power is defined by the relationship between the output dc power and load power. Thus, the rate of change of power in the equivalent dc-link capacitor is simplified as:

$$\Delta P_{dc} = \frac{C_{eq}}{2}\frac{dV_{dc}^2(t)}{dt} = \frac{3}{2}V_m I_m - \frac{V_{dc}^2(t)}{R} \tag{5.16}$$

where R is known as the load resistance of the rectifier. From equation (5.16), the capacitor current in equation (5.9) is calculated as

$$I_{Ceq}(t) = I_{dc}(t) - I_{load}(t)$$
$$C_{eq}\frac{dV_{dc}(t)}{dt} = \frac{3V_m I_m}{V_{dc}(t)} - \frac{2V_{dc}(t)}{R} \tag{5.17}$$

From equation (5.17), the proportional gain of the outer loop voltage control is designed according to the open-loop transfer function of the voltage control as shown in Fig. 5.5. Therefore, the open-loop transfer function L(s) is written as

$$L(s) = K_p\left(1 + \frac{1}{T_i s}\right) \times \frac{3V_m}{V_{dc}C_{eq}} \times \frac{1}{s + 6V_m/V_{dc}C_{eq}R} \tag{5.18}$$

where the parameter of K_p is selected based on the pole cancellation for a stable system as given in the following:

$$K_p = \frac{V_{dc}C_{eq}R}{6V_m} \tag{5.19}$$

The root locus of the energy store model is shown in Fig. 5.6(a). Based on the pole trajectory in Fig. 5.6, the gain K_p is selected as $0.16A^{-1}$ and T_i is 0.02s. With this control parameter, the pole of the close-loop system is placed at the left-half-plane and it gives a stable close-loop system.

With control parameter $K_p = 0.16A^{-1}$ and $T_i = 0.02$s, the voltage control shows a good tracking performance as shown in Fig. 5.7. From Fig. 5.7, one can see that the steady-state error of the dc voltage is zero at time $t = 0.065$s, which shows that the controller tracked the desired reference value.

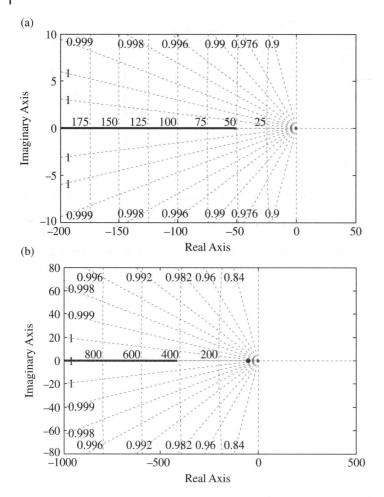

Figure 5.6 Root locus of the respective system in an (a) open-loop system and (b) close-loop system with $K_p = 0.16$ (frequency unit: rad/s).

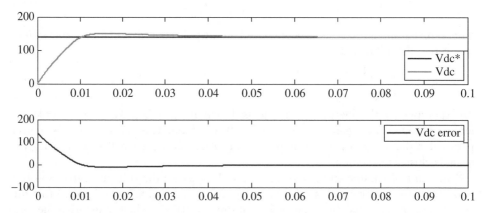

Figure 5.7 Obtained results of voltage control in Fig. 5.2.

5.3.2 Current Control

The average current control (ACC) and hysteresis current control (HCC) are often used for current error compensation [2–4]. Both the above current control techniques are used to reduce the harmonic content at the grid side. In the ACC, the current error is shaped with a carrier waveform, whereas the current error of the HCC is suppressed within the hysteresis band limit as shown in Fig. 5.8. However, the HCC provides a better dynamic behavior than the ACC method when there is a step change in the reference current [3].

This section provides a study of the HCC method with the decoupling current control. The switching function of the bidirectional switches is given in the following equation:

$$SW_s(t) = \begin{cases} 1, & \text{if } I_s^*(t) - I_s(t) < -h \\ 0, & \text{else } I_s^*(t) - I_s(t) > +h \end{cases} \tag{5.20}$$

With the HCC method, the dc voltages across the capacitors are not equal due to non-uniform charging. Furthermore, an unbalanced grid can inject zero sequence current components into the neutral point of the dc-link capacitors. Therefore, a decoupling current control is employed to eliminate the zero sequence current from entering into the neutral point as shown in Fig. 5.2.

Assuming the three-phase supply operates an under-balanced condition, the summation of the phase voltages of the grid and the pole voltages of the rectifier are

$$\begin{cases} V_{an}(t) + V_{bn}(t) + V_{cn}(t) = 0 \\ V_{am}(t) + V_{bm}(t) + V_{cm}(t) = 0 \end{cases} \tag{5.21}$$

On substituting equation (5.21) into (5.1),

$$\frac{3V_{mn}(t)}{L} = \frac{dI_a(t)}{dt} + \frac{dI_b(t)}{dt} + \frac{dI_c(t)}{dt} \tag{5.22}$$

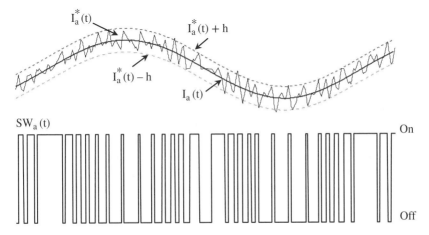

Figure 5.8 Hysteresis current control in the controller of Fig. 5.2.

From equation (5.22), it is clear that an additional zero sequence current is flowing through the virtual ground. Based on this assumption, a zero sequence current is added in the current control loop to compensate the zero sequence components in the floating ground. This is expressed as

$$\frac{V_{mn}(t)}{L} = \frac{dI_o(t)}{dt} \tag{5.23}$$

Hence, the zero sequence current from equation (5.33) is decoupled with the reference current as given in the following expression:

$$\begin{cases} I_a^*(t) = I_a(t) - I_o(t) \\ I_b^*(t) = I_b(t) - I_o(t) \\ I_c^*(t) = I_c(t) - I_o(t) \end{cases} \tag{5.24}$$

where $L = L_a = L_b = L_c$ of the input filter inductance and $I_o(t)$ is written as

$$I_o(t) = \frac{-V_{dc}(t)}{6L} \int \begin{bmatrix} (1 - S_a(t))\text{sign}(I_a(t)) \\ +(1 - S_b(t))\text{sign}(I_b(t)) \\ +(1 - S_c(t))\text{sign}(I_c(t)) \end{bmatrix} dt \tag{5.25}$$

Thus, the unbalanced capacitor voltage is successfully eliminated by the new reference currents in (5.34) under the balanced and unbalanced grid conditions.

5.3.3 Validation

The operational behavior of the three-phase unity power factor rectifier under several operating conditions is verified in MATLAB Simulink with a SimCoupler PSIM simulator. The obtained results show the reliability and robustness of the proposed power balanced control strategy with the decoupling current control based on a hysteresis fix band under three different operating conditions. Three cases are simulated from balanced to an extremely unbalanced grid condition as shown in Figs. 5.9–5.14.

Case 1 illustrates the input and output performance of the unity power factor rectifier under a balanced and clean grid condition: (Fig. 5.9 and Fig. 5.10).

A balanced grid operation (Fig 5.9) results in low THD content in the input grid current and zero phase shifts on the main phase voltage. Besides, the dynamic behavior (load change) of the converter shows that the dc voltage is tracked well even during a short interval of the voltage dip. With a stable dc-link voltage, the voltage across each capacitor is balanced before and after the load change as shown in Fig. 5.10.

Case 2 illustrates the input and output performance of the unity power factor rectifier under a ±10% unbalanced and distorted grid condition: (Fig. 5.11 and Fig. 5.12).

To visualize the feasibility of the proposed power balanced control strategy, two operating modes under an unbalanced grid condition are analyzed. In Case 2, the inner loop current control limits the harmonic current to minimum. The quality of the grid current in Case 2 is similar to a balanced grid operation as shown in Fig. 5.9(c), (d) and Fig. 5.11(c), (d). Under Case 2 operating conditions, the dynamic performance of the

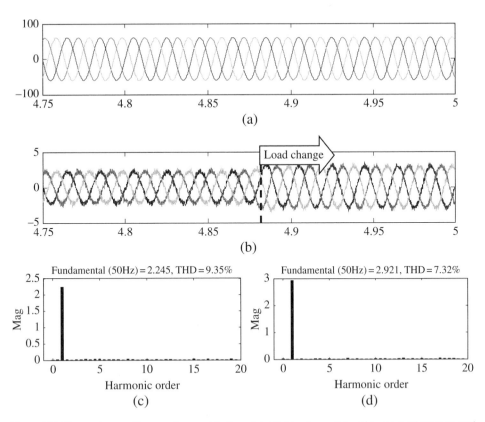

Figure 5.9 Three-phase voltage and current in Case 1. (a) Grid voltage, (b) grid current, (c) THD on grid current before load change, and (d) THD on grid current after load change at t = 4.875s.

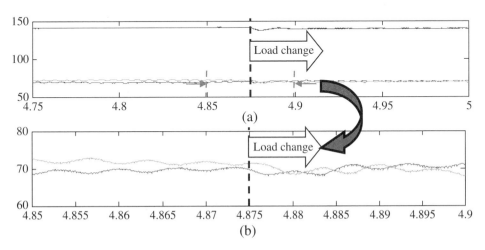

Figure 5.10 DC bus in Case 1. (a) dc-link voltage and voltage across capacitor C1 and C2 and (b) balance capacitor voltage before and after load change at t = 4.875s.

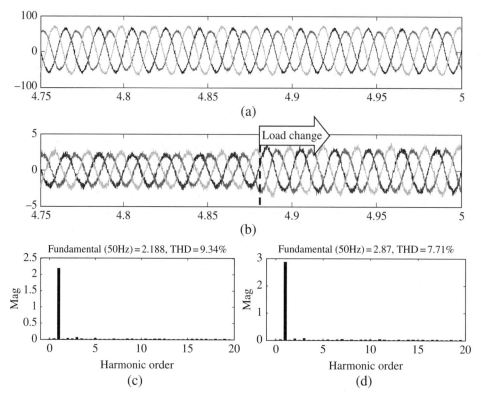

Figure 5.11 Three-phase voltage and current in Case 2. (a) Grid voltage, (b) Grid current, (c) THD on grid current before load change, and (d) THD on grid current after load change at t = 4.875s.

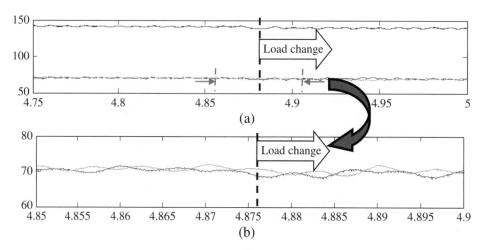

Figure 5.12 DC bus in Case 2. (a) dc-link voltage and voltage across capacitor C1 and C2 and (b) balance capacitor voltage before and after load change at t = 4.875s.

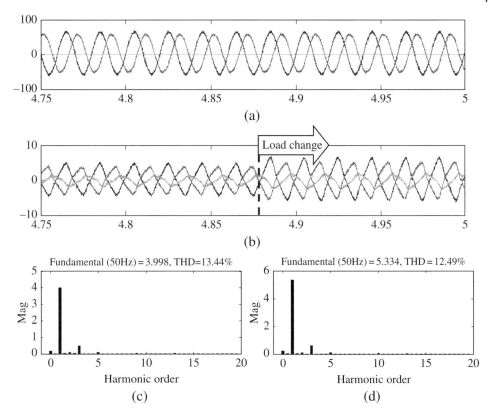

Figure 5.13 Three-phase voltage and current in Case 3. (a) Grid voltage, (b) grid current, (c) THD on grid current before load change, and (d) THD on grid current after load change at t = 4.875s.

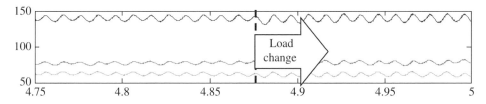

Figure 5.14 Voltage across each capacitor and dc-link voltage in Case 3 before and after load change at t = 4.875.

dc-link voltage is well tracked and both capacitor voltages are balanced as shown in Fig. 5.12.

Case 3 illustrates the input and output performance of the unity power factor rectifier under a $\pm10\%$ unbalanced with phase c short to ground and distorted grid condition: (Fig. 5.13 and Fig. 5.14).

Under extreme unbalanced grid conditions, the proposed controller is able to regulate two line currents in phase with their respective phase voltages while the remaining phase is shorted to the ground. Besides, the THD current in Case 3 is high as compared with

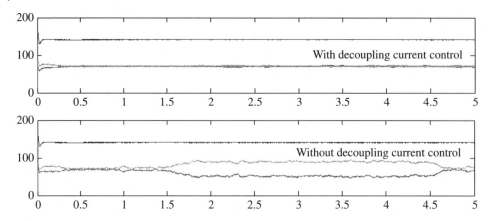

Figure 5.15 dc-link voltage and voltage across each capacitor in the dc bus with and without the decoupling current control.

Cases 1 and 2. This is due to the occurrence of the dominant second order harmonic content in the dc link. This as well causes distortion in the reference current wave in the inner loop. Furthermore, dc voltage across each capacitor is not balanced due to the existence of the negative and zero sequence line current, which are heavily injected into the neutral point. However, the proposed controller regulates the power closer to the unity power factor by which one can extract the optimum active power from the grid.

Fig. 5.15 shows the performance of the voltage-balancing property with and without the decoupling current control. Due to the non-identical parameters of the dc-link capacitors, the voltage across each capacitor will be unequal due to non-uniform charging and discharging. In contrast with the additional decoupling current control, both the capacitor voltages are balanced during the steady-state condition. The parameters of the simulated results are given in Table 5.1.

5.4 Experimental Verification

To evaluate the performance of the proposed controller, a low power laboratory prototype will be constructed based on the given parameters as stated in Table 5.1. The following experimental results are measured according to the above explained cases (1, 2, and 3). The voltage of phase c is set to very low voltage instead of short circuit to ground to replicate an extremely unbalanced grid operating condition. This is due to the safety requirement for the laboratory. The control algorithm is built into dSPACE RTI1103 with the MATLAB Simulink program and the sampling time of the proposed controller is set to 60 μs.

From the obtained experimental results, one can notice that the proposed controller is working well for the three operating grid conditions. The obtained results for Cases 2 and 3 are constructed based on the three isolated single-phase transformers.

Table 5.1 Parameters rating for both simulation and experiment converter test.

	Simulated parameter				Experimental parameter		
	Grid condition				Grid condition		
Description	Balanced	Unbalanced	Extremely unbalanced	Description	Balanced	Unbalanced	Extremely unbalanced
Grid voltage	42.43 Vrms, 42.43 Vrms, 42.43 Vrms	42.43 Vrms, 38.18 Vrms, 46.67 Vrms	42.43 Vrms, 38.18 Vrms, 0 Vrms	Grid voltage	42.43 Vrms, 42.43 Vrms, 42.43 Vrms	42.43 Vrms, 38.18 Vrms, 46.67 Vrms	42.43 Vrms, 38.18 Vrms, 2.14 Vrms
5th-harmonic voltage	0	0.77 Vrms, 0.76 Vrms, 0.93 Vrms	0.77 Vrms, 0.76 Vrms, 0 Vrms				
Noise voltage (random variable)	0	3.5 Vpeak	3.5 Vpeak				
Frequency	50 Hz	50 Hz	50 Hz	Frequency	50 Hz	50 Hz	50 Hz
Filter nductor	5 mH	5 mH	5 mH	Filter inductor	5 mH ± 5%	5 mH ± 5%	5 mH ± 5%
Filter resistor	0.2 Ω	0.2 Ω	0.2 Ω	Filter resistor	0.2 Ω ± 1%	0.2 Ω ± 1%	0.2 Ω ± 1%
DC capacitor C1	1050 µF	1050 µF	1050 µF	DC capacitor C1	1000 µF ± 20%	1000 µF ± 20%	1000 µF ± 20%
ESR resistor C1	110 mΩ	110 mΩ	110 mΩ	ESR resistor C1	100 mΩ	100 mΩ	100 mΩ
DC capacitor C2	920 µF	920 µF	920 µF	DC capacitor C2	1000 µF ± 20%	1000 µF ± 20%	1000 µF ± 20%
ESR resistor C2	80 mΩ	80 mΩ	80 mΩ	ESR resistor C2	100 mΩ	100 mΩ	100 mΩ

Case 1. Balanced and clean grid condition.

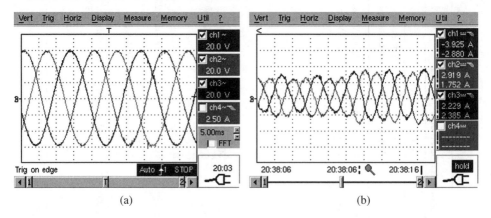

(a)   (b)

Figure 5.16 Three-phase voltage and current for Case 1. (a) Grid voltage and (b) grid current.

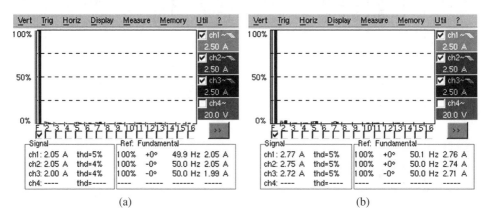

(a)   (b)

Figure 5.17 FFT of the three-phase grid current for Case 1. (a) Before load change and (b) after load change.

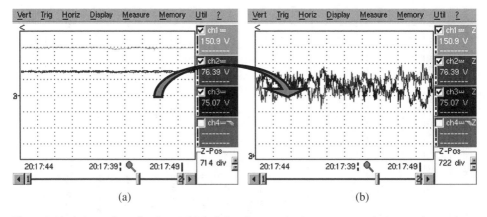

(a)   (b)

Figure 5.18 dc bus voltage for Case 1. (a) dc-link voltage and voltage across each dc capacitor and (b) voltage across both dc capacitors.

Figure 5.19 Power factor measurement is defined by the phase "a" voltage and current for Case 1.

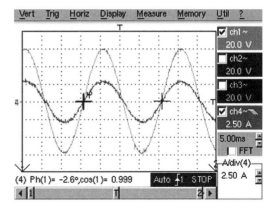

Case 2. Unbalanced and distorted grid condition with ±10% of the nominal phase voltage.

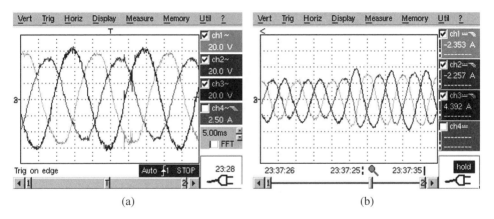

(a)   (b)

Figure 5.20 Three-phase voltage and current for Case 2. (a) Grid voltage and (b) grid current.

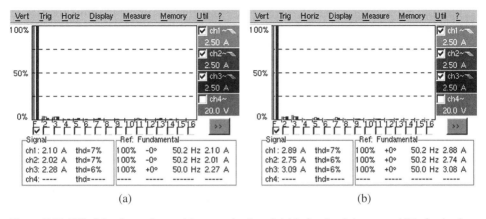

(a)   (b)

Figure 5.21 FFT of the three-phase grid current for Case 2. (a) Before load change and (b) after load change.

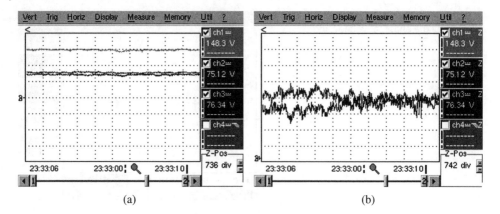

(a)   (b)

Figure 5.22 dc bus voltage for Case 2. (a) dc-link voltage and voltage across each dc capacitor and (b) voltage across both dc capacitors.

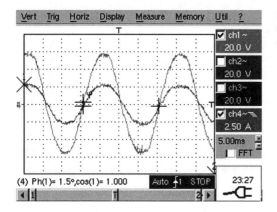

Figure 5.23 Power factor measurement is defined by the phase "a" voltage and current for Case 2.

Case 3. Unbalanced and distorted grid condition with Phase c down.

The fast Fourier transform of the three-phase load currents under the aforementioned test conditions are showed in Figs. 5.17, 5.21, and 5.25. The voltage across capacitors in Cases 1, 2, and 3 are balanced with reduced zero sequence current through the neutral-point of the star-connected bidirectional switches as shown in Fig. 5.18, Fig 5.22, and Fig. 5.26, respectively. The phase currents in the grid for Cases 1, 2, and 3 are shaped closely to a sinusoidal waveform with hysteresis current regulator as shown in Figs. 5.16, Fig. 5.20, and Fig. 5.24, respectively. However, the phase currents in the grid under the balanced condition in Fig. 5.16(b) are slightly unbalanced. This is due to the limitation of the sampling time of the dSPACE controller board, which results in an error in current within the limits of fixed hysteresis band. Hence, the peak current of the grid can cause a small overshoot current error. However, the THD of the three-phase currents in Cases 1 and 2 are still considerably low and the THD is limited in the range of 3% to 7%. The power factor of the load under these test conditions is presented in Figs. 5.19, 5.23, and 5.27. The power factor in Case 3 is smaller than in both the other test conditions.

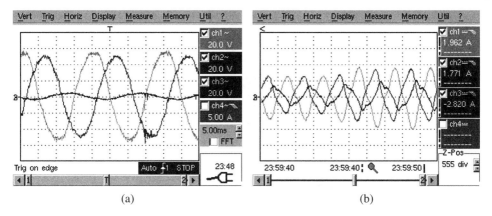

Figure 5.24 Three-phase voltage and current for Case 3. (a) Grid voltage and (b) grid current.

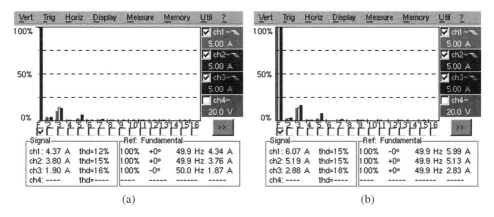

Figure 5.25 FFT of the three-phase grid current for Case 3. (a) Before load change and (b) after load change.

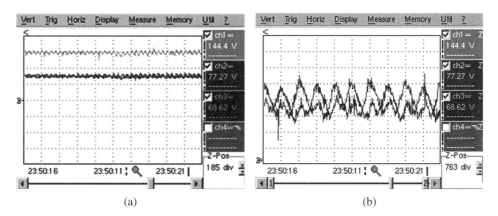

Figure 5.26 DC bus voltage for Case 3. (a) dc-link voltage and voltage across each dc capacitor and (b) voltage across both dc capacitors.

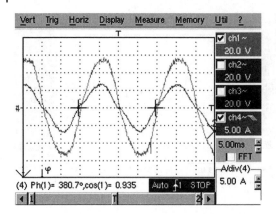

Figure 5.27 Power factor measurement is defined by the phase "a" voltage and current for Case 3.

By implementing the proposed power balance control strategy with decoupled current control, a good current compensation is obtained, while the voltage across the dc-link capacitors is balanced, and unity power factor operation is achieved.

5.5 Summary

The objective of this chapter was to introduce the power balanced control strategy with decoupling current control in a unidirectional three-level rectifier. With this proposed control method, the rectifier does not require any bulky and expensive passive filters to eliminate the input current harmonic while yielding low output dc ripple voltage. Hence, the current compensation effort in the L filter (first order filter) is low due to the synthesizing of the zero pole voltage through the aid of three switches.

The experiment results in this chapter proved that the proposed current control can be an alternative solution without the need for using complicated control algorithms. Based on the obtained experiment results, the grid current has a sinusoidal waveform with low harmonic distortion contents. Unity power factor is also achieved even after a sudden load change is performed. Hence, the three-level rectifier behaves like a symmetrical resistive load. Moreover, the voltages distributed across the two dc capacitors are almost equal. This is achieved with the compensation from the zero sequence current control.

Such a power balanced control method with the zero sequence current control for the multilevel rectifier can be an excellently retrofit for the grids connected or the PMSGs connected in wind turbine applications.

References

1 A. I. Maswood, A. K. Yusop, and M. Azizur Rahman, "A novel suppressed-link rectifier-inverter topology with near unity power factor," *Power Electronics, IEEE Transactions on*, vol. 17, pp. 692–700, 2002.

2 A. I. Maswood and L. Fangrui, "A unity power factor front-end rectifier with hysteresis current control, " *Energy Conversion, IEEE Transactions on*, vol. 21, pp. 69–76, 2006.

3 L. Dalessandro, U. Drofenik, S. D. Round, and J. W. Kolar, "A novel hysteresis current control for three-phase three-level PWM rectifiers, " in *Applied Power Electronics Conference and Exposition, 2005, APEC 2005, Twentieth Annual IEEE*, pp. 501–507, vol. 1, 2005.

4 A. I. Maswood, E. Al-Ammar, and F. Liu, "Average and hysteresis current-controlled three-phase three-level unity power factor rectifier operation and performance," *Power Electronics, IET*, vol. 4, pp. 752–758, 2011.

6

Bidirectional and Unidirectional Five-Level Multiple-Pole Multilevel Rectifiers

Gabriel H. P. Ooi

6.1 Introduction

The maintenance of a high quality of utility power supply is very important for the ac bus when a single-phase or three-phase nonlinear electronic load is installed. Nonlinear electronic loads like rectifier systems are frequently carried out with the power factor correction (PFC) to make sure the input grid voltage and current are operating at the same phase angle, as well as achieving low input current distortion. The power electronic supply for a high power electrical drive is usually implemented through two stages of energy conversion. The first stage of energy conversion is the main ac voltage converted into the dc voltage level and then the load voltage can be adapted with the dc–dc converter with or without galvanic isolation for the ac electrical drive of the inverter side. Often only three conductors are connected to the first stage of the ac–dc conversion without any neutral conductor, which can be known as the three-phase three-wire rectifier system.

In some applications, a three-phase three-wire single stage energy conversion system has been implemented for medium and high power electrical drives such as the vienna rectifier. However, the boost inductors at the input side of this rectifier occupy a significant amount of space. Although the voltage stress across the power devices is about half that of the dc-link voltage, which allows the high switching frequency rectifier to operate at a higher power density as compared with the conventional two-level rectifier or the diode-bridge rectifier. But in terms of low switching frequency operations for the high power inductive component with minimum cost implementation, the volume and weight of the rectifier are required to optimize.

Therefore, the proposed new bidirectional and unidirectional five-level multilevel rectifier based on a multiple-pole approach is presented in this chapter to optimize the inductive component with the low switching frequency operations. The derivation of a multiple-pole concept has been discussed [1].

6.2 Circuit Configuration

This section presents the operating principle of two different five-level pulse-width modulation (5L PWM) ac–dc inverters based on the multiple-pole approach for

Advanced Multilevel Converters and Applications in Grid Integration, First Edition.
Edited by Ali I. Maswood and Hossein Dehghani Tafti.
© 2019 John Wiley & Sons Ltd. Published 2019 by John Wiley & Sons Ltd.

bidirectional and unidirectional rectifiers. The 5L PWM inverter can be retrofit as a load for the 5L ac–dc–ac drive with a unity power factor (UPF) operation. The derivation of the proposed multiple-pole multilevel diode-clamped rectifier (M^2DCR) and multiple-pole multilevel switch clamped rectifier (M^2SCR) topologies are explained in the following section.

6.2.1 Bidirectional Front-End Five-Level/Multiple-Pole Multilevel Diode-Clamped Rectifier

The proposed bidirectional five-level multiple-pole multilevel diode-clamped rectifier ($5L-M^2DCR$) topology used in an ac–dc–ac drive with a five-level multiple-pole multilevel diode-clamped inverter ($5L-M^2DCI$) topology as a dynamic load is presented in Fig. 6.1. This back-to-back topology requires only eight power diodes in each phase to achieve the same input and output quality as the classical five-level diode-clamped converter presented in [2]. However, when the number of cells in this proposed topology increases, a total number of $6(n-3)$ diode components are reduced.

A five-level voltage stepped waveform of the input rectifier side is achieved with the same switching positions of $5L-M^2DCI$ topology as presented in [1]. The incremental input voltage step level can be classified into different sectors and switching states as shown in Table 6.1.

6.2.2 Unidirectional Front-End Five-Level/Multiple-Pole Multilevel Switch-Clamped Rectifier

A bidirectional front-end rectifier is not necessarily needed for the load application such as propulsion, compressor, or others containing a non-regenerative braking system. Thus, a proposed unidirectional multilevel rectifier is reconfigured and modified from Fig. 6.1 by arranging and replacing the semiconductor devices to form a unidirectional power flow.

By observing the current flow through the two identical insulated gate bipolar transistors (IGBTs) and the diode (i.e., Sa1, Sa2, and Da1 of phase "a") connected at the

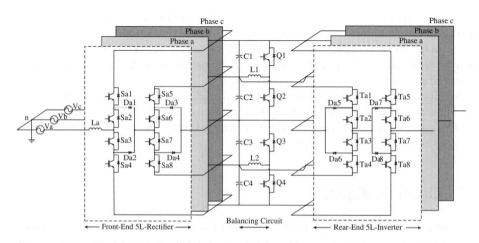

Figure 6.1 Proposed bidirectional $5L-M^2DCR$ at the front-end side of the ac–dc–ac drive.

Table 6.1 Switching Logic for Respective IGBT in Front-End 5L-M²DCR Topology.

Vom / Switch	Vdc/2 (Sector II)	Vdc/4 (Sector I & III)	0 (Sector I, III, IV & VI)	−Vdc/4 (Sector IV & VI)	−Vdc/2 (Sector V)
Sa1, Ta1	1	0	0	0	0
Sa2, Ta2	1	1	1	1	0
Sa3, Ta3	0	1	1	1	1
Sa4, Ta4	0	0	0	0	1
Sa5, Ta5	1	1	0	0	0
Sa6, Ta6	1	1	1	0	0
Sa7, Ta7	0	0	1	1	1
Sa8, Ta8	0	0	0	1	1

T-junction of the upper phase "a" with the unidirectional power flow operation, the current flow through the respective device is determined by the switching states as stated in Table 6.1, as well as the polarity of the grid current (e.g., When Da1 is conducting, Sa1 = 0 and Sa2 = 1, current can either flow through Da1 or Da2 depending on the polarity of the grid current. When Da1 is off, Sa1 = Sa2 = 1, the current will charge to the filter dc capacitor by passing through Sa1 and Sa2). Based on this switching current operation, Sa1 and Sa2 can be replaced with the diode and Da1 is replaced by IGBT to form a unidirectional power flow. Hence, the unidirectional 5L-rectifier in Fig. 6.2 is named as "the multiple-pole multilevel switch-clamped rectifier (M²SCR)."

In short, each phase leg of the proposed unidirectional 5L-M²SCR consists of two cells with four IGBTs and eight diodes to achieve a five-level input pole voltage. In Fig. 6.2, two switching devices (Sa3 and Sa4) of the inner cell are connected and clamped directly to the neutral point of the four dc-link capacitors, while both the other switching devices (Sa1 and Sa2) of the outer cell are clamped to the output terminal of the inner cell. With this unidirectional rectifier configuration, a higher power density is achieved due to the lower switching and conduction losses.

6.3 Modulation Scheme

The hardware implementation of the front-end rectifier and rear-end inverter of the 5L ac–dc–ac drives are operated independently with the level shifted pulse width modulation (LSPWM) technique. The LS-PWM of the front-end rectifier and rear-end inverter are set at a 1 kHz switching frequency operation.

The switching function of the LSPWM technique for the 5L rectifier and inverter topologies is expressed as:

$$\begin{cases} S_{a1}(t) = T_{a1}(t) = 2m_a \sin \omega t - 1 \\ S_{a2}(t) = T_{a2}(t) = 2m_a \sin \omega t + 2 \\ S_{a5}(t) = T_{a5}(t) = 2m_a \sin \omega t \\ S_{a6}(t) = T_{a6}(t) = 2m_a \sin \omega t + 1 \end{cases} \tag{6.1}$$

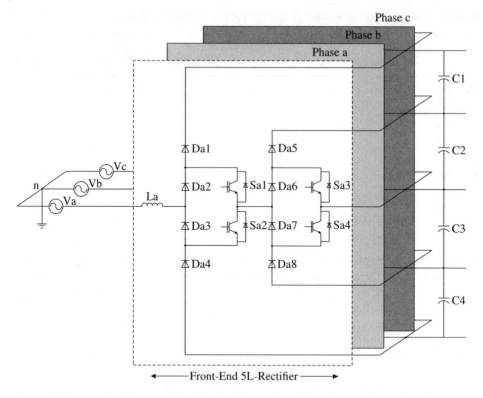

Figure 6.2 Proposed unidirectional five-level multiple-pole multilevel switch-clamped rectifier (5L-M²SCR).

where m_a is the ratio of twice the fundamental component of the pole voltage to the dc-link voltage.

Under the conditions of steady-state and balanced capacitors voltage in the dc bus, the general expression of the incremental output pole voltage is written as:

$$V_{xm}(t) = \frac{V_{dc}(t)}{n-1}\left(\sum_{i=1}^{n-1} S_{xi} - \frac{n-1}{2}\right) \tag{6.2}$$

where x represents phases "a," "b," and "c" and n is the number of the voltage level. S_{xi} is the switching state of each switching device depicted in the rectifier side.

The voltage transfer ratio of the both front-end and rear-end converters between the dc bus voltage to the input and output voltage are defined as:

$$\begin{cases} M_{x,rectifier}(t) \approx \dfrac{V_{dc}(t)}{V_{x,L-L}(t)}, & M_{x,rectifier}(t) > 1 \\[3mm] M_{x,inverter}(t) \approx \dfrac{2V_{xm}(t)}{V_{dc}(t)}, & M_{x,rectifier}(t) \leq 1 \end{cases} \tag{6.3}$$

$V_{x,L-L}(t)$ is the line-to-line voltage measured from the grid side and $V_{xm}(t)$ is the output pole voltage referred to the inverter side.

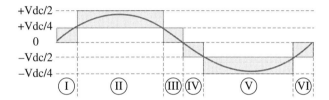

Figure 6.3 Incremental input pole voltage stepped waveform with the appropriate sector occurring.

In general, a high modulation index ($M_{x,rectifier}(t) > 1$) of the front-end rectifier is required to mitigate high input current distortion and achieve good dc-link voltage tracking due to its boosting effect in nature. Meanwhile, the modulation index of the rear-end inverter is usually operated at the linear region ($M_{x,inverter}(t) < 1$) to prevent any higher order harmonic components incurred in the ac load. But the high modulation index of the inverter side will occur when an ac machine is loaded and under the startup condition.

Thus, a low switching frequency can be used for the 5L rectifier to achieve a low ripple current and better power conversion efficiency due to the five-level voltage step. The ripple current is expressed in Eq. (6.4) and is based on equations (6.2) and (6.3).

$$\Delta I_{Lx}(t) \approx \frac{k}{L_x F_s} \left\{ \begin{array}{c} \dfrac{3V_{x,L-L}(t) - 3\sqrt{3}V_{mn}(t)}{3\sqrt{3}} \\[2ex] - \dfrac{M_{x,rectifier}(t)V_{x,L-L}(t)}{(n-1)\sqrt{3}} \left[\displaystyle\sum_{i=1}^{n-1} S_{xi,d} - \dfrac{n-1}{2} \right] \end{array} \right\} \tag{6.4}$$

F_s is the switching frequency of the rectifier and $V_{mn}(t)$ is the virtual ground voltage referred from node m to node n in Fig. 6.1. $S_{xi,d}$ is the switching state with the respective sectors shown in Fig. 6.3. k is the duty ratio of the switching state selection occurring in the sector (Fig. 6.3), and this is expressed as:

$$\begin{cases} 0 \le [k = 2\sin\omega t] \le 1 & 0 \le \omega t \le \pi/6 \\ 0 \le [k = 2(\sin\omega t - 1/2)] \le 1 & \pi/6 \le \omega t \le \pi/2 \end{cases} \tag{6.5}$$

The reduction of the input inductance value for the 5L rectifier topologies can be estimated with the duty cycle and the switching states with respect to the sectors shown in Fig. 6.3. According to equation (6.4), the maximum peak value of the input current ripple is determined using the differential equation at $\omega t = 30°$ (assume $V_{mn} = V_{dc} = 1$p.u). Thus, the critical inductance value is estimated as follows:

$$L_{x,max} \approx \frac{4(V_{xn} - V_{mn}) - V_{dc}}{4\Delta I_{Lx}F_s} \tag{6.6}$$

V_{xn} is the peak value of the grid phase voltage.

6.4 Design Considerations

Comparative evaluation of the design and implementation of the high power factor rectifiers with the 5L incremental voltage stepped waveform is discussed. The comparative study of the proposed topologies will cover the device rating, components count, and the control algorithm as given in the following section.

6.4.1 Device Voltage Stresses

Voltage and current stresses are the dominant factors considered in the hardware development process, so that the converter can achieve optimum performance and a higher reliability. Proper selection of the device rating for the proposed 5L rectifier topologies are determined based on the local and global stress analysis.

The voltage and current stress expressions for the respective front-end rectifiers are derived with the switching function in equation (6.1) and based on the following assumptions: (a) high power factor, (b) current and voltage ripple-free, (c) constant switching frequency, (d) balanced electrolytic capacitors voltage in the dc-link, and € zero voltage dropped across boost inductor, L_x.

The maximum voltage stress across the power devices of the unidirectional 5L-M^{2}SCR and bidirectional 5L-M^{2}DCR topologies are expressed, respectively, in equations (6.7) and (6.8).

$$\begin{cases} V_{Da1} = \dfrac{3V_{dc}}{4} \\[2mm] V_{Da2} = V_{Sa1} = \dfrac{3V_{dc}}{8} \\[2mm] V_{Da5} = V_{Da6} = V_{Sa3} = \dfrac{V_{dc}}{4} \end{cases} \tag{6.7}$$

$$\begin{cases} V_{Sa1} = \dfrac{3V_{dc}}{4} \\[2mm] V_{Sa2} = \dfrac{V_{dc}}{2} \\[2mm] V_{Sa5} = V_{Sa6} = V_{Da1} = V_{Da3} = \dfrac{V_{dc}}{4} \end{cases} \tag{6.8}$$

The maximum voltage stress expressed in equations (6.7) and (6.8) is for the power devices in the upper phase leg; hence the respective complimentary power devices in the lower phase leg are also determined using the same expressions.

6.4.2 Device Current Stresses

The average current stress is analyzed over one period of the fundamental frequency based on the assumed factors (a) to (e). For simplification, the average current stress is approximated as follows based on the respective switching function in (6.1).

$$\begin{cases} I_{Sa\langle Outer\ Cell\rangle}(t) = I_{Da\langle Outer\ Cell\rangle}(t) = \dfrac{1}{2\pi} \displaystyle\int_0^{2\pi} I_a \sin(\omega t) S_{a\langle Outer\ Cell\rangle} d\omega t \\[4mm] I_{Sa\langle Inner\ Cell\rangle}(t) = I_{Da\langle Inner\ Cell\rangle}(t) = \dfrac{1}{2\pi} \displaystyle\int_0^{2\pi} I_a \sin(\omega t) \left[2 - \dfrac{4V_{am1}}{V_{dc}} \right] S_{a\langle Inner\ Cell\rangle} d\omega t \end{cases}$$

$$\tag{6.9}$$

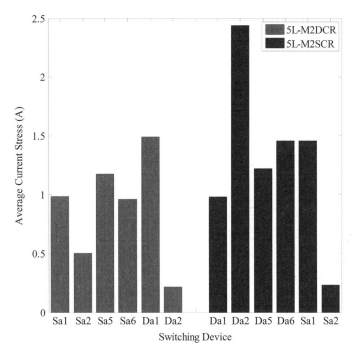

Figure 6.4 Average current stress level for each device of the respective front-end rectifier topologies.

V_{am1} is the peak value of the fundamental component of the grid phase voltage. With equation (6.9), the final expression of the average current and RMS current stresses flowing through each power device in the upper phase-leg of the proposed rectifier topologies is shown in Fig. 6.4 according to the expression defined in Tables 6.2 and 6.3.

The data obtained in Fig. 6.4 has been verified using the analytical expression (6.9). It is proven that during the negative cycle of the conduction period, the reverse current through the semiconductor switches in the bidirectional 5L-M^{2}DCR will be canceled out from the current stress during the positive cycle period. Unlike the case for unidirectional 5L-M^{2}SCR, the current only flows through the diodes during the positive cycle conduction period. Hence, the final net average current stress of 5L-M^{2}DCR in Fig. 6.4 is found to be lower compared with 5L-M^{2}SCR.

6.5 Comparative Evaluation

6.5.1 Input Current Shaping

The performances of the grid current response for the proposed front-end unidirectional and bidirectional rectifiers are carried out with the simulation as shown in Fig. 6.5(a) and (b), respectively. The harmonic current distortions are obtained based on the switching frequency (F_s) and the voltage transfer ratio ($M_{x,rectifier}$).

Fig. 6.5 shows that both the proposed front-end rectifiers have the highest current distortions when the $M_{x,rectifier}$ is as low as 1. The harmonic current distortion can be minimized in this case by increasing the F_s to 5 kHz and above.

Table 6.2 Expression of Average Current and RMS Current Stresses for 5L-M²DCR under Unity Power Factor Operation.

	Average Current Stress	RMS Current Stress
Da1 Da2	$$\frac{V_{am1}\,i_a}{V_{dc}\pi}\left\{\frac{V_{dc}}{V_{am1}}\left[1+\cos\left(\sin^{-1}\frac{V_{dc}}{4V_{am1}}\right)\right]-2\left[\pi+\sin\left(2\sin^{-1}\frac{V_{dc}}{4V_{am1}}\right)-2\sin^{-1}\left(\frac{V_{dc}}{4V_{am1}}\right)\right]\right\}$$	$$i_a\sqrt{\frac{V_{am1}}{V_{dc}\pi}\left[\frac{V_{dc}}{V_{am1}}\left[\frac{1}{2}\sin^{-1}\left(\frac{V_{dc}}{4V_{am1}}\right)-\frac{V_{dc}}{8V_{am1}}\cos\left(\sin^{-1}\frac{V_{dc}}{4V_{am1}}\right)\right]+\pi-2\sin^{-1}\left(\frac{V_{dc}}{4V_{am1}}\right)+\sin\left(2\sin^{-1}\frac{V_{dc}}{4V_{am1}}\right)+\frac{2}{3}\cos\left(3\sin^{-1}\frac{V_{dc}}{4V_{am1}}\right)-6\cos\left(\sin^{-1}\frac{V_{dc}}{4V_{am1}}\right)\right]}$$
Da3 Da4	$$\frac{V_{am1}\,i_a}{V_{dc}\pi}\left\{\frac{V_{dc}}{V_{am1}}\left[1-\frac{1}{2}\cos\left(\sin^{-1}\frac{V_{dc}}{4V_{am1}}\right)\right]-2\sin^{-1}\left(\frac{V_{dc}}{4V_{am1}}\right)\right\}$$	$$i_a\sqrt{\frac{V_{am1}}{V_{dc}\pi}\left[\frac{V_{dc}}{2V_{am1}}\left[\sin^{-1}\left(\frac{V_{dc}}{4V_{am1}}\right)-\frac{V_{dc}}{4V_{am1}}\cos\left(\sin^{-1}\frac{V_{dc}}{4V_{am1}}\right)\right]-\frac{16}{3}\sin^4\left(\frac{1}{2}\sin^{-1}\frac{V_{dc}}{4V_{am1}}\right)\cdot\left[2+\cos\left(\sin^{-1}\frac{V_{dc}}{4V_{am1}}\right)\right]\right]}$$
Sa1 Sa4	$$\frac{V_{am1}\,i_a}{V_{dc}\pi}\left\{\pi+\sin\left(2\sin^{-1}\frac{V_{dc}}{4V_{am1}}\right)-2\sin^{-1}\left(\frac{V_{dc}}{4V_{am1}}\right)-\frac{V_{dc}}{V_{am1}}\cos\left(\sin^{-1}\frac{V_{dc}}{4V_{am1}}\right)\right\}$$	$$i_a\sqrt{\frac{V_{am1}}{V_{dc}\pi}\left[\frac{V_{dc}}{2V_{am1}}\left[\sin^{-1}\left(\frac{V_{dc}}{4V_{am1}}\right)-\frac{\pi}{2}\right]-\cos\left(\sin^{-1}\frac{V_{dc}}{4V_{am1}}\right)\cdot\left[\frac{V_{dc}}{4V_{am}}-\frac{6V_{am}}{V_{dc}}\right]-\frac{1}{3}\cos\left(3\sin^{-1}\frac{V_{dc}}{4V_{am1}}\right)\right]}$$

$$i_a = \frac{V_{am1}}{V_{dc}\pi}\left\{ \left[\pi+\sin\left(2\sin^{-1}\frac{V_{dc}}{4V_{am1}}\right)\right]\left[1+\frac{V_{dc}}{2V_{am1}}\right] - \left[2+\frac{V_{dc}}{2V_{am1}}\right]\sin^{-1}\frac{V_{dc}}{4V_{am1}} - \left[3+\frac{2V_{dc}}{V_{am1}}+\frac{1}{8}\left(\frac{V_{dc}}{V_{am1}}\right)^2\right]\cos\left(\sin^{-1}\frac{V_{dc}}{4V_{am1}}\right) + \frac{1}{3}\cos\left(3\sin^{-1}\frac{V_{dc}}{4V_{am1}}\right) \right\}$$

Sa2
Sa3

$$\frac{V_{am1}i_a}{V_{dc}\pi}\left\{ 2\pi + 2\sin\left(2\sin^{-1}\frac{V_{dc}}{4V_{am1}}\right) - 4\sin^{-1}\left(\frac{V_{dc}}{4V_{am1}}\right) - \frac{5V_{dc}}{V_{am1}}\cos\left(\sin^{-1}\frac{V_{dc}}{4V_{am1}}\right) - \frac{V_{dc}}{V_{am1}} \right\}$$

$$i_a = \frac{V_{am1}}{V_{dc}\pi}\left\{ \frac{16}{3}\sin^4\left(\frac{1}{2}\sin^{-1}\frac{V_{dc}}{4V_{am1}}\right)\cdot\left[\cos\left(\sin^{-1}\frac{V_{dc}}{4V_{am1}}\right)+2\right] + \frac{1}{3}\cos\left(3\sin^{-1}\frac{V_{dc}}{4V_{am1}}\right) - 3\cos\left(\sin^{-1}\frac{V_{dc}}{4V_{am1}}\right) + \frac{V_{dc}}{2V_{am1}}\left[\pi+\sin\left(2\sin^{-1}\frac{V_{dc}}{4V_{am1}}\right)-2\sin^{-1}\frac{V_{dc}}{4V_{am1}}\right] \right\}$$

Sa5
Sa8

$$\frac{V_{am1}i_a}{V_{dc}\pi}\left\{ 4\sin^{-1}\left(\frac{V_{dc}}{4V_{am1}}\right) + 2\left[2-\frac{V_{dc}}{4V_{am1}}\right]\cdot\cos\left(\sin^{-1}\frac{V_{dc}}{4V_{am1}}\right) - \pi - \sin\left(2\sin^{-1}\frac{V_{dc}}{4V_{am1}}\right) \right\}$$

$$i_a = \frac{V_{am1}}{V_{dc}\pi}\left\{ \frac{1}{3}\cos\left(3\sin^{-1}\frac{V_{dc}}{4V_{am1}}\right) - 3\cos\left(\sin^{-1}\frac{V_{dc}}{4V_{am1}}\right) + \frac{V_{dc}}{2V_{am1}}\left[\pi+\sin\left(2\sin^{-1}\frac{V_{dc}}{4V_{am1}}\right)-\sin^{-1}\frac{V_{dc}}{4V_{am1}}\right] - \frac{V_{dc}}{2V_{am1}}\cos\left(\sin^{-1}\frac{V_{dc}}{4V_{am1}}\right) \right\}$$

Sa6
Sa7

$$\frac{V_{am1}i_a}{V_{dc}\pi}\left\{ 6\sin^{-1}\left(\frac{V_{dc}}{4V_{am1}}\right) + \frac{1}{2}\left[9-\frac{V_{dc}}{V_{am1}}\right]\cdot\cos\left(\sin^{-1}\frac{V_{dc}}{4V_{am1}}\right) - \sin\left(2\sin^{-1}\frac{V_{dc}}{4V_{am1}}\right) - \pi - 1 \right\}$$

Table 6.3 Expression of Average Current and RMS Current Stresses for 5L-M^{2}SCR under Unity Power Factor Operation.

	Average Current Stress	RMS Current Stress
Da1 Da4	$\dfrac{V_{am1}\,i_a}{V_{dc}\pi}\left\{\pi+\sin\left(2\sin^{-1}\dfrac{V_{dc}}{4V_{am1}}\right)-2\sin^{-1}\left(\dfrac{V_{dc}}{4V_{am1}}\right)-\dfrac{V_{dc}}{V_{am1}}\cos\left(\sin^{-1}\dfrac{V_{dc}}{4V_{am1}}\right)\right\}$	$i_a\sqrt{\dfrac{V_{am1}}{V_{dc}\pi}\left[\dfrac{V_{dc}}{2V_{am1}}\left(\sin^{-1}\dfrac{V_{dc}}{4V_{am1}}-\dfrac{\pi}{2}-\dfrac{V_{dc}}{4V_{am1}}\cos\left(\sin^{-1}\dfrac{V_{dc}}{4V_{am1}}\right)\right)+3\cos\left(\sin^{-1}\dfrac{V_{dc}}{4V_{am1}}\right)-\dfrac{1}{3}\cos\left(3\sin^{-1}\dfrac{V_{dc}}{4V_{am1}}\right)\right]}$
Da2 Da3	$\dfrac{i_a}{\pi}$	$\dfrac{i_a}{2}$
Da5 Da8	$\dfrac{V_{am1}\,i_a}{V_{dc}\pi}\left\{4\sin^{-1}\dfrac{V_{dc}}{4V_{am1}}+\dfrac{7V_{dc}}{4V_{am1}}\cos\left(\sin^{-1}\dfrac{V_{dc}}{4V_{am1}}\right)-\sin\left(2\sin^{-1}\dfrac{V_{dc}}{4V_{am1}}\right)-\pi\right\}$	$i_a\sqrt{\dfrac{V_{am1}}{V_{dc}\pi}\left[\dfrac{V_{dc}}{2V_{am1}}\left(\pi+\sin\left(2\sin^{-1}\dfrac{V_{dc}}{4V_{am1}}\right)-2\sin^{-1}\dfrac{V_{dc}}{4V_{am1}}\right)+\dfrac{1}{3}\cos\left(3\sin^{-1}\dfrac{V_{dc}}{4V_{am1}}\right)-3\cos\left(\sin^{-1}\dfrac{V_{dc}}{4V_{am1}}\right)+\dfrac{16}{3}\sin^4\left(\dfrac{1}{2}\sin^{-1}\dfrac{V_{dc}}{4V_{am1}}\right)\cdot\left[2+\cos\left(\sin^{-1}\dfrac{V_{dc}}{4V_{am1}}\right)\right]\right]}$

Da6
Da7

$$\frac{V_{am1}\, i_a}{V_{dc}\pi}\left\{2\sin^{-1}\frac{V_{dc}}{4V_{am1}} - \sin\left(2\sin^{-1}\frac{V_{dc}}{4V_{am1}}\right) - \pi + \frac{V_{dc}}{V_{am1}}\left[1+\cos\left(\sin^{-1}\frac{V_{dc}}{4V_{am1}}\right)\right]\right\}$$

$$i_a\left\{\frac{V_{am1}}{V_{dc}\pi}\,\frac{V_{dc}}{2V_{am1}}\left[\pi+\sin^{-1}\frac{V_{dc}}{4V_{am1}}+\sin\left(2\sin^{-1}\frac{V_{dc}}{4V_{am1}}\right)-2\sin^{-1}\frac{V_{dc}}{4V_{am1}}-\frac{V_{dc}}{4V_{am1}}\cos\left(\sin^{-1}\frac{V_{dc}}{4V_{am1}}\right)+\frac{1}{3}\cos\left(3\sin^{-1}\frac{V_{dc}}{4V_{am1}}\right)-3\cos\left(\sin^{-1}\frac{V_{dc}}{4V_{am1}}\right)\right]\right\}$$

Sa1
Sa2

$$\frac{V_{am1}\, i_a}{V_{dc}\pi}\left\{2\sin^{-1}\frac{V_{dc}}{4V_{am1}} - \sin\left(2\sin^{-1}\frac{V_{dc}}{4V_{am1}}\right) - \pi + \frac{V_{dc}}{V_{am1}}\left[1+\cos\left(\sin^{-1}\frac{V_{dc}}{4V_{am1}}\right)\right]\right\}$$

$$i_a\left\{\frac{V_{am1}}{V_{dc}\pi}\,\frac{V_{dc}}{2V_{am1}}\left[\pi+\sin^{-1}\frac{V_{dc}}{4V_{am1}}+\sin\left(2\sin^{-1}\frac{V_{dc}}{4V_{am1}}\right)-2\sin^{-1}\frac{V_{dc}}{4V_{am1}}-\frac{V_{dc}}{4V_{am1}}\cos\left(\sin^{-1}\frac{V_{dc}}{4V_{am1}}\right)+\frac{1}{3}\cos\left(3\sin^{-1}\frac{V_{dc}}{4V_{am1}}\right)-3\cos\left(\sin^{-1}\frac{V_{dc}}{4V_{am1}}\right)\right]\right\}$$

Sa3
Sa4

$$\frac{V_{am1}\, i_a}{V_{dc}\pi}\left\{1 - 2\sin^{-1}\left(\frac{V_{dc}}{4V_{am1}}\right) - \frac{V_{dc}}{2V_{am1}}\cos\left(\sin^{-1}\frac{V_{dc}}{4V_{am1}}\right)\right\}$$

$$i_a\left\{\frac{V_{am1}}{V_{dc}\pi}\left[\frac{V_{dc}}{2V_{am1}}\sin^{-1}\frac{V_{dc}}{4V_{am1}}-\frac{V_{dc}}{4V_{am1}}\cos\left(\sin^{-1}\frac{V_{dc}}{4V_{am1}}\right)-\frac{16}{3}\sin^4\left(\frac{1}{2}\sin^{-1}\frac{V_{dc}}{4V_{am1}}\right)\cdot\left[\cos\left(\sin^{-1}\frac{V_{dc}}{4V_{am1}}\right)+2\right]\right]\right\}$$

Note: M_a is the phase a modulation index, which is expressed as $M_a = (2V_{am1}/V_{dc})\sin\omega t$.

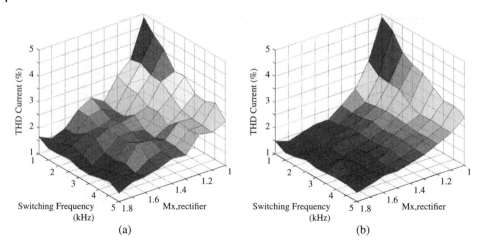

Figure 6.5 Harmonic current distortion characteristics of the respective front-end 5L-rectifier. (a) Unidirectional 5L-M^{2}SCR and (b) bidirectional 5L-M^{2}DCR.

However, the grid current distortion does not deviate much with any range of switching frequency utilized when the voltage transfer ratio $M_{x,rectifier}$ is more than 1.3. This can be proved by substituting equations (6.6) into (6.4), where the F_s is found to be nullified.

The harmonic current distortion varies nonlinearly for the unidirectional 5L-M^{2}SCR as shown in Fig. 6.5(a), while the results obtained for bidirectional 5L-M^{2}DCR in Fig. 6.5(b) has a smoother curve. This is caused by the forward blocking voltage where the current flows from the grid to the dc-link in one direction for the 5L-M^{2}SCR topology.

Even though there is a slight difference in the performances, both proposed front-end rectifiers have met the adopted standard current total harmonic distortion (THD) requirements of 6% below even at a low switching frequency. Hence, the proposed low switching frequency rectifiers have also been found to be attractive for relatively high power applications.

6.5.2 Components Count

The number of the components count in the front-end rectifiers is shown in Table 6.4. One can observe that the number of diodes used in the proposed bidirectional rectifier topology is reduced. This yields better efficiency because of the low conduction loss. For unidirectional rectifiers, the proposed topology reduces six MOSFET/IGBT devices. In addition to that, the required six isolated gate drivers are also eliminated.

Moreover, the proposed rectifier topology also reduces the cost of implementation, as well as the size of the converters. The overall components reduction not only consists of the active switches, but also the number of necessary control circuit, and the size of the heat sink. This can also improve the complexity of the control gating signals through less isolated power supply requirements.

Table 6.4 Number of Components Used in Front-End Rectifier Topologies.

	Topologies				
	Unidirectional Rectifier			Bidirectional Rectifier	
Devices	Classical 5L-MDCI (ref. to [3] of Fig. 6)	Classical 5L-MDCI (ref. to [3] of Fig. 7)	Proposed 5L-M^{2}SCR	Classical 5L-MDCI	Proposed 5L-M^{2}DCR
Diode	24	24	24	18	12
IGBT/MOSFET	18	18	12	24	24
Capacitors	4	4	4	4	4
Complementary switch	0	9	0	12	12
Isolated gate driver	18	18	12	24	24
Cost (USD)	$7472.72	$8943.56	$5698.07	$6616.08	$7410.76
Efficiency (%)	77.37	80.44	85.42	80.99	82.30
Weight (kg)	8.01	9.63	6.17	7.14	7.91

6.6 Control Strategy

The proposed control algorithm with the PFC technique is shown in Fig. 6.6. Two control loops, that is, synchronous-reference-frame (SRF) current control and constant switching frequency modulation are implemented to regulate the dc-link voltage and mitigate the current distortions. Due to the simplicity of the control strategy, a low cost integrated control circuit can be designed. The balancing control for the dc–dc balancing circuit is presented in [2, 4].

The UPF controller for the front-end five-level rectifiers (M^{2}DCR or M^{2}SCR) designed in Fig. 6.6 is based on the SRF current control with the LSPWM technique. The detailed analysis of the outer-loop dc-link voltage control and inner-loop current control are both presented in [5]. The SRF controller provides a good dynamic response to achieve high quality input sinusoidal current with constant UPF performance.

The open-loop transfer function of the dc-link voltage control under a steady-state condition is expressed as follows to achieve a stable control system:

$$L(s) = \frac{K_p s + K_I}{s} \cdot \frac{L_x I_d}{C_{eq} V_{dc}} \cdot \frac{\left(\sqrt{3} V_p / L_x I_d\right) - s}{s + \left(I_{dc} / C_{eq} V_{dc}\right)} \tag{6.10}$$

K_p and K_I are the PI parameters of the dc voltage control loop. C_{eq} and L_x are the equivalent capacitance value of the dc bus and the input (phase "a," "b," and "c") filter inductance value, respectively. V_p is the rms value of the grid phase voltage, and V_{dc} is the mean value of the dc-link voltage.

I_d is the peak value of the reference current that is obtained from the summation of I_{dc} (output of dc voltage control) and feed-forward current (output of K_f).

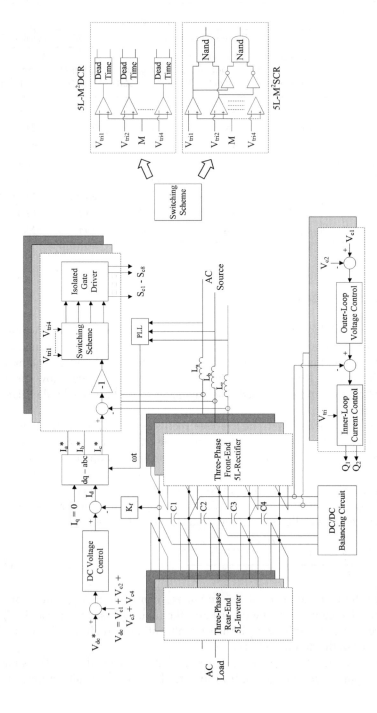

Figure 6.6 Proposed front-end rectifier controller for 5L-ac–dc–ac drive based on synchronous-reference-frame (SRF) current control.

The feed-forward current control loop under a steady-state condition is derived based on the power-balance principle, which is expressed as the following:

$$K_f = \frac{V_{dc}}{\sqrt{3}V_p} \tag{6.11}$$

6.7 Experimental Verification

The ac–dc–ac hardware prototypes in Fig. 6.7 are constructed based on the proposed circuit diagrams of Figs. 6.1 and 6.2 with the controller loop as shown in Fig. 6.6. The controller is implemented utilizing the dSPACE RTI1103 controller board. The experimental results are obtained based on the chosen system parameters values shown in Table 6.5.

The input current shape of the two proposed front-end rectifiers is achieved through the application of proper gating signals based on the LSPWM. The front-end bidirectional 5L-M^{2}DCR consists of complementary switches; therefore, additional dead time circuits are needed to prevent a short circuit. Unlike the 5L-M^{2}DCR, the unidirectional 5L-M^{2}SCR does not have any complementary switches. Hence, the gating signals for the 5L-M^{2}SCR are generated with the four logic NAND gates as shown in Fig. 6.6.

The experimental results obtained in Fig. 6.8(a) and (b) show the input voltage and current quality of front-end unidirectional 5L-M^{2}SCR and bidirectional 5L-M^{2}DCR, respectively. Both proposed topologies achieve low input current distortions (as shown in Fig. 6.8(c)) yielding a 0.99 power factor. The input pole voltage in Fig. 6.8 (b) is slightly

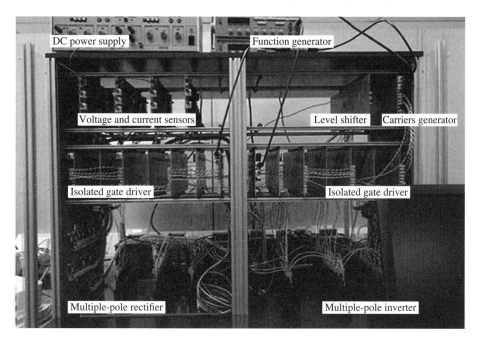

Figure 6.7 Experimental setup on ac–dc–ac converter with different configurations based on bidirectional and unidirectional 5L unity power factor rectifiers.

Table 6.5 Parameter Setting for Experimental ac/dc/ac Converters.

System parameters	Values
Input grid voltage	60 Vrms (50 Hz)
dc-link voltage	200 Vdc
Input inductors (Lx)	5 mH
Mx, inverter	0.8
Switching frequency	1 kHz

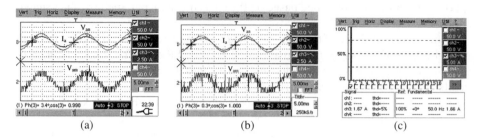

Figure 6.8 Input voltage and current waveform (upper trace: grid phase voltage, and current, lower trace: rectifier pole voltage) based on respective front-end multilevel rectifier topologies. (a) 5L-M²SCR, (b) 5L-M²DCR, and (c) THD of the grid current of Fig. 10.8(a), the THD of the grid current is also similar result for 5L-M²DCR.

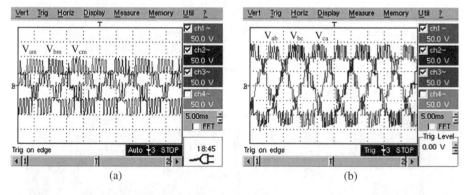

Figure 6.9 Output voltage of rear-end 5L-M²DCI topology in both ac–dc–ac configuration. (a) Output pole voltage and (b) output line-to-line voltage.

distorted as compared with Fig. 6.8 (a). This is due to the instantaneous high reverse peak I_{dc} current fed into the close-loop controller. Thus, the current error signal is affected and the input pole voltage is synthesized from the positive voltage step to the negative voltage step or vice versa.

The same rear-end 5L-M²DCI topology is implemented for both front-end uni- and bidirectional topologies. The experimental results of the output pole voltages and output line-to-line voltages are obtained and shown in Fig. 6.9(a) and (b), respectively.

6.8 Summary

A new generation of front-end unidirectional 5L-M^{2}SCR and bidirectional 5L-M^{2}DCR topologies are introduced to reduce the number of semiconductor devices usually required in conventional converters. The experimental results shown have proven the theory and analysis and also verified the feasibility of the proposed ac–dc–ac topologies.

Excellent performance and low input current distortion with a high power factor is achieved even while operating at a low switching frequency of 1 kHz. This is without the use of any bulky LC passive filter. Low voltage/current stress and low switching losses due to the reduced components count lead to better converter efficiency. The size of input reactors required is also reduced, thanks to the higher number of voltage levels while operating at a low switching frequency. This is at a small cost of an additional balancing capacitor voltage circuitry to balance the capacitor's voltage in the dc-link.

Alternatively, dc-link-balancing strategies such as modulation schemes or control algorithms can be implemented to replace the additional balancing circuit. These will provide a more cost-effective and energy-efficient solution for a higher-level ac–dc–ac drive and can especially be suitable for renewable energy conversion where high efficiency is paramount.

References

1 O. H. P. Gabriel, A. I. Maswood, and A. Venkataraman, "Multiple-poles multilevel diode-clamped inverter (M2DCI) topology for alternative multilevel converter," in *IPEC, 2012 Conference on Power & Energy*, pp. 497–502, 2012.

2 N. Hatti, Y. Kondo, and H. Akagi, "Five-level diode-clamped PWM converters connected back-to-back for motor drives," *Industry Applications, IEEE Transactions on*, vol. 44, pp. 1268–1276, 2008.

3 K. A. Corzine and J. R. Baker, "Reduced-parts-count multilevel rectifiers," *Industrial Electronics, IEEE Transactions on*, vol. 49, pp. 766–774, 2002.

4 C. Newton and M. Sumner, "Novel technique for maintaining balanced internal DC link voltages in diode clamped five-level inverters," *Electric Power Applications, IEE Proceedings -*, vol. 146, pp. 341–349, 1999.

5 A. I. Maswood and L. Fangrui, "A unity-power-factor converter using the synchronous-reference-frame-based hysteresis current control," *Industry Applications, IEEE Transactions on*, vol. 43, pp. 593–599, 2007.

7

Five-Level Multiple-Pole Multilevel Vienna Rectifier

Gabriel H. P. Ooi and Ali I. Maswood

7.1 Introduction

This chapter introduces a new unidirectional three-phase five-level rectifier topology that requires only a minimum of six identical insulated gate bipolar transistor (IGBT) devices as shown in Fig. 7.1. The concept of this rectifier topology with a reduced number of operating devices is derived and modified from the unidirectional multiple-pole multilevel switch-clamped rectifier topology presented in Chapter 6. Hence, this rectifier is named multiple-pole multilevel vienna rectifier (M^2VR) topology. The M^2VR topology achieves a high power factor in the main line requiring only a small inductor with the five-level input voltage stepped waveform. The reliability of this converter is improved by employing a lesser number of components as compared with the unidirectional five level (5L) rectifier topology in Chapter 6, which does not require any dead time circuitries and additional isolated gate drivers. However, an alternative method for reducing the losses in the unidirectional 5L rectifier is required to investigate to reduce the number of the components count.

Several methods for reducing the stress and the switching loss across the semiconductor devices have been proposed based on: (1) the low conduction period control [1, 2], (2) the resonant dc-link circuit [3], and (3) the transformer-assisted pulse-width modulation (PWM) zero voltage switching technique [4].

The proposed method (1) has the advantage of reducing the losses and stresses on the semiconductor devices, but the converter would still need a large input reactor for the lower conduction angles [5]. The proposed methods (2) and (3) achieve lower switching losses due to the zero voltage and zero current switching. However, the disadvantage of both methods will increase the complexity of the control scheme and the overall weight of the converters.

A new switching scheme in this chapter is proposed to achieve lower losses by remaining the minimum number of switches in unidirectional 5L rectifier topologies in Fig. 7.1. Two types of switching schemes—PWM techniques and low conduction period control [1, 2] are applied to the outer cell and inner cell switches, respectively, to achieve low switching and conduction losses as compared with the classical switching method. PWM techniques such as the level-shifted PWM (LS-PWM) or the phase-shifted PWM

Advanced Multilevel Converters and Applications in Grid Integration, First Edition.
Edited by Ali I. Maswood and Hossein Dehghani Tafti.
© 2019 John Wiley & Sons Ltd. Published 2019 by John Wiley & Sons Ltd.

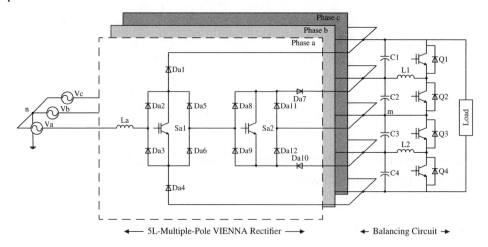

Figure 7.1 The proposed unidirectional five-level/multiple-pole multilevel vienna rectifier (5L-M²VR) with a balancing circuit.

(PS-PWM) can be applied to this topology. The performances of both the PWM techniques are evaluated in the following section.

7.2 Operating Principle

A conventional three-phase vienna rectifier topology consists of three switches. Each of the switches is connected across the four adjacent diodes to enable a bidirectional current flow between the grid and the neutral-point clamp of the two dc-link capacitors. Zero voltage conduction is achieved across the switching device due to the configuration of the vienna topology. Thus, three voltage levels ($+Vdc/2$, 0 and $-Vdc/2$) are synthesized.

The proposed five-level rectifier topology (5L-M²VR) in Fig. 7.1 is configured with two vienna rectifier cells in each phase, which are constructed based on the multiple-pole hierarchy. According to the pole diagram concept (Fig. 7.2), alternative structures of the bidirectional switches are configured as shown in Fig. 7.2. Therefore, the five-level stepped input voltage waveform is obtained with the proposed M²VR configuration.

Two types of switching schemes are utilized to modulate the gate signals for both the outer and the inner cells, in order to achieve a low current distortion. The switches (S_{a2}, S_{b2}, and S_{c2}) in the outer cell of the rectifier are operated with the PWM techniques as shown in Fig. 7.4, while the switches (S_{a1}, S_{b1}, and S_{c1}) in the inner cell of the rectifier start conducting at each zero crossing point of the grid voltage with a switching angle (α) of 30° as shown in Fig. 7.3.

The approximated maximum current (peak value of the current) flow through the power device can be estimated using a differential equation together with the summation of the initial value. This initial value is determined by the maximum input current obtained from a previous group of 30° switching angles, which is explained in [6].

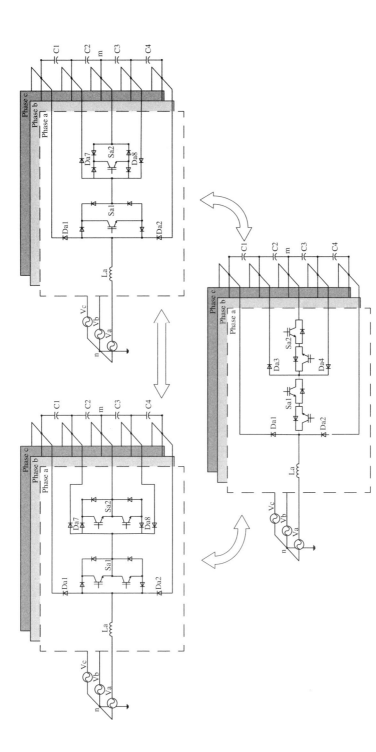

Figure 7.2 Alternative configuration of the proposed unidirectional 5L multiple-pole multilevel rectifier topologies based on the various types of the four quadrant switches.

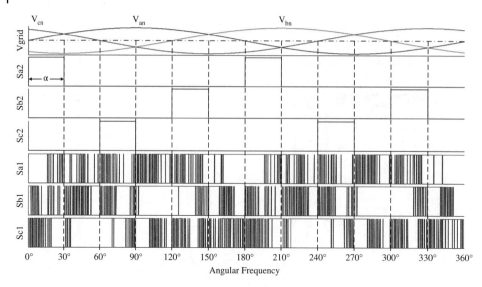

Figure 7.3 Two proposed switching schemes: the switching angle firing method for Sx1 and LS-PWM techniques through the logic gate for Sx2.

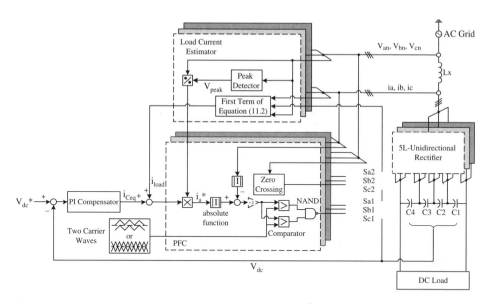

Figure 7.4 Proposed power balanced control technique for 5L-M²VR topology.

7.3 Design Considerations

7.3.1 Device Current Stress

The average current and RMS current stresses are estimated using the local and global stress analysis as presented in Chapter 6. Theoretical analysis of the switching function of the local current stress of all of the power devices mainly depends on the switching states selection of the outer cell switch, while the average value of the global current stress estimation on particular devices like Sa2, Da11, and Da12 in the inner cell are formed by the integration limit within the switching angle period, α over one main period ($\omega t = 2\pi$) of the grid.

The expression of the average and RMS current stress on each power device based on the integral limit with the switching function analysis is expressed in Tables 7.1 and 7.2. The final expression of the current stress is extended in term of the modulation amplitude and with or without the switching angle depending on the conduction period of the power device. The derivation of the switching function for the switching device has been detailed in Chapter 3. The amplitude of the carrier waveform in LS-PWM technique is normalized at 1 p.u, which is set at 0 to 1 p.u for the upper triangle wave and 0 to -1 p.u for the lower triangle wave. Similarly for the PS-PWM, both carrier waves are normalized at -1 to 1 p.u. With the LS-PWM through the logic gate, the final switching function expression can be written in equation (7.1). Similarly, a PS-PWM can be expressed with the same derivation concept as presented in Chapter 3.

$$S_{a1}(\omega t) = 1 - m_a \sin \omega t \tag{7.1}$$

The remaining current stress analysis for the PS-PWM technique will not be further discussed in this chapter.

7.3.2 Device Voltage Stress

The maximum voltage across the switches can be determined based on the switching conditions. Thus, the maximum voltage stress for $S\times2$ is obtained when $S\times1$ is switched

Table 7.1 Average Current Stress Based on Integral Limit with the Switching Function Expression.

Power Device	Average Current Stress Expression	Final Expression of the Average Current Stress
Da1	$\dfrac{1}{2\pi}\displaystyle\int_0^\pi i_a \sin \omega t[1 - S_{a1}(\omega t)]d\omega t$	$\dfrac{m_a i_a}{4}$
Da2	$\dfrac{1}{2\pi}\displaystyle\int_0^\pi i_a \sin \omega t\, d\omega t$	$\dfrac{i_a}{\pi}$
Da5	$\dfrac{1}{2\pi}\displaystyle\int_0^\pi i_a \sin \omega t \cdot S_{a1}(\omega t)d\omega t$	$\dfrac{i_a}{2\pi}\left\{2 - \dfrac{\pi m_a}{2}\right\}$
Da7	$\dfrac{1}{2\pi}\displaystyle\int_\alpha^\pi i_a \sin \omega t \cdot S_{a1}(\omega t)d\omega t$	$\dfrac{i_a}{2\pi}\left\{1 + \cos\alpha - \dfrac{1}{2}m_a\left[\begin{array}{c}\pi - \alpha \\ + \sin\alpha\cos\alpha\end{array}\right]\right\}$
Da8	$\dfrac{1}{2\pi}\displaystyle\int_0^\pi i_a \sin \omega t \cdot S_{a1}(\omega t)d\omega t$	$\dfrac{i_a}{2\pi}\left\{2 - \dfrac{\pi m_a}{2}\right\}$
Da11	$\dfrac{1}{2\pi}\displaystyle\int_0^\alpha i_a \sin \omega t \cdot S_{a1}(\omega t)d\omega t$	$\dfrac{i_a}{2\pi}\left\{1 - \cos\alpha + \dfrac{1}{4}m_a[\sin 2\alpha - 2\alpha]\right\}$
Sa1	$\dfrac{1}{\pi}\displaystyle\int_0^\pi i_a \sin \omega t \cdot S_{a1}(\omega t)d\omega t$	$\dfrac{i_a}{\pi}\left\{2 - \dfrac{\pi m_a}{2}\right\}$
Sa2	$\dfrac{1}{\pi}\displaystyle\int_0^\alpha i_a \sin \omega t \cdot S_{a1}(\omega t)d\omega t$	$\dfrac{i_a}{\pi}\left\{1 - \cos\alpha + \dfrac{1}{4}m_a[\sin 2\alpha - 2\alpha]\right\}$

Table 7.2 RMS Current Stress Based on Integral Limit with the Switching Function Expression.

Power Device	Average Current Stress Expression	Final Expression of the Average Current Stress
Da1	$i_a\sqrt{\dfrac{1}{2\pi}\displaystyle\int_0^\pi \sin^2\omega t[1-S_{a1}(\omega t)]d\omega t}$	$i_a\sqrt{\dfrac{2m_a}{3\pi}}$
Da2	$i_a\sqrt{\dfrac{1}{2\pi}\displaystyle\int_0^\pi \sin^2\omega t\, d\omega t}$	$\dfrac{i_a}{2}$
Da5	$i_a\sqrt{\dfrac{1}{2\pi}\displaystyle\int_0^\pi \sin^2\omega t\cdot S_{a1}(\omega t)d\omega t}$	$i_a\sqrt{\dfrac{1}{4}-\dfrac{2m_a}{3\pi}}$
Da7	$i_a\sqrt{\dfrac{1}{2\pi}\displaystyle\int_\alpha^\pi \sin^2\omega t\cdot S_{a1}(\omega t)d\omega t}$	$i_a\sqrt{\dfrac{1}{2\pi}\left\{\begin{array}{l}\dfrac{4}{3}m_a(\cos\alpha-2)\cos^4\left(\dfrac{\alpha}{2}\right)\\[4pt]+\dfrac{1}{2}[\pi+\sin\alpha\cos\alpha-\alpha]\end{array}\right\}}$
Da8	$i_a\sqrt{\dfrac{1}{2\pi}\displaystyle\int_0^\pi \sin^2\omega t\cdot S_{a1}(\omega t)d\omega t}$	$i_a\sqrt{\dfrac{1}{4}-\dfrac{2m_a}{3\pi}}$
Da11	$i_a\sqrt{\dfrac{1}{2\pi}\displaystyle\int_0^\alpha \sin^2\omega t\cdot S_{a1}(\omega t)d\omega t}$	$i_a\sqrt{\dfrac{1}{2\pi}\left\{\begin{array}{l}\dfrac{1}{2}[\alpha-\sin\alpha\cos\alpha]\\[4pt]-\dfrac{4}{3}m_a\sin^4\left(\dfrac{\alpha}{2}\right)[\cos\alpha+2]\end{array}\right\}}$
Sa1	$i_a\sqrt{\dfrac{1}{\pi}\displaystyle\int_0^\pi \sin^2\omega t\cdot S_{a1}(\omega t)d\omega t}$	$i_a\sqrt{\dfrac{1}{2}-\dfrac{4m_a}{3\pi}}$
Sa2	$i_a\sqrt{\dfrac{1}{\pi}\displaystyle\int_0^\alpha \sin^2\omega t\cdot S_{a1}(\omega t)d\omega t}$	$i_a\sqrt{\dfrac{1}{\pi}\left\{\begin{array}{l}\dfrac{1}{2}[\alpha-\sin\alpha\cos\alpha]\\[4pt]-\dfrac{4}{3}m_a\sin^4\left(\dfrac{\alpha}{2}\right)[\cos\alpha+2]\end{array}\right\}}$

off and vice versa. Therefore, the maximum voltage across the inner cell switches (S×2) and the outer cell switches (S×1) will be $V_{dc}/4$ and $V_{dc}/2$, respectively.

7.4 Control Strategy

The proposed power balance control technique outlined in Chapter 5 is applied for this 5L-M²VR topology. Two control loops (dc-link voltage control and carrier-based current control) in Fig. 7.4 are derived according to the power balance principle, and they are briefly explained in the following subsections.

7.4.1 Power Balance Principle

A high input power factor can be achieved under both balanced and unbalanced grid conditions with the simplest power balance control technique based on the dynamic current control. According to the power balance principle, the reference current is equivalent to the summation of load current estimator (the first current component) and the equivalent capacitor current (the second current component) is as given below:

$$
I_a^*(t) = \begin{bmatrix} i_{load}(t) \\ +i_{Ceq}(t) \end{bmatrix} \sin \omega t = \left[\underbrace{\frac{\sum (V_{xn}(t) - L_x \frac{dI_x(t)}{dt} - R_x I_x(t)) I_x(t)}{V_{dc}(t)}}_{\text{First current component}} + \underbrace{C_{eq} \frac{dV_{dc}(t)}{dt}}_{\text{Second current component}} \right] \sin \omega t \tag{7.2}
$$

"x" is denoted as the phase "a," "b," and "c" for the grid phase voltage $V_{xn}(t)$ and input current $i_x(t)$. C_{eq} is the equivalent capacitance value of the dc-link capacitors.

The equivalent capacitor current (i_{Ceq}) in equation (7.2) is obtained from the dc-link voltage control. The unit template of the supply voltage, $\sin(\omega t)$, in equation (7.2) for the reference current phase "a" is formulated as follows:

$$
V_{peak} = \left[\left(\frac{2}{3} V_{an}(t) - \frac{1}{3} V_{bn}(t) - \frac{1}{3} V_{cn}(t) \right)^2 + \left(\frac{1}{\sqrt{3}} V_{bn}(t) - \frac{1}{\sqrt{3}} V_{cn}(t) \right)^2 \right]^{1/2} \tag{7.3}
$$

$$
\sin \omega t = \frac{V_{an}(t)}{V_{peak}} \tag{7.4}
$$

7.4.2 dc-Link Voltage Control

The second current component depends on the variation of the dc-link voltage. Therefore, the energy storage model can also be derived based on the power balance principle. With this model, a simple PI control is implemented to regulate the dc-link voltage in the outer control loop. Thus, the open-loop transfer function of the energy storage model is written as:

$$
L(s) = K_p \left(1 + \frac{1}{T_i s} \right) \times \frac{3V_m}{V_{dc} C_{eq}} \times \frac{1}{s + 6V_m/V_{dc} C_{eq} R_{load}} \tag{7.5}
$$

Based on equation (7.5), the proportional gain K_p is selected according to the pole cancellation in order to achieve a stable output dc voltage.

7.4.3 Current Control

Two types of low switching modulation schemes are applied to the switches (Sx1 and Sx2) in the 5L-M^{2}VR topology. For the outer cell switches (Sx1), the PWM techniques are implemented by comparing the two triangular carrier waves of 1 kHz frequency with the current error. Thus, the two modulated signals are generated. However, only a single gate signal drives the switch (Sx1); therefore, the two modulated signals are fed into the logic NAND gate.

As mentioned earlier, the inner cell switches (S×2) are controlled by the switching angle at the zero crossing point of the grid voltage. In order to achieve optimal performances for the 5L-M^{2}VR, it is very important to define the range of the switching angle for S×2 due to the combination with the PWM techniques applied for S×1.

An input power factor of 0.99 is achieved with a switching angle of more than 10° as shown in Fig. 7.5(a). Besides that, to achieve low current distortion, the range of the switching angle is between 20° and 30°, which can be seen in Fig. 7.5(b). The comparison results of both the PWM techniques (LS-PWM and PS-PWM) are presented in Fig. 7.5(a) and (b) are obtained based on the same constant modulation index. These results have clearly shown that the optimum performances can be achieved by the combination of LS-PWM with the switching angle of 30° or PS-PWM with 25°.

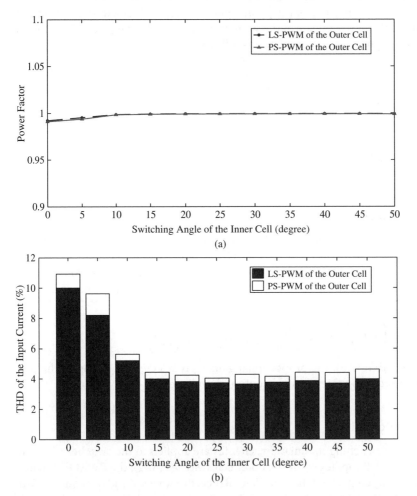

Figure 7.5 Input characteristic performance of the novel rectifier with the proposed switching technique based on 1 kHz PWM with the respective switching angle range between 0° and 50°. (a) Input power factor and (b) THD of the input grid current.

In the closed-loop control strategy (Fig. 7.4), the inner cell switches driven by the fixed switching angle are unable to control the dc-link voltage. Due to the 5L-M²VR configuration, only the outer cells are directly connected between the input grid and the output dc-bus terminal. Hence, the dc-link voltage is regulated with the outer loop voltage control involving only outer cell switches.

Such a power balance control technique does not include any balancing control for the dc-link capacitors voltage. Thus, an additional dc–dc balancing circuit is required to balance the capacitor voltage in the dc link. The voltage balancing dc–dc converter is not discussed in this chapter. The voltage balancing control strategy and analysis are detailed in [6, 7].

7.5 Validation

The results for the proposed 5L-M²VR topology with the proposed controller scheme are obtained with the aid of the MATLAB simulator. The results are shown in Figs. 7.6 to 7.8 with the implementation of the LS-PWM technique for Sx1 and a fixed switching angle of 30° for Sx2.

In Fig. 7.6, the input characteristic of 5L-M²VR (Fig. 7.1) results in a good approximation of current ripple reduction with the time varying DT_s, which is expressed in equation (7.6):

$$\Delta I_a(t) = \int_0^{DT_s} \frac{V_{an}(t) - V_{am}(t) - V_{mn}(t)}{L_a} \, dt \tag{7.6}$$

With this approximation, the rate of change of the current depends on the variable duty cycle ($0 \leq D \leq 1$) of PWM signals. Due to the advantage of a higher-level stepped voltage waveform, a lower switching frequency of PWM signals for Sx1 is sufficient to improve the quality of the line current, together with a fixed switching angle of 30° conducted by Sx2. As the number of the synthesized voltage steps increase, the shape of the input voltage is more sinusoidal in nature. This results in a drastic reduction in the size of the input line inductance with lower switching and conduction losses.

The input line current total harmonic distortion (THD) obtained from the FFT analysis tool is below 6% as shown in Fig. 7.8. Having such a lower THD current results in a higher power factor (cos φ) near to unity (Fig. 7.7); THD and power factor are related

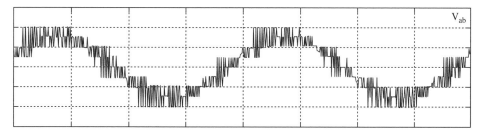

Figure 7.6 Input rectifier line-to-line voltage (voltage referred from node "a" to node "b") refers to Fig. 7.1.

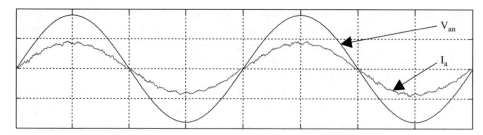

Figure 7.7 Grid phase voltage (voltage referred from node "a" to node "n") and line current of phase "a" with a power factor of 0.99.

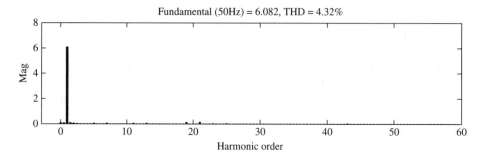

Figure 7.8 Input line current THD of phase "a".

as shown in the equation (7.7):

$$PF = \frac{\cos \varphi_1}{\sqrt{1 - (THD_i)^2}} \tag{7.7}$$

As a result, the displacement factor ($\cos \varphi_1$) in (7.7) is approximately equal to 1 due to the significantly low distortion in the input current and the voltage waveform.

7.6 Summary

In this chapter, a high power factor five-level rectifier topology (5L-M^{2}VR) based on the concepts of a three-phase three-switch three-level rectifier (vienna rectifier) is discussed. The proposed 5L-M^{2}VR utilizes lesser number of components as compared with the conventional unidirectional diode-clamped rectifier.

The advantages of this topology over conventional topologies are discussed in the following:

(a) Good performance is achieved even while operating at a lower switching frequency requiring no complicated EMI filter.
(b) A low-cost integrated control unit for two semi-conductor switches in each phase.
(c) Wide bandwidth control for achieving low THD current and high power factor.
(d) The elimination of the short circuit fault due to the failure of the switching devices results in a higher reliability of the converter.

References

1 E. M. Mehl and I. Barbi, "An improved high-power factor and low-cost three-phase rectifier," *Industry Applications, IEEE Transactions on*, vol. 33, pp. 485–492, 1997.

2 A. I. Maswood and L. Fangrui, "A novel unity power factor input stage for AC drive application," *Power Electronics, IEEE Transactions on*, vol. 20, pp. 839–846, 2005.

3 H. Jin and N. Mohan, "Parallel resonant DC link circuit–a novel zero switching loss topology with minimum voltage stresses," *Power Electronics, IEEE Transactions on*, vol. 6, pp. 687–694, 1991.

4 Y. Xiaoming and I. Barbi, "Zero-voltage switching for the neutral-point-clamped (NPC) inverter," *Industrial Electronics, IEEE Transactions on*, vol. 49, pp. 800–808, 2002.

5 A. I. Maswood and L. Fangrui, "A unity power factor front-end rectifier with hysteresis current control," *Energy Conversion, IEEE Transactions on*, vol. 21, pp. 69–76, 2006.

6 C. Newton and M. Sumner, "Novel technique for maintaining balanced internal DC link voltages in diode clamped five-level inverters," *Electric Power Applications, IEE Proceedings* vol. 146, pp. 341–349, 1999.

7 N. Hatti, Y. Kondo, and H. Akagi, "Five-level diode-clamped PWM converters connected back-to-back for motor drives," *Industry Applications, IEEE Transactions on* vol. 44, pp. 1268–1276, 2008.

8

Five-Level Multiple-Pole Multilevel Rectifier with Reduced Components

Gabriel H. P. Ooi

8.1 Introduction

Most of the commercially available multilevel inverters require a bulky phase-shifted transformer with multiple bridge rectifiers connected at the front-end side [1]. How-ever, the volume and the weight of such a configuration are large and heavy. In addition, more losses are experienced in the transformer during low utilization due to its core resistance. Several new transformerless multilevel rectifier topologies with low switching frequency operation have been reported in the literature. Recent developments on high incremental voltage level rectifier topologies with a reduced number of components are found, namely: (i) packed U cells multilevel converter (PUC) [2], (ii) reduced-part-count diode-clamped rectifier (RPC-DCR) [3], and (iii) hybrid diode-clamped and flying capacitor rectifier (DCLP-FC) [4]. The mentioned low-cost topologies have achieved efficiency and also proven that the filter size can be drastically reduced even with the low switching frequency operation as discussed in Chapter 4.

However, each of these topologies has its limitations and disadvantages. For instance, a complex control algorithm is required to balance the flying capacitors of the three-phase star-configured PUC topology. While in the case of the RPC-DCR topology, only two switches are reduced in each phase-leg but the total component counts are not signifi-cantly optimized. Hence, a huge number of gate drivers and isolated gate supplies are still required. As for the DCLP-FC topology, a good arrangement of the hybrid approach is introduced to reduce 50% of the switching devices as compared with both conventional diode-clamped and flying capacitor rectifiers. Nevertheless, a total of 13 dc capacitors are still needed for this three-phase topology to synthesize a five-level phase voltage stepped waveform. Due to the involvement of the dc electrolytic capacitors, the lifetime of the power converter will eventually be affected by the thermal aging.

This chapter presents a components count optimization for the proposed five-level multiple-pole multilevel unity power factor rectifier (5L-M^{2}UPFR) topology in Fig. 8.1 with a simple design and a cost-effective control algorithm based on the proposed observer control technique. This observer control technique is implemented with the in-phase quantity current control, which can eliminate some of the physical measure-ment sensors in the feedback control loop and provide excellent dynamic response for a two-phase operation in the grid. As a result, more compact size and high reliability

Advanced Multilevel Converters and Applications in Grid Integration, First Edition.
Edited by Ali I. Maswood and Hossein Dehghani Tafti.

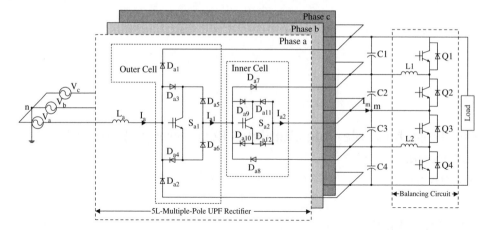

Figure 8.1 Proposed unidirectional five-level multiple-pole multilevel unity power factor rectifier (5L-M²UPFR) with a balancing circuit.

of the three-phase five-level/multiple-pole multilevel unity power factor rectifier is achieved.

8.2 Operation Principle

A proposed 5L-M²UPFR topology shown in Fig. 8.1 is constructed using two three-level (3L) vienna rectifier cells in each phase leg. The multiple-pole structure is similar to that in Fig. 7.1 and the multiple-pole concept is presented in Chapter 6. The output terminals of both vienna rectifier cells are connected to the respective dc capacitors with the aid of a balancing circuit.

Since each cell is characterized by a three-level input voltage stepped waveform, the overall performance for a five-level incremental voltage step is synthesized based on the switching state selection and the direction of the corresponding grid phase current stated in Table 8.1. Therefore, the expression for the input pole voltages of the proposed five-level multiple-pole multilevel unity power factor rectifier (5L-M²UPFR) is written as follows:

$$\begin{cases} V_{am}(t) = \dfrac{V_{dc}(t)}{4}[2 - S_{a1}(t) - S_{a2}(t)] \cdot sign(I_a(t)) \\[2mm] V_{bm}(t) = \dfrac{V_{dc}(t)}{4}[2 - S_{b1}(t) - S_{b2}(t)] \cdot sign(I_b(t)) \\[2mm] V_{cm}(t) = \dfrac{V_{dc}(t)}{4}[2 - S_{c1}(t) - S_{c2}(t)] \cdot sign(I_c(t)) \end{cases} \tag{8.1}$$

where $V_{dc}(t)$ is the dc-link voltage and sign(.) indicates the directional flow of the respective phase current according to the condition given in (8.2). For simplicity, the equations in this paper expressed with the phase "a" term are applied for the other phases "b" and "c" as well.

$$sign(I_a(t)) = \begin{cases} 1 & \text{for } I_a(t) \geq 0 \\ -1 & \text{for } I_a(t) < 0 \end{cases} \tag{8.2}$$

Table 8.1 Switching States for Corresponding Phase "A" Voltage Level.

States	Switching		sign(Ia)	Input Pole Voltage Level, Vam
	Sa1	Sa2		
1	0	0	+	Vdc/2
2	1	0	+	Vdc/4
3	1	1	+ or −	0
4	1	0	−	−Vdc/4
5	0	0	−	−Vdc/2

sign(Ia) is the phase "a" current direction flow from the grid through the switching devices.

The current through the four quadrant switch of the outer vienna rectifier cell is expressed in equation (8.3) based on the respective switching states in Table 8.1.

$$\begin{cases} I_{a1}(t) = I_a(t)S_{a1}(t) \\ I_{b1}(t) = I_b(t)S_{b1}(t) \\ I_{c1}(t) = I_c(t)S_{c1}(t) \end{cases} \tag{8.3}$$

The output bidirectional current ($I_{a2}(t)$, $I_{b2}(t)$, and $I_{c2}(t)$) of the inner cell depends on the current expression in equation (8.3) and the corresponding switching states of the inner cell switch. Hence, the neutral-point-clamped current $I_m(t)$ through node "m" written in equation (8.4) is the summation of the three-phase output bidirectional current.

$$\begin{aligned} I_m(t) &= I_{a1}(t)S_{a2}(t) + I_{b1}(t)S_{b2}(t) + I_{c1}(t)S_{c2}(t) \\ &= I_a(t)S_{a1}(t)S_{a2}(t) + I_b(t)S_{b1}(t)S_{b2}(t) + I_c(t)S_{c1}(t)S_{c2}(t) \end{aligned} \tag{8.4}$$

The bidirectional current through the outer and inner switches (Sa1 and Sa2) of the 5L-M²UPFR expressed in (8.5) is always positive over the fundamental period due to the floating diodes ($D_{a3} − D_{a6}$ and $D_{a9} − D_{a12}$).

$$\begin{cases} I_{Sa1}(t) = |I_a(t)S_{a1}(t)| \\ I_{Sa2}(t) = |I_a(t)S_{a1}(t)S_{a2}(t)| \end{cases} \tag{8.5}$$

The instantaneous current through the respective diode elements of both outer and inner cells are determined by the current flow over the fundamental period and expressed as follows:

Outer Cell 5L-M²UPFR:

a) During positive half cycle

$$\begin{cases} I_{Da1}(t) = I_a(t)[1 − S_{a1}(t)] \\ I_{Da2}(t) = 0 \\ I_{Da3}(t) = I_{Da6}(t) = I_a(t)S_{a1}(t) \\ I_{Da4}(t) = I_{Da5}(t) = 0 \end{cases} \tag{8.6}$$

b) During negative half cycle

$$
\begin{cases}
I_{Da1}(t) = 0 \\
I_{Da2}(t) = I_a(t)[1 - S_{a1}(t)] \\
I_{Da3}(t) = I_{Da6}(t) = 0 \\
I_{Da4}(t) = I_{Da5}(t) = I_a(t)S_{a1}(t)
\end{cases} \tag{8.7}
$$

Inner Cell 5L-M^{2}UPFR:

a) During positive half cycle

$$
\begin{cases}
I_{Da7}(t) = I_a(t)S_{a1}(t)[1 - S_{a2}(t)] \\
I_{Da8}(t) = 0 \\
I_{Da9}(t) = I_{Da12}(t) = I_a(t)S_{a1}(t)S_{a1}(t) \\
I_{Da10}(t) = I_{Da11}(t) = 0
\end{cases} \tag{8.8}
$$

b) During negative half cycle

$$
\begin{cases}
I_{Da1}(t) = 0 \\
I_{Da2}(t) = I_a(t)S_{a1}(t)[1 - S_{a2}(t)] \\
I_{Da3}(t) = I_{Da6}(t) = 0 \\
I_{Da4}(t) = I_{Da5}(t) = I_a(t)S_{a1}(t)S_{a1}(t)
\end{cases} \tag{8.9}
$$

8.3 Modulation Scheme

The gating signals for the switches Sa1 and Sa2 based on the switching states given in Table 8.1 are obtained by comparing the absolute function of the modulation signal with the triangular carriers as shown in Fig. 8.2. The switching states are achieved according to the condition given in (8.10).

$$
S_{a1}(t) = \begin{cases} 1 & \text{if } M_a(t) \leq V_{tri1}(t) \\ 0 & \text{otherwise} \end{cases}
$$

$$
S_{a2}(t) = \begin{cases} 1 & \text{if } M_a(t) \leq V_{tri2}(t) \\ 0 & \text{otherwise} \end{cases} \tag{8.10}
$$

where $V_{tri1}(t)$ and $V_{tri2}(t)$ are the triangular carriers and $M_a(t)$ is the phase "a" modulation signal.

The switching function of the respective phase "a" switches is expressed in (8.11), which describes the corresponding duty cycle distribution following the condition in (8.10) when the modulation signal crosses the edge of the triangular carriers.

$$
\begin{cases}
S_{a1}(t) = 2[1 - m \sin \omega t] \\
S_{a2}(t) = 1 - 2m \sin \omega t
\end{cases} \tag{8.11}
$$

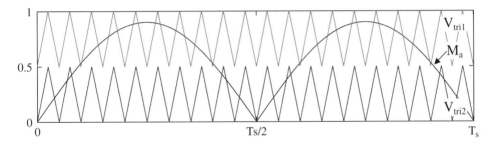

Figure 8.2 Switching scheme for the 5L-M²UPFR topology.

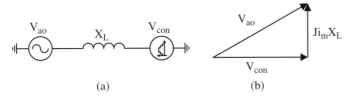

(a) (b)

Figure 8.3 Space vector calculation for the amplitude modulation. (a) Schematic diagram and (b) phasor diagram.

where m is the modulation amplitude obtained according to the space vector diagram in Fig. 8.3 and is expressed as follows:

$$m = \frac{2\sqrt{V_m^2 - \left(\dfrac{2\omega LP_{dc}}{3V_m}\right)^2}}{V_{dc}} \tag{8.12}$$

The space vector calculation for the amplitude modulation is presented in Fig. 8.3, in which V_m is the amplitude of the grid voltage, L is the input line inductance, ω is the angular frequency of the grid side expressed in terms of rad/s and P_{dc} is the output power of the rectifier.

8.4 Control Strategy

The proposed control algorithm is implemented by employing the MATLAB Simulink toolbox with the configuration parameter of fix time step based on Runge-Kutta ODE4 and the sampling time of the control algorithm is set at 90μs. The close-loop control system based on the proposed observer technique and power factor correction control loop is discussed in the following subsection:

8.4.1 Unity Power Factor Control

The proposed unity power factor control with both grid voltage and load current observers is shown in Fig. 8.4. The basic structure of this controller is constructed using the synchronous-reference-frame (SRF) current control [5]. However, the d-q transformation in the SRF control requires a complicated phase lock loop (PLL)

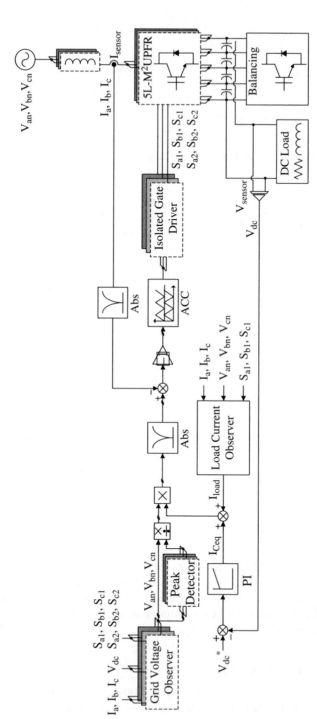

Figure 8.4 Block diagram of the unity power factor controller with the proposed grid phase voltage and load current observers.

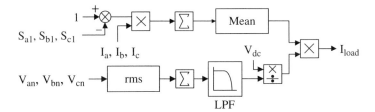

Figure 8.5 Load current observer of Fig. 8.4 based on equation (8.14).

design that limits the control bandwidth. Therefore, the in-phase quantities current control technique in the proposed control is implemented to eliminate the mentioned disadvantages of the d-q (direct-quadrature) transformation. Moreover, the two-phase operation is allowed with this proposed control that provides higher reliability in a three-phase power supply system [6].

This control method consists of a dc-link voltage control at the outer loop and current control at the inner loop. The outer loop control computes the dc equivalent capacitors current and regulates the output dc-link voltage. Meanwhile, the load current is formulated from the power balanced principle as shown in the following:

$$V_{dc}I_{dc}(t) = \sum \left[V_{sn}(t) - L_s\frac{dI_s(t)}{dt} - R_sI_s(t) \right] I_s(t) + C_{eq}V_{dc}(t)\frac{dV_{dc}(t)}{dt} \qquad (8.13)$$

The actual load current measurement is replaced with the load current control loop of Fig. 8.5. The feed-forward observer technique is implemented to achieve a better dc-link voltage tracking response during the load change. The load current expression in (8.14) can be estimated from equation (8.13) assuming that the average power in the energy storage elements is zero.

$$I_{load} = \frac{V_{dc}(I_{Da1} + I_{Db1} + I_{Dc1})}{V_{an} + V_{bn} + V_{cn}} \qquad (8.14)$$

V_{an}, V_{bn}, and V_{cn} are the rms values of the grid phase voltage. V_{dc} and I_{dc} are the mean values of the dc-link voltage and output rectifier current, respectively. Since six IGBTs are used in this rectifier configuration, the I_{dc} current can be easily obtained by using the feedback gating signals and the summation of the diodes current (I_{Da1}, I_{Db1}, and I_{Dc1}) from equation (8.6) during the positive half period.

8.4.2 Voltage Control

In order to compute the equivalent dc capacitor current in the second term of (8.13), the dc-link voltage is regulated with a simple PI controller which is expressed as follows:

$$I_{Ceq} = K_p\frac{T_i s + 1}{T_i s}[V_{dc}^*(t) - V_{dc}(t)] \qquad (8.15)$$

where K_p is the proportional gain of the dc-link voltage regulator and T_i is the settling time of the dc-link voltage tracking. The implemented algorithm for the calculation of the grid maximum voltage is presented in Fig. 8.6. Alpha–beta equivalent vectors of the abc voltage are calculated. Subsequently, the maximum amplitude is computed.

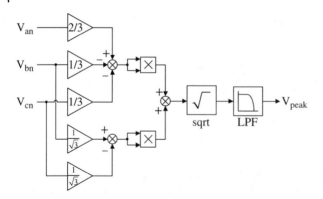

Figure 8.6 Peak detector of the grid voltage for the reference sinusoidal wave.

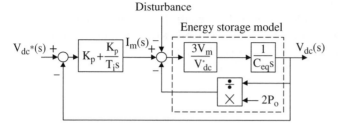

Figure 8.7 Energy storage model of the output rectifier.

The approximation value of K_p is obtained from the energy storage model as shown in Fig. 8.7. The derivation of this model is based on the power balanced principle to determine the output equivalent dc capacitor current.

During the time interval from (t_0) to (t_1) of the charging period of the capacitors, the voltage variation due to the charge of energy storage, ΔE_{dc}, is shown in the following expression:

$$\Delta E_{dc} = \frac{C_{eq}}{2}[V_{dc}^2(t_1) - V_{dc}^2(t_0)] \tag{8.16}$$

The power of the C_{eq}, ΔP_{dc} in (8.17) is obtained from the derivative of the energy stored.

$$\Delta P_{dc} = \frac{C_{eq}}{2}\frac{dV_{dc}^2(t)}{dt} = \frac{3}{2}V_m I_m - P_o(t) \tag{8.17}$$

where V_m and I_m are the amplitude of the grid voltage and grid current, assuming the losses are neglected for the balanced three-phase system in this case.

According to stability criteria, the proportional gain of the control system expressed in (8.18) is derived from the open-loop transfer function of Fig. 8.7 with the pole cancellation.

$$K_p = \frac{V_{dc}^3 C_{eq}}{6P_o V_m} \tag{8.18}$$

where C_{eq} is the equivalent dc-link capacitors and P_o is the average output load power.

8.4.3 Current Control

The grid current control technique of the active rectifier can be classified into four main categories such as space vector modulation (SVM) [7], fixed hysteresis band current control (FHBCC) [5, 8, 9], variable hysteresis band current control (VHBCC) [10], and average current control (ACC) [9, 11]. The SVM scheme requires a higher computational effort due to the complex sector control algorithm required for the higher voltage stepped level rectifier topology [12]. Both HBCC and ACC can overcome the stated problems of the SVM scheme. However, the FHBCC scheme exhibits the disadvantage of the variable switching frequency that complicates the design of the input inductance filter. The comparison performance of the FHBCC and ACC method is detailed in [9].

The carrier based ACC scheme is applied for the proposed 5L-M^2UPFR and allows the desired voltage space vector to be modulated using simple analog comparators. By doing so, the lower cost implementation and the lesser computational effort needed are achieved.

The input current shaping of 5L-M^2UPFR depends on the carrier-based modulation scheme in Fig. 8.2 and the condition given in equation (8.10). Hence, the current error, which is the difference between the measured current and the sinusoidal reference current template, will be the modulation signal $M_a(t)$ of Fig. 8.2.

The sinusoidal reference current template is realized from the grid voltage and the peak detector as shown in Fig. 8.4. The peak detector is designed based on the mathematical analysis of the space vector diagram as shown in Figs. 8.3 and 8.6.

8.4.4 Grid Voltage Observer

Several observer techniques have been proposed for various types of rectifier configuration [13–15]. Besides the advantage of eliminating the sensors needed, the observer technique reduces the size of the converter and provides a lower production cost as well. A PWM rectifier without voltage measurements is presented by Ohnuki *et al.* in [14]. Even though the information of the three-phase grid currents are sufficient to derive and estimate the AC and DC voltages, large dc-link voltage ripples are experienced during the computational process, and hence, causing the input current THD to be considerably high. As a result, a bulky input inductor is required to filter the current distortion.

In [13], a three-level vienna rectifier without current sensors is proposed. The power factor control is designed based on the phase angle difference between the grid and pole voltages of the rectifier, which is determined for calculating the modulation index space vector. Although this method provides a good dynamic response during load change, the unbalanced grid condition is not considered in this case.

The grid voltage observer of the proposed controller shown in Fig. 8.8 is modified from the single-phase two-stages PWM rectifier in [15]. The high accuracy of the grid phase voltage is estimated with three-phase grid currents and dc-link voltage measurements using the proportional + resonant control method. Low input current THD is achieved with the proposed observer and good dynamic response is performed as well for the case of a single phase grid voltage down during the operation.

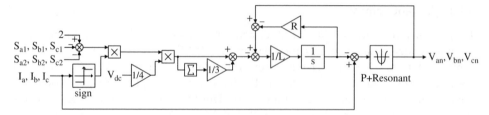

Figure 8.8 Grid voltage observer of Fig. 8.4 based on equations (8.19) and (8.20).

The voltage across the input phase "a" inductor is expressed as follows:

$$V_{La}(t) + V_{Ra}(t) = V_{an}(t) - V_{am}(t) - V_{mn}(t)$$

$$= V_{an}(t) - \frac{V_{dc}(t)}{12} \left\{ \begin{array}{l} 2[2 - S_{a1}(t) - S_{a2}(t)] \cdot \text{sign}(I_a(t)) \\ -[2 - S_{b1}(t) - S_{b2}(t)] \cdot \text{sign}(I_b(t)) \\ -[2 - S_{c1}(t) - S_{c2}(t)] \cdot \text{sign}(I_c(t)) \end{array} \right\} \tag{8.19}$$

$V_{mn}(t)$ is the virtual ground voltage of node "m" referred to the ground terminal of node n. $V_{Ra}(t)$ is the voltage drop across the inductor core resistor of phase "a."

The grid phase voltage can be calculated by expanding the voltage expression in (8.19) and it is given by

$$\tilde{I}_a(t) = \frac{1}{L_a} \int \left\{ \begin{array}{c} V_{an}(t) - R_a \tilde{I}_a(t) \\ -\frac{V_{dc}(t)}{12} \left\{ \begin{array}{l} 2[2 - S_{a1}(t) - S_{a2}(t)] \cdot \text{sign}(I_a(t)) \\ -[2 - S_{b1}(t) - S_{b2}(t)] \cdot \text{sign}(I_b(t)) \\ -[2 - S_{c1}(t) - S_{c2}(t)] \cdot \text{sign}(I_c(t)) \end{array} \right\} \end{array} \right\} dt \tag{8.20}$$

From equation (8.20), a proportional + resonant controller is used for the grid phase voltage estimation and is expressed in equation (8.21):

$$V_{an}(t) = [I_a(t) - \tilde{I}_a(t)] \cdot \left[K_p + \frac{K_r s}{s^2 + \omega^2} \right] \tag{8.21}$$

ω is the angular frequency of the grid side ($\omega = 314$ rad/s) and the proportional gain and resonant gain is chosen to be $K_p = 3.25$ and $K_r = 125$, respectively.

8.5 Design Considerations

In order to maintain a minimum safety requirement for the rectifier operation in the grid, an analytical approximation of the device stress level is calculated based on the presented analytical method as detailed in Chapters 6 and 7. To select the components rating for a 5L-M²UPFR topology, a worst case operating condition is considered. The voltage stress level across the power device is selected according to the dc voltage reference set by the feedback control loop in the control algorithm and the current stress of the power device can be determined by the minimum modulation indices as given in equation (8.12) for the minimum worst case scenario. The analytical approximation of the minimum voltage and current stresses is presented in the following subsection.

8.5.1 Device Current Stresses

Several assumptions have been made for the current stress analysis: (i) pure sinusoidal grid current without any ripple content, (ii) unity power factor, (iii) zero voltage drop across the inductors, (iv) constant switching frequency, (v) balanced input grid condition and (vi) the negligence of any losses in the balancing circuit. Hence, the average and RMS of the approximated global current stress are expressed in equations (8.22) to (8.23).

$$
\begin{aligned}
I_{Sai\langle avg\rangle} &= \frac{1}{2\pi} \int_\alpha^\beta \left[\frac{1}{t_p} \sum_{k=t_{on}/t_p}^{\gamma_{ai}} I_a(kt_p, \omega t) \right] d\omega t \\
&\approx \frac{1}{2\pi} \int_\alpha^\beta \left[\frac{1}{t_p} \int_0^{\gamma_{ai}t_p} I_a(\omega t)\, dt_\mu \right] d\omega t
\end{aligned}
\tag{8.22}
$$

$$
\begin{aligned}
I_{Sai\langle avg\rangle} &= \sqrt{ \frac{1}{2\pi} \int_\alpha^\beta \left[\frac{1}{t_p} \sum_{k=t_{on}/t_p}^{\gamma_{ai}} I_a^2(kt_p, \omega t) \right] d\omega t } \\
&\approx \sqrt{ \frac{1}{2\pi} \int_\alpha^\beta \left[\frac{1}{t_p} \int_0^{\gamma_{ai}t_p} I_a^2(\omega t)\, dt_\mu \right] d\omega t }
\end{aligned}
\tag{8.23}
$$

γ_{ai} is the switching state function of the instantaneous grid current. k is the duty cycle during the turn on period. α and β are the switching times occurring within one fundamental period. Similarly, the current stress for the diode elements can also be expressed using the same expressions in (8.22) and (8.23).

Based on the current stresses expression, the final expression of the current stresses for each power device is written as:

Average current stress:

$$
\begin{cases}
I_{Sa1\langle avg\rangle} = \dfrac{i_a}{\pi} \left\{ 2 + 2\cos\left(\sin^{-1}\dfrac{1}{2m_a}\right) + m_a \left[2\sin^{-1}\dfrac{1}{2m_a} - \pi - \sin\left(2\sin^{-1}\dfrac{1}{2m_a}\right) \right] \right\} \\[2mm]
I_{Sa2\langle avg\rangle} = \dfrac{i_a}{\pi} \left\{ 2 - 2m_a\sin^{-1}\dfrac{1}{2m_a} - \cos\left(\sin^{-1}\dfrac{1}{2m_a}\right) \right\} \\[2mm]
I_{Da1\langle avg\rangle} = I_{Da2\langle avg\rangle} = \dfrac{i_a}{\pi} \left\{ \dfrac{m_a}{2}\left[\pi + \sin\left(2\sin^{-1}\dfrac{1}{2m_a}\right) - 2\sin^{-1}\dfrac{1}{2m_a} \right] \right. \\[2mm]
\hspace{6cm} \left. - \cos\left(\sin^{-1}\dfrac{1}{2m_a}\right) \right\} \\[2mm]
I_{Da3\langle avg\rangle} = I_{Da4\langle avg\rangle} = I_{Da5\langle avg\rangle} = I_{Da6\langle avg\rangle} = \dfrac{I_{Sa1\langle avg\rangle}}{2}
\end{cases}
$$

$$
\begin{cases}
I_{Da7\langle avg\rangle} = I_{Da8\langle avg\rangle} = \dfrac{i_a}{\pi} \left\{ \dfrac{3}{2}\cos\left(\sin^{-1}\dfrac{1}{2m_a}\right) \right. \\[2mm]
\hspace{3cm} \left. - \dfrac{m_a}{2}\left[\pi + \sin\left(2\sin^{-1}\dfrac{1}{2m_a}\right) - 4\sin^{-1}\dfrac{1}{2m_a} \right] \right\} \\[2mm]
I_{Da9\langle avg\rangle} = I_{Da10\langle avg\rangle} = I_{Da11\langle avg\rangle} = I_{Da12\langle avg\rangle} = \dfrac{I_{Sa2\langle avg\rangle}}{2}
\end{cases}
\tag{8.24}
$$

RMS current stress:

$$
\left\{
\begin{aligned}
I_{Sa1\langle RMS\rangle} &= i_a \frac{1}{\pi} \sqrt{
\begin{aligned}
&\left\{
\begin{aligned}
&\pi + \sin\left(2\sin^{-1}\frac{1}{2m_a}\right) - \sin^{-1}\frac{1}{2m_a} - \frac{1}{2m_a}\cos\left(\sin^{-1}\frac{1}{2m_a}\right) \\
&+\frac{m_a}{3}\left[\cos\left(3\sin^{-1}\frac{1}{2m_a}\right) - 9\cos\left(\sin^{-1}\frac{1}{2m_a}\right)\right]
\end{aligned}
\right\}
\end{aligned}
} \\[3mm]
I_{Sa2\langle RMS\rangle} &= i_a \frac{1}{\pi} \sqrt{
\left\{
\begin{aligned}
&\sin^{-1}\frac{1}{2m_a} - \frac{1}{2m_a}\cos\left(\sin^{-1}\frac{1}{2m_a}\right) \\
&-\frac{16}{3}m_a\sin^4\left(\frac{1}{2}\sin^{-1}\frac{1}{2m_a}\right)\cdot\left[\cos\left(\sin^{-1}\frac{1}{2m_a}\right) + 2\right]
\end{aligned}
\right\}
}
\end{aligned}
\right.
$$

$$
\left\{
\begin{aligned}
I_{Da1\langle RMS\rangle} &= I_{Da2\langle RMS\rangle} = i_a \frac{1}{\pi}\sqrt{
\left\{
\begin{aligned}
&\frac{m_a}{6}\left[9\cos\left(\sin^{-1}\frac{1}{2m_a}\right) - \cos\left(3\sin^{-1}\frac{1}{2m_a}\right)\right] \\
&-\frac{1}{4}\left[\pi + \sin\left(2\sin^{-1}\frac{1}{2m_a}\right) - 2\sin^{-1}\frac{1}{2m_a}\right]
\end{aligned}
\right\}
} \\[3mm]
I_{Da3\langle RMS\rangle} &= I_{Da4\langle RMS\rangle} = I_{Da5\langle RMS\rangle} = I_{Da6\langle RMS\rangle} = \frac{I_{Sa1\langle RMS\rangle}}{\sqrt{2}} \\[3mm]
I_{Da7\langle RMS\rangle} &= I_{Da8\langle RMS\rangle} \\[1mm]
&= i_a \frac{1}{\pi}\sqrt{
\left\{
\begin{aligned}
&\frac{1}{2}\left[\pi + \sin\left(2\sin^{-1}\frac{1}{2m_a}\right) - 2\sin^{-1}\frac{1}{2m_a}\right] \\
&+\frac{m_a}{6}\left[
\begin{aligned}
&\cos\left(3\sin^{-1}\frac{1}{2m_a}\right) - 9\cos\left(\sin^{-1}\frac{1}{2m_a}\right) \\
&+16\sin^4\left(\frac{1}{2}\sin^{-1}\frac{1}{2m_a}\right)\cdot\left[\cos\left(\sin^{-1}\frac{1}{2m_a}\right) + 2\right]
\end{aligned}
\right]
\end{aligned}
\right\}
} \\[3mm]
I_{Da9\langle RMS\rangle} &= I_{Da10\langle RMS\rangle} = I_{Da11\langle RMS\rangle} = I_{Da12\langle RMS\rangle} = \frac{I_{Sa2\langle RMS\rangle}}{\sqrt{2}} \\[3mm]
I_{C1\langle RMS\rangle} &= I_{C4\langle RMS\rangle} = \sqrt{3\left(I^2_{Da1\langle RMS\rangle} - 3I^2_{Da1\langle avg\rangle}\right)} \\[3mm]
I_{C2\langle RMS\rangle} &= I_{C3\langle RMS\rangle} = \frac{I_{C1\langle RMS\rangle}}{2}
\end{aligned}
\right.
$$

$$(8.25)$$

8.5.2 Device Voltage Stress

Voltage stress analysis is essential to prevent the power devices from being damaged and affecting the reliability of the converter. Thus, minimum safety requirements and smooth operations can be assured with the proper blocking capability. Two conditions are assumed for the voltage stress analysis, which are: (i) the negligence of overvoltage caused by the parasitic inductance due to low switching frequency operations, and (ii) the negligence of the transient output dc voltage caused by the inrush current since a pre-charging circuit is typically employed in the dc-link for practical cases.

The overall voltage stress for the outer and inner cell of 5L-M^{2}UPFR is approximately V_{dc} and $3V_{dc}/4$ across the respective clamped dc-link capacitors to the negative dc rail. Therefore, the maximum voltage stress across each active component during the steady-state operation is expressed in the following:

Switching elements:

a)

$$\begin{cases} V_{Sa1} = \dfrac{V_{dc}}{3} \\ V_{Sa2} = \dfrac{V_{dc}}{4} \end{cases} \tag{8.26}$$

b)

$$\begin{cases} V_{Da1} = V_{Da2} = V_{dc} \\ V_{Da3} = V_{Da4} = V_{Da5} = V_{Da6} = \dfrac{V_{dc}}{3} \\ V_{Da7} = V_{Da8} = \dfrac{V_{dc}}{2} \\ V_{Da9} = V_{Da10} = V_{Da11} = V_{Da12} = \dfrac{V_{dc}}{4} \end{cases} \tag{8.27}$$

8.6   Validation

The laboratory prototype is developed with the control algorithm implemented in the dSPACE DS1103 development controller board to verify the performance of 5L-M^{2}UPFR. Both simulation and experimental results have proven the feasibility of the proposed rectifier topology with the controller based on the grid phase voltage and load current observers.

The dynamic response of the grid phase voltage estimation during the unbalanced grid condition is verified with both the estimation and the actual measurement as shown in Fig. 8.9. In addition to that, the peak value of the grid phase voltage obtained by the peak detector is stable in Fig. 8.10 even during a sudden grid phase voltage change.

8.7   Experimental Verification

The experimental results in Fig. 8.11 show the input characteristic performance of the new 5L-M^{2}UPFR with the proposed controller under a balanced grid supply condition. The power factor for the balanced grid supply is high and low THD current is achieved with a low input inductance filter.

On top of that, the experimental results in Fig. 8.12 show the feasibility study of the converter response during two phase operations with the same in-phase quantity current control. Even though one of the phase voltages is down during the operation, comparatively low THD current in Fig.8.12(c) is achieved with the same input inductance filter. With the supported experimental results, the proposed controller proved that a high reliability of the three-phase power supply unit is achievable under an extreme unbalanced grid condition.

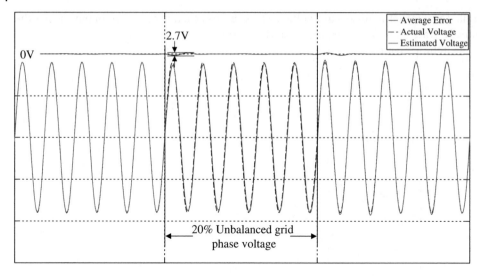

Figure 8.9 Grid phase voltage of the estimation method and actual measurement.

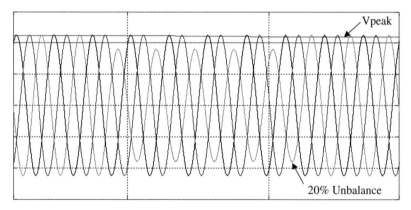

Figure 8.10 Peak voltage measurement and three-phase grid phase voltage with one phase 20% unbalance.

Based on both experimental results obtained for balanced and unbalanced grid operations, a five-level incremental stepped waveform is synthesized with the proposed ACC scheme based on the low switching frequency carrier. The proposed dynamic control with the observers reduces the cost implementation and the complexity of the algorithm.

8.8 Summary

A new cost-effective 5L-M²UPFR topology is introduced in this chapter to achieve a high power factor and low current distortion with a drastic reduction in the total number of switching devices and sensors. Moreover, no isolated gate power supply is required for the rectifier side since there are no complementary switches used in this proposed rectifier topology. The great advantage of this proposed rectifier is that a low grid current

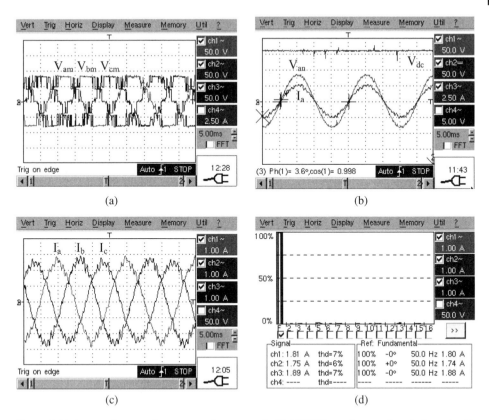

Figure 8.11 Input voltage and current performance of a 5L-M²UPFR topology under a balanced grid condition. (a) Input pole voltage, (b) dc-link voltage (upper trace), grid phase voltage and line current (lower trace), (c) input line current, and (d) THD of line current.

THD is achievable with a constant low switching frequency operation and a reduced input inductance filter size.

Both simulation and experimental results proved the dynamic response of the proposed in-phase quantities current control using the observer technique. Therefore, the proposed controller provides a higher reliability of the three-phase power supply even during the extreme unbalanced grid condition. Since grid voltage and load current observers are designed in the control loop, the reduction of sensors can avoid technical issues such as sensor failure and measurement errors. In addition to, light weight and high power density can be achieved for the proposed 5L-M²UPFR topology.

Several technical challenges and improvements need to be considered for future development such as dc balancing and dc voltage ripple content that occur during the extreme unbalance grid condition. However, the dc voltage ripple can be filtered out by replacing a larger filter capacitor. Thanks to the use of the in-phase quantities current control using the observer technique, a 0.85 power factor with a low current distortion of 7% is achievable. The proposed rectifier with the control technique is recommended for high power application in ac–dc–ac drives for energy efficiency.

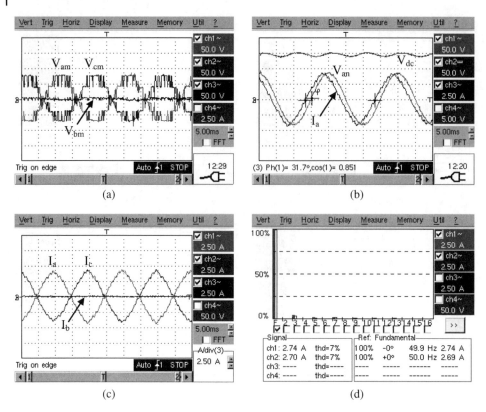

Figure 8.12 Input voltage and current performance of a 5L-M²UPFR topology under an extreme unbalanced grid condition. (a) Input pole voltage, (b) dc-link voltage (upper trace), grid phase voltage and line current (lower trace), input line current, and (d) THD of line current.

References

1 S. Kouro, M. Malinowski, K. Gopakumar, J. Pou, L. G. Franquelo, W. Bin, *et al.*, "Recent advances and industrial applications of multilevel converters," *Industrial Electronics, IEEE Transactions on*, vol. 57, pp. 2553–2580, 2010.
2 Y. Ounejjar, K. Al-Haddad, Gre, x, and L. A. goire, "Packed U cells multilevel converter topology: Theoretical study and experimental validation," *Industrial Electronics, IEEE Transactions on*, vol. 58, pp. 1294–1306, 2011.
3 K.A. Corzine and J.R. Baker, "Reduced-parts-count multilevel rectifiers," *Industrial Electronics, IEEE Transactions on*, vol. 49, pp. 766–774, 2002.
4 J.I. Itoh, Y. Noge, and T. Adachi, "A novel five-level three-phase PWM rectifier with reduced switch count," *Power Electronics, IEEE Transactions on*, vol. 26, pp. 2221–2228, 2011.
5 A.I. Maswood and L. Fangrui, "A unity-power-factor converter using the synchronous-reference-frame-based hysteresis current control," *Industry Applications, IEEE Transactions on*, vol. 43, pp. 593–599, 2007.

6 J. Minibock and J.W. Kolar, "Novel concept for mains voltage proportional input current shaping of a VIENNA rectifier eliminating controller multipliers," *Industrial Electronics, IEEE Transactions on*, vol. 52, pp. 162–170, 2005.

7 I. Barbi and F.A.B. Batista, "Space vector modulation for two-level unidirectional PWM rectifiers," *Power Electronics, IEEE Transactions on*, vol. 25, pp. 178–187, 2010.

8 A.I. Maswood and L. Fangrui, "A unity power factor front-end rectifier with hysteresis current control," *Energy Conversion, IEEE Transactions on*, vol. 21, pp. 69–76, 2006.

9 A.I. Maswood, E. Al-Ammar, and F. Liu, "Average and hysteresis current-controlled three-phase three-level unity power factor rectifier operation and performance," *Power Electronics, IET*, vol. 4, pp. 752–758, 2011.

10 L. Fangrui and A.I. Maswood, "A novel variable hysteresis band current control of three-phase three-level unity PF rectifier with constant switching frequency," *Power Electronics, IEEE Transactions on*, vol. 21, pp. 1727–1734, 2006.

11 S. Jong-Won and C. Bo-Hyung, "Digitally implemented average current-mode control in discontinuous conduction mode PFC rectifier," *Power Electronics, IEEE Transactions on*, vol. 27, pp. 3363–3373, 2012.

12 P. Ide, N. Froehleke, and H. Grotstollen, "Investigation of low cost control schemes for a selected 3-level switched mode rectifier," in *Telecommunications Energy Conference, 1997. INTELEC 97., 19th International*, pp. 413–418, 1997.

13 B. Wang, G. Venkataramanan, and A. Bendre, "Unity power factor control for three-phase three-level rectifiers without current sensors," *Industry Applications, IEEE Transactions on*, vol. 43, pp. 1341–1348, 2007.

14 T. Ohnuki, O. Miyashita, P. Lataire, and G. Maggetto, "Control of a three-phase PWM rectifier using estimated AC-side and DC-side voltages," *Power Electronics, IEEE Transactions on*, vol. 14, pp. 222–226, 1999.

15 R. Ghosh and G. Narayanan, "Generalized feedforward control of single-phase PWM rectifiers using disturbance observers," *Industrial Electronics, IEEE Transactions on*, vol. 54, pp. 984–993, 2007.

9

Four-Quadrant Reduced Modular Cell Rectifier
Ziyou Lim

9.1 Introduction

Multilevel converters are popularly recognized for their perceived merits such as possessing a higher power handling capability as well as delivering excellent power quality with the intelligent synthesis of m-level (mL) waveforms. Such attractive benefits have helped multilevel converters emerge as the most viable industrial solutions for medium to high voltage drives [1, 2], renewable energy systems [3, 4], power transmission systems [5, 6], and electrical transportations [7, 8].

Producing a low harmonic profile with high incremental stepped voltage waveforms has proved to be advantageous at all times. However, several complications arise in these established multilevel topologies, especially the increase in component counts, production cost, and converter size. Besides complicating the converter design, the overall reliability and system efficiency are also greatly affected. An extensive research is further carried out on multilevel converters by enhancing the performance with alternative new topologies.

Several multilevel topologies thus far reported in the recent literature are developed to synthesize higher mL voltages with reduced component counts. But most of these converters are employed as high power inverters to perform dc–ac electrical conversion. Even though higher resolution waveforms and better efficiency are achieved with these multilevel inverters [9–13], none of them is retrofittable as a three-phase front-end active rectifier in the ac/dc/ac system due to the multiple isolated dc sources configuration involved in the designs.

Multilevel rectifiers with reduced part counts have been proposed previously as well, as shown in Fig. 9.1, which are known as (I) unidirectional flying capacitor rectifier [14], (II) unidirectional diode-clamped rectifier [15], (III) reduced-parts-count multilevel rectifier [16], and (IV) hybrid PWM rectifier [17]. The active switch counts of these high-power rectifiers are reduced because of the replacement made with the power diodes. As a result, these multilevel rectifiers are two quadrants operated as the positive switched voltage waveforms are produced only with the positive polarity of input currents and vice versa for the negative cycle. In this case, the applicability of (I) to (IV) is restricted to non-regenerative systems where bidirectional power flow is not required.

Therefore, a modular cell rectifier with reduced flying capacitors and active switches (RFCR) is proposed using a hybrid carrier-based PWM named "hybrid phase-shifted

Advanced Multilevel Converters and Applications in Grid Integration, First Edition.
Edited by Ali I. Maswood and Hossein Dehghani Tafti.
© 2019 John Wiley & Sons Ltd. Published 2019 by John Wiley & Sons Ltd.

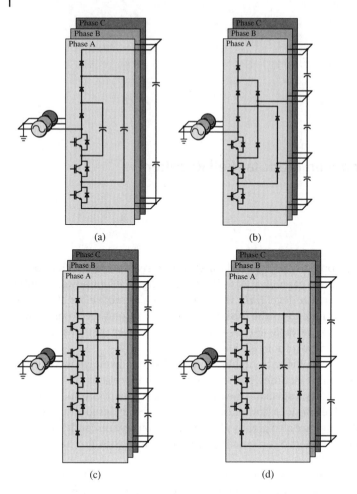

(a) (b)

(c) (d)

Figure 9.1 Recent three-phase multilevel rectifier with reduced parts topologies (I) to (IV). (a) 4L unidirectional flying capacitor rectifier. (b) 4L unidirectional diode-clamped rectifier. (c) 4L reduced-parts-count rectifier. (d) 5L hybrid PWM rectifier.

carrier phase-disposition PWM (PSC-PD-PWM)". The proposed topological design is achieved by clamping a series of modular cells onto a low operational frequency cell. By doing so, high flexibility in imbricating the cells is retained in the RFCR with the modularity approach incorporated. In addition to that, the RFCR can synthesize higher mL voltages with the component counts significantly reduced without affecting its full four-quadrant operations. Moreover, the single dc-link configured RFCR eliminates the need for an additional bulky transformer at all times. Hence, lower cost and compact size are achieved with the RFCR.

Similarities in the converter design are found between the proposed RFCR and the classical multilevel flying capacitor converter (MFCC) developed by Meynard *et al.* [18]. The MFCC is commercially available in ALSTOM Power Conversion for medium voltage (MV) up to 4.16 kV drives [1] and has also operated as the input chopper for the European railways T13 locomotives [19]. Typically, the voltage waveforms of the classical MFCC are synthesized using the commonly known phase-shifted carrier PWM (PSC-PWM). Nevertheless, several factors such as high switching frequency operation and large flying capacitor energy make the MFCC unfeasible for higher voltage applications.

Therefore, a comparative study is evaluated between the proposed RFCR and the classical flying capacitor rectifier (MFCR) to highlight the enhanced performances achieved based on the seven-level (7L) stepped voltage levels. The proven merits of the proposed RFCR make it attractively suitable for higher MV applications with space constraint. The feasibility of the near unity power factor operated RFCR is verified with a proposed closed-loop flying capacitor voltage balancing control using hybrid PSC-PD-PWM under various dynamic conditions.

9.2 Circuit Configuration

The general configuration of the proposed transformerless three-phase modular cell rectifier with reduced component counts (RFCR) is shown in Fig. 9.2. Each modular cell is constructed using a pair of semiconductor switches S_x and its complementary S_x' with a capacitor p clamped between the cells. The number of active switches $i_{switches}$ and flying capacitors $p_{capacitors}$ can be determined according to the number of modular cells (n_{cells}) utilized. The pair of switches S_{x1} and S_{x1}' in Cell 1 nearest to the dc-link operate at a low frequency (LF) of 50 Hz depending on the modulating frequency (f_o) while the remaining switches in Cell 2 to Cell n operate at constant frequency matching with the carrier frequency (f_c).

The proposed design of clamping the series-connected modular cells (Cell 2 to Cell n) onto the LF operated cell (Cell 1) allows the RFCR to synthesize higher mL voltages flexibly by extending the n_{cells}. Thus, the stepped voltage levels are now doubly increased with every incremental modular cell based on (9.1) while retaining the transformerless operation with its single dc-link structure.

$$(mL)\text{RFCR} = n_{cells} \times 2 - 1 \tag{9.1}$$

Even though a similarity is found between the proposed RFCR and the classical MFCR, the component count, cost, and size of the converter are significantly reduced with the proposed modularity approach. Since the semiconductor switches are reduced with the introduction of LF switches instead of replacing with the power diodes like those of (I) to (IV), RFCR can continue to perform the four-quadrant operations that are suitable for those regenerative drive systems. Hence, a higher efficiency is achieved while the harmonic profile is improved as well.

9.3 Operating Principle

Having three modular cells being clamped onto the low frequency operated Cell 1 results in a single-phase seven-level (7L) reduced flying capacitor rectifier (RFCR) in Fig. 9.3(a). A minimum of 8 $i_{switches}$ (S_{x1} to S_{x4} and S_{x1}' to S_{x4}') and 3 $p_{capacitors}$ (C_x1 to C_x3) are required to construct each phase-leg depending on the device voltage ratings considered for the converter design. The number of component counts is clearly observed to

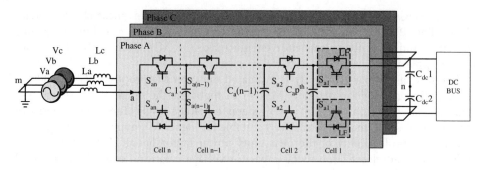

Figure 9.2 Proposed transformerless three-phase *n*-cell reduced flying capacitor rectifier (RFCR) topology.

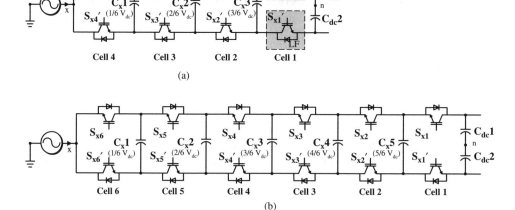

Figure 9.3 (a) Proposed 4-Cell 7L-RFCR. (b) Classical 6-Cell 7L-MFCR.

be reduced when compared with the classical 7L-MFCR (12 i_{switches} and 5 $p_{\text{capacitors}}$ per phase-leg) in Fig. 9.3(b). The difference would be even greater when the converter is designed for the three-phase systems.

A total combination of $2^4 = 16$ valid operating states from "0000" to "1111" as listed in Table 9.1 are required to drive the four pairs of semiconductor switches. There are two additional switching states generated for the zero voltage level and three redundant states for each $\pm V_{\text{dc}}/3$ and $\pm V_{\text{dc}}/6$ voltage level. Apart from that, one can observe that all of the positive voltage levels are produced only when S_{x1} conducts while the negative voltage levels are synthesized when S_{x1}' conducts. Hence, only the two switches S_{x1} and S_{x1}' in Cell 1 are always performing at a fundamental frequency of 50 Hz.

The seven-level stepped voltage waveforms are synthesized at $\pm V_{\text{dc}}/6$, $\pm V_{\text{dc}}/3$, $\pm V_{\text{dc}}/2$, and zero level accordingly, based on the given switching states in Table 9.1 and the flying capacitor voltages in Table 9.2. The flying capacitor voltages $V_{\text{Cx}}1$, $V_{\text{Cx}}2$, and $V_{\text{Cx}}3$ of 4-cell 7L-RFCR are maintained at $V_{\text{dc}}/6$, $V_{\text{dc}}/3$, and $V_{\text{dc}}/2$, respectively. The voltage

Table 9.1 3Cell 7L-RFCR Voltage Level and Corresponding Switching States

Switching States $(S_{x1}, S_{x2}, S_{x3}, S_{x4})$	Phase Leg x Voltage V_{xn}
(0000)	$-V_{dc}/2$
(0001) (0010) (0100)	$-V_{dc}/3$
(0011) (0101) (0110)	$-V_{dc}/6$
(0111) (1000)	0
(1001) (1010) (1100)	$+V_{dc}/6$
(1011) (1101) (1110)	$+V_{dc}/3$
(1111)	$+V_{dc}/2$

Where x represents phases a, b, and c, and Logic "1" and "0" represent "on" and "off" state, respectively.

Table 9.2 Capacitor Voltages for Classical MFCR and Proposed RFCR

Capacitors Clamped at pth Position	Phase x Capacitor Voltages $V_{Cx,p}{}^{th}$			
	MFCR		RFCR	
	4-Cells (5-Levels)	6-Cells (7-Levels)	3-Cells (5-Levels)	4-Cells (7-Levels)
1	$V_{dc}/4$	$V_{dc}/6$	$V_{dc}/4$	$V_{dc}/6$
2	$2V_{dc}/4$	$2V_{dc}/6$	$2V_{dc}/4$	$2V_{dc}/6$
3	$3V_{dc}/4$	$3V_{dc}/6$	$\sim$	$3V_{dc}/6$
4	$\sim$	$4V_{dc}/6$	$\sim$	$\sim$
5	$\sim$	$5V_{dc}/6$	$\sim$	$\sim$

Where x represents phases a, b, and c, and pth represents the placing position number of the flying capacitor being clamped.

across the corresponding pth flying capacitor of n-cell mL-RFCR can be determined using (9.2).

$$V_{Cx(p^{th})}^{RFCR} = \frac{p^{th}}{mL - 1} \times V_{dc} \tag{9.2}$$

It is necessary to observe the two key factors in Table 9.2, which are the numbers of the flying capacitors and the corresponding capacitor voltage ratings. The proposed 4-cell 7L-RFCR does not only reduce the total number of flying capacitors but also greatly decreases the required voltages across the flying capacitors as compared with the same number of modular cells implemented in the classical 4-cell 5L-MFCR.

9.4 Design Considerations

The following analysis in this section is expressed to determine the appropriate switching device ratings as well as designing both the input filter inductors and the flying capacitors. It is imperative to have the parameters calculated when designing the converter so that the reliability and the efficiency can be optimized. Hence, this section provides the theoretical expressions for obtaining the voltage stress, current stress, input inductance, and flying capacitance values.

9.4.1 Devices Voltage Stress

The synthesized phase voltages and line-to-line voltages for RFCR are determined using (9.3) and (9.4), respectively, based on the given switching states in Table 9.1 and the corresponding flying capacitor voltages determined from (9.2).

$$V_{xn}(t) = V_{dc}(t)(S_{x1}(t) - 0.5) + \sum_{k=2}^{n_{cells}} V_{Cx((P_{capacitors}+2)-k)}(t)\left(S_{x(k)}(t) - S_{x(k-1)}(t)\right) \tag{9.3}$$

$$\begin{bmatrix} Vab \\ Vbc \\ Vca \end{bmatrix}_{mL} = V_{dc}(t)\begin{bmatrix} (S_{a1}(t) - S_{b1}(t)) \\ (S_{b1}(t) - S_{c1}(t)) \\ (S_{c1}(t) - S_{a1}(t)) \end{bmatrix}$$

$$+ \sum_{m=2}^{n_{cells}} V_{Cx((P_{capacitors}+2)-m)}(t)\begin{bmatrix} (S_{a(m)}(t) - S_{a(m-1)}(t) - S_{b(m)}(t) + S_{b(m-1)}(t)) \\ (S_{b(m)}(t) - S_{b(m-1)}(t) - S_{c(m)}(t) + S_{c(m-1)}(t)) \\ (S_{c(m)}(t) - S_{c(m-1)}(t) - S_{a(m)}(t) + S_{a(m-1)}(t)) \end{bmatrix}_{x\in\{a,b,c\}} \tag{9.4}$$

9.4.2 Voltage Stress Analysis

The voltage stress across the switching devices of the proposed RFCR as expressed in (9.5) is prominently affected by the instantaneous flying capacitor voltages.

$$V_{Sx(cell\ n^{th})\atop RFCI}(t) \approx \begin{cases} -\left(\dfrac{(mL-1)-p^{th}}{mL-1}\right) V_{dc} \times S_{x(cell\ n^{th})}(t) & \text{if cell } n^{th} = 1 \\[3mm] -\left(\dfrac{p^{th}-(p^{th}-1)}{mL-1}\right) V_{dc} \times S_{x(cell\ n^{th})}(t) & \text{if cell } n^{th} \geq 2 \end{cases} \tag{9.5}$$

According to (9.2) and (9.5), it should be noted that the blocking voltage of the two LF switches S_{x1} and S_{x1}' in Cell 1 must be capable of withstanding $V_{dc}/2$ for *n*-cell *mL*-RFCR at all times. The same voltage stress expression (9.5) applies to the respective complementary switches as well.

The maximum current stress must be determined since it is one of the dominant factors considered for designing the converter with proper selection of the device rating. The average current stress experienced in each active switches of the proposed RFCR is approximated in the following (9.6) over one fundamental period.

$$I_{Sx1}(t) \approx \frac{\hat{I}_x}{2\pi} \int_{0-\theta_x}^{\pi-\theta_x} \sin(\omega t - \delta + \theta_x)d\omega t$$

$$I_{Sx(\text{cell } n^{th})}(t) \approx \frac{\hat{I}_x}{2\pi} \left[\begin{array}{l} \int_{0-\phi_{(n-1)}-\theta_x}^{\pi-\phi_{(n-1)}-\theta_x} \left[\begin{array}{l} \sin(\omega t - \delta + \phi_{(n-1)} + \theta_x)\times \\ m_x \sin(\omega t + \phi_{(n-1)} + \theta_x) \end{array} \right] d\omega t \\ + \int_{\pi-\phi_{(n-1)}-\theta_x}^{2\pi-\phi_{(n-1)}-\theta_x} \left[\begin{array}{l} \sin(\omega t - \delta + \phi_{(n-1)} + \theta_x)\times \\ (m_x \sin(\omega t + \phi_{(n-1)} + \theta_x) + 1) \end{array} \right] d\omega t \end{array} \right]_{n \geq 2} \tag{9.6}$$

where $Mx_{\text{sine<rectifier>}}$ is the voltage transfer ratio of the system between the dc-link voltage and the respective input phase voltage as expressed in (9.7):

$$Mx_{\text{sine}\langle\text{rectifier}\rangle}(t) = \frac{V_{dc}(t)}{2\sqrt{2}V_{xm(rms)}(t)} \geq 1 \tag{9.7}$$

The average current stress is determined based on the power factor between the grid voltage and current in radians (δ) as well as the peak magnitude of the phase current ($\hat{I}_x$) assuming that the ripple free current is sinusoidal. Since the active switches S_{x1} and S_{x1}' of Cell 1 are conducting only during half of the fundamental period, there is low to no reverse current through the pair of LF switches depending on the power factor. Therefore, the net average current stress of the Cell 1 switches is found higher as compared with the remaining other switches.

It is proven that the average current stress of the switches in the Cell 2 to Cell n is equivalently the same even though the carriers are being phase-shifted evenly by ϕ_n in radians. The maximum average current stress analysis is expressed for all three phases considering the respective reference phase angle (θ_x).

9.4.3 Design of Input Inductors

Low input current total harmonic distortion (THD) is an important concern for the grid connected system which must be strictly complied with the stringent regulations to guarantee excellent power quality and maintain high grid reliability. Thus, it is essential to obtain the critical input inductance value according to (9.8) so that the ripple current is limited within the permissible range (ΔI_{Lx}).

$$L_{x\langle\text{critical}\rangle} \approx \frac{k\left(\begin{array}{l} \sqrt{2}V_{xm(rms)} \times \sin(\omega t) \\ -\dfrac{Sec_n \times m_{x\langle\text{rectifier}\rangle} \times 2\sqrt{2}V_{xm(rms)}}{mL-1} \end{array} \right)}{\Delta I_{Lx} \times f_{Os}} \tag{9.8}$$

The input inductance is determined based on the duty ratio (k) of the switching states with respect to the sectors (Sec_n) in Fig. 9.4, while it is inversely proportional to the overall switching frequency (f_{Os}). The sectors I ($Sec_I = 1$) to III ($Sec_{III} = 3$) from 0 to $\pi/2$ radians defined in (9.9) are considered for the analysis due to the phase symmetry.

$$
\begin{array}{lll}
Sec_I & k = 3\sin(\omega t) & 0 \leq \omega t \leq 0.1082\pi \\
Sec_{II} & k = 3\left(\sin(\omega t) - \dfrac{1}{3}\right) & 0.1082\pi \leq \omega t \leq 0.2323\pi \\
Sec_{III} & k = 3\left(\sin(\omega t) - \dfrac{2}{3}\right) & 0.2323\pi \leq \omega t \leq 0.5\pi
\end{array} \tag{9.9}
$$

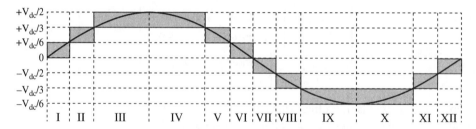

Figure 9.4 Incremental input voltage waveforms of the proposed 4-cell 7L-RFCR.

The ripple current is observed to be higher when the controlled dc-link voltage increases at a 50% duty ratio (k = 0.5) assuming the input filter inductance value and output load are fixed. Consequently, the ripple current can be reduced by increasing the f_{Os} or controlling the dc-link voltage in (9.7) to attain a near unity modulation index value ($m_{x\langle rectifier\rangle} \approx 1$). Apart from that, the ripple current is also greatly reduced due to the incremental stepped voltage waveforms according to the $Sec_n / (mL - 1)$ term in expression (9.8). Hence, the maximum current rating including the peak ripple current must be taken into consideration when designing the input inductor windings on top of the f_{Os}, which affects the number of turns.

9.4.4 Design of Flying Capacitors

Since the proposed *n*-cell *mL*-RFCR is constructed with flying capacitors clamped between the modular cells, the critical flying capacitance value is estimated using (9.10):

$$C_{x,p^{th}\langle critical\rangle} \approx \frac{4 \times m_{x\langle rectifier\rangle} \times \cos(\varphi) \times \bar{I}_{dc}}{3 \times \Delta V_{Cx,p^{th}} \times f_{Os}} \tag{9.10}$$

The flying capacitance is inversely proportional to the maximum permissible flying capacitor ripple voltage ($\Delta V_{Cx,p}{}^{th}$) and the overall switching frequency (f_{Os}). The critical flying capacitance value is determined based on the modulation index ($m_{x\langle rectifier\rangle}$), the average dc-link current ($\bar{I}_{dc}$) affected by the output load as well as the power factor angle (φ) between the input voltage and the input current.

High flying capacitor ripple voltage is experienced when the rectifier operates at near unity power factor ($\cos(\varphi) = 1$). However, the ripple voltage can be ensured to be low, similarly, by controlling the V_{dc} and the f_{Os}. Selecting the flying capacitors appropriately is important to achieve a safety operating environment by ensuring that both maximum ripple voltage and peak current are lesser than the rated voltage and current ratings given by the manufacturer's datasheet.

9.5 Control Strategy

The closed-loop control is proposed for the 4-cell 7L-RFCR in this section as shown in Fig. 9.5. It is constructed based on the two control schemes (synchronous-reference-frame current control and flying capacitor voltage balancing control), which are comprehensively described in the following sections.

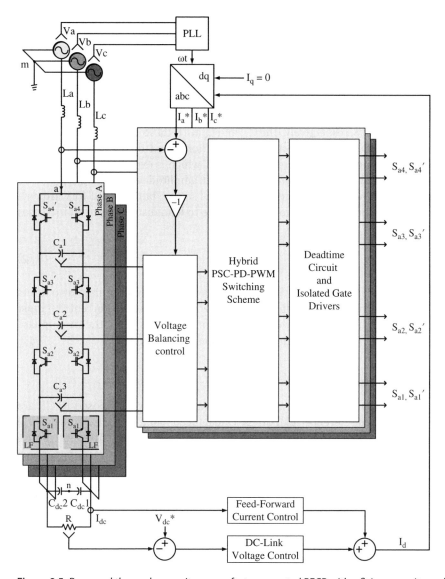

Figure 9.5 Proposed three-phase unity power factor operated RFCR with a flying capacitor voltage balancing control scheme using hybrid PSC-PD-PWM.

9.5.1 Synchronous-Reference-Frame Current Control

The unity power factor controller for the 4-cell 7L-RFCR proposed in Fig. 9.5 is constructed using the synchronous reference frame current (SRFC) control as detailed in [20]. The SRFC controller provides a good dynamic response and shapes input current with a high resolution quality while maintaining the unity power factor operation. The SRFC consists of a dc-link voltage control, a feed-forward current control, and a reference current generator (dq0 Park's transformation).

In order to achieve a stable control system, the outer-loop dc-link voltage control is designed following the open-loop transfer function as expressed in (9.11):

$$L(s) = \frac{K_p s + K_i}{s} \cdot \frac{L_x I_d}{C_{eq} V_{dc}} \cdot \frac{\left(\sqrt{3} V_p / L_x I_d\right) - s}{s + \left(I_{dc} / C_{eq} V_{dc}\right)} \tag{9.11}$$

The proportional gain (K_p) and integral gain (K_i) of the dc-link voltage control loop are selected at 0.16 and 8, respectively. The C_{eq} and L_x are total equivalent dc-link capacitance and the filter inductance, respectively. An additional feed-forward current parameter (K_f) is implemented in the control loop as well, which is expressed in (9.12) based on the power-balanced principle where V_p is the peak amplitude of the instantaneous ac supply voltage. Even though it requires an additional sensor, but the feed-forward current control allows a better dc-link voltage response to load disturbance and increases the system stability.

$$K_f = \frac{2 V_{dc}}{3 V_p} \tag{9.12}$$

The active current (I_d) fed into the SRFC control is achieved by summing the outputs of the dc-link feed-forward and feedback controls while the reactive current (I_q) must remain null to attain near unity power factor performance. As such, the three-phase reference currents (I_x^*) are now obtained using Park's transformation.

9.5.2 Flying Capacitor Voltage Balancing Control

It is necessary to ensure that the flying capacitor voltages of the proposed RFCR are tracked to their desired reference at all times. Unbalanced flying capacitor voltages will cause an increased voltage stress on the power semiconductors and affect the entire system's stability. This issue can be resolved easily by oversizing the device ratings, which is not a cost-effective solution.

Alternatively, a flying capacitor voltage balancing control is developed using the simple proportional gain. It is clearly observed in (9.13) that the current through the respective flying capacitor is affected by the switching states of the two adjacent switches.

$$C_{x,p^{th}} \times \frac{dV_{Cx,p^{th}}}{dt} = I_{Cx,p^{th}}(t) = \left(S_{x(n-p^{th}+1)} - S_{x(n-p^{th})}\right) I_x \tag{9.13}$$

The flying capacitor current in (9.13) can be expressed in a local average form over a switching period by replacing the switching states with the duty ratio. The locally average capacitor voltage balancing dynamic is given in the following:

$$\frac{\Delta \overline{V}_{Cx,p^{th}}}{\Delta t} = \frac{\left(\Delta d_{x(n-p^{th}+1)} - \Delta d_{x(n-p^{th})}\right) \overline{I}_x}{C_{x,p^{th}}} \tag{9.14}$$

Hence, the flying capacitor voltages can be regulated by varying the duty ratios of the switching pulses according to the respective capacitor voltage errors ($\varepsilon_{x(n)}$) between the desired reference voltage in Table 9.2 and the instantaneous measured values. These voltage errors are minimized using the proportional (P) controller with a gain of 0.4.

$$\Delta d_{x(n)} = \text{sign}(I_x)(\varepsilon_{x(n-1)} - \varepsilon_{x(n)})P \tag{9.15}$$

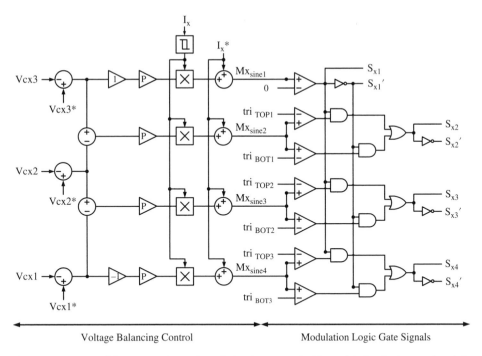

Figure 9.6 Proposed flying capacitor voltage balancing control with modulation logic gate signals for 4-cell 7L-RFCR.

The duty cycle varies depending on the magnitude of the instantaneous input current, which is defined as 1 and −1 for the respective positive and negative cycle using the sign term.

$$Mx_{sine(n)} = I_x^* + \Delta d_{x(n)} \tag{9.16}$$

The variation of the duty ratio will be added to the three-phase reference current I_x^* (output of the Park transformation) to obtain the modulation sine templates ($Mx_{sine(n)}$). The proposed flying capacitor voltage balancing control scheme for the 4-cell 7L-RFCR is shown in Fig. 9.6.

9.5.3 Hybrid Carrier-Based Pulse-Width Modulation Schemes

The proposed RFCR is operated using the hybrid phase-shifted carrier phase-disposition PWM (PSC-PD-PWM) in Fig. 5.4(b). The alternative hybrid PSC-POD-PWM can also be applied for synthesizing the input voltages of the RFCR. However, the hybrid PSC-PD-PWM is chosen in this case over the other proposed PSC-POD-PWM due to its improved voltage harmonic profile.

Similar to all cases, (mL-1) number of triangular carriers (n_{tri}) would be needed for any mL modulation scheme. Thus, a total of six n_{tri} are required for the proposed 7L-RFCR, which are divided equally into two sets. One set is level-shifted above the zero reference magnitude ($tri_{TOP(1,2,3)}$) and the other set is level-shifted below the zero reference magnitude ($tri_{BOT(1,2,3)}$). The top and bottom carriers are evenly phase-shifted 120° apart.

The final gating signals for the switching devices are achieved directly by comparing the $Mx_{sine(n)}$ with the respective top and bottom phase-shifted carriers using simple comparators and logic gates as shown in Fig. 9.6. Since there are only three carriers being phase-shifted at the top and the bottom level in this case compared with the six phase-shifted carriers in PSC-PWM, the f_{Os} is also reduced by 50%.

9.6 Comparative Evaluation of Classical MFCR and Proposed RFCR

The performances of the proposed four quadrants of RFCR are compared against those of the classical MFCR in terms of the input current harmonic distortions as well as the distributed semiconductor losses as discussed in the following sections. Significant improvements can be observed in the proposed RFCR over the classical MFCR.

9.6.1 Input Current Shaping Performance

Shaping the input current to near pure sinusoidal waveforms is always desired to be achieved for the grid-connected rectifier. The performance of the grid current response for the proposed RFCR is evaluated in Fig. 9.7 based on different parameters such as (i) output power, (ii) input filter inductance, (iii) controlled V_{dc}, and (iv) f_{Os}. Generally, the input current THD is observed high for the light-load conditions despite any f_{Os} and $Mx_{sine<rectifier>}$ operations. This is because the fundamental current harmonic amplitude is low with the accompanied ripple harmonics. Thus, the input current THD can be greatly reduced with the increasing magnitude of the fundamental current harmonic, which can be achieved by increasing the output load power or the $Mx_{sine<rectifier>}$ with higher controlled V_{dc}.

Alternatively, the ripple current can also be further decreased with larger input filter inductance. Apart from that, the synthesis of higher mL voltages reduces the ripple current as well. It is clearly proven in Fig. 9.7 that the input current THD of the proposed 7L-RFCR is approximately the same as the classical 7L-MFCR and is much lower than the classical 5L-MFCR. Therefore, the input inductance of the proposed RFCR can be further reduced while achieving good input current shaping with reduced component counts.

9.6.2 Power Semiconductor Device Losses

The total distributed losses consisting of both a total switching loss and total conduction loss of per phase-leg discrete components are shown in Fig. 9.8 based on the different parameters (i) to (iv) as mentioned previously. The power converters achieve better efficiency with near unity $Mx_{sine<rectifier>}$ operation since higher losses are experienced when the controlled V_{dc} reference is increased ($Mx_{sine<rectifier>} > 1$) as well as for larger output loads. The total distributed losses are also reduced with the increasing input filter inductance due to the lower input current THD since conduction loss is dominantly affected by current performance.

The results in Fig. 9.8 clearly show that the proposed 7L-RFCR experienced a similar quantum of loss as the classical 5L-MFCR and a much lower level compared with the

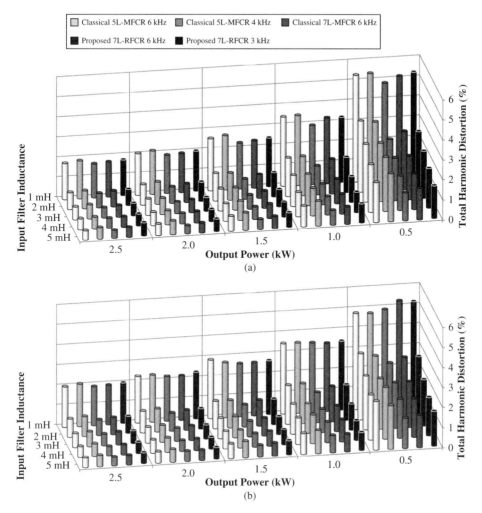

Figure 9.7 Comparison of input current total harmonic distortion based on different dc-link voltage reference. (a) 170 Vdc (1.0 Mx$_{sine}$). (b) 187 Vdc (0.9 Mx$_{sine}$).

classical 7L-MFCR. The total distributed losses in the RFCR configuration are notably decreased because of the reduced component counts that lead to the lower dominant conduction loss and the reduced flying capacitor voltages. Hence, the proposed RFCR features a higher efficiency than the classical MFCR.

9.7 Experimental Verification

The comparative studies between the proposed RFCR using hybrid PSC-PD-PWM and the classical MFCR under PSC-PWM are carried out using PSIM v9.0.4 Simcouple and MATLAB R2010b Simulink®. The hardware prototype is developed to verify the discussed performances with the aid of dSPACE RT1103 Real-Time Workshop (RTW). All presented results are based on the detailed settings given in Table 9.3.

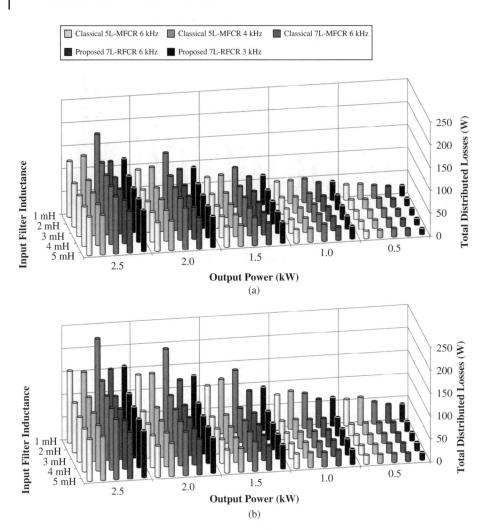

Figure 9.8 Comparison of per phase-leg semiconductor losses based on different dc-link voltage reference. (a) 170 Vdc (1.0 Mx$_{sine}$). (b) 187 Vdc (0.9 Mx$_{sine}$).

Table 9.3 Configuration Settings for Simulation and Hardware Prototype of 4-Cell 7L-RFCR

System Parameters	Settings
Reference sinusoidal frequency (f_o)	50 Hz
Triangular carrier frequency (f_c)	1 kHz
ac voltage source	60 Vrms
Output dc-link voltage reference	170 Vdc
Flying capacitors ($C_{x,p}{}^{th}$)	2200 μF
Filter – inductors	5 mH
Load – Resistors	28.9 Ω (1kW)

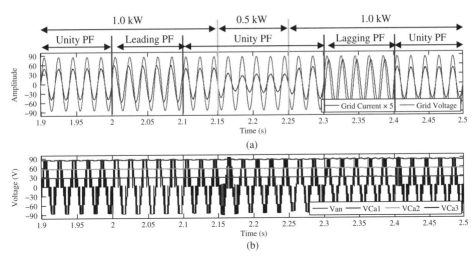

Figure 9.9 Simulated dynamic performance of 4-cell 7L-RFCR with PF variation and load change. (a) Grid voltage and current (×5). (b) Input rectifier pole voltage and flying capacitor voltages.

The dynamic performances of the 4-cell 7L-RFCR are evaluated with a series of step changes in the power factor (PF), modulation index ($Mx_{sine<rectifier>}$), and load parameters to verify the feasibility of the proposed voltage balancing control.

The proposed 7L-RFCR is simulated with leading PF (0.8 PF) between t = 2s and 2.1s, lagging PF (0.8 PF) between t = 2.3s and 2.4s as well as unity PF for the remaining period as shown in Fig. 9.9(a) while having the dc-link voltage (V_{dc}) controlled to an output voltage command reference of 170 V. Besides that, the output load is step changed from 1 kW to 0.5 kW at t = 2.15s and back from 0.5 kW to 1 kW at t = 2.25s. However, it can be clearly observed in Fig. 9.9(b) that the flying capacitor voltages $V_{Ca}1$, $V_{Ca}2$, and $V_{Ca}3$ remain balanced at their respective desired voltage levels (28.33 V, 59.67 V, and 85 V) despite the variations in the PF and the load-changed condition. The simulated dynamic performances of 4-cell 7L-RFCR in Fig. 9.9 are also validated with the experimental results.

The unity power factor controlled performance of the 4-cell 7L-RFCR following the control scheme in Fig. 9.5 using hybrid PSC-PD-PWM can be clearly observed in Fig. 9.10. A sinusoidal grid current shaping waveform is achieved while yielding a near unity power factor (0.999 PF). The dc-link voltage V_{dc} is also tracked to the output voltage command reference at 170 V. Besides that, the flying capacitor voltages $V_{Ca}1$, $V_{Ca}2$, and $V_{Ca}3$ in Fig. 9.11 are well balanced at 28.3 V, 59.7 V, and 85.1 V, respectively. Hence, the input pole voltage, Van, in Fig. 9.10 is synthesized correspondingly to the seven levels.

Moreover, the proposed 4-cell 7L-RFCR is also evaluated with the leading (0.872 PF) power factor in Figs. 9.12 and 9.13 as well as with the lagging (0.808 PF) power factor in Figs. 9.14 and 9.15. Despite the variations in the power factor, the dc-link voltages in Figs. 9.16, 9.18, and 9.20 are well maintained at 170 V constantly. However, there is a difference between the input current magnitudes in Figs. 9.12 and 9.14 that are caused by the angle displacement, which affects the input seven-level pole voltage waveforms

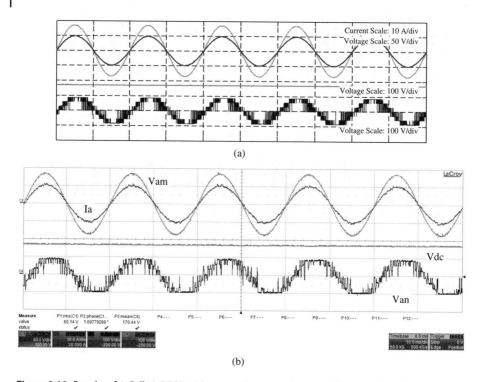

(a)

(b)

Figure 9.10 Results of 4-Cell 7L-RFCR with near unity power factor performance based on input grid voltage (Vam), input grid current (Ia), and output dc-link voltage (V_{dc}). (a) Simulation results. (b) Experimental results.

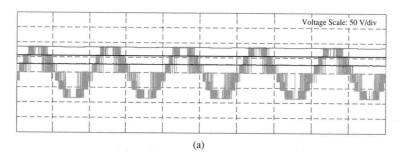

(a)

Figure 9.11 Results of 4-Cell 7L-RFCR with near unity power factor performance based on input pole voltage (Van) and flying capacitor voltages ($V_{Ca}1$, $V_{Ca}2$, and $V_{Ca}3$). (a) Simulation results. (b) Experimental results.

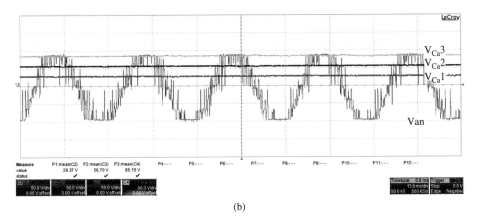

(b)

Figure 9.11 (*Continued*)

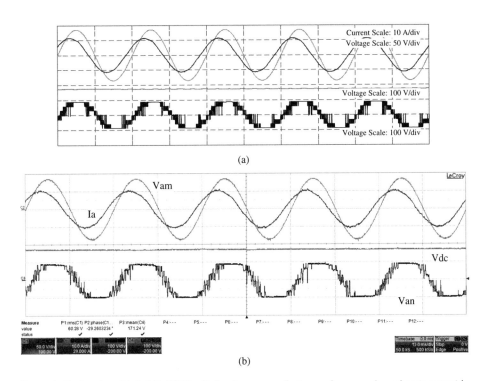

(a)

(b)

Figure 9.12 Results of 4-Cell 7L-RFCR with leading power factor performance based on input grid voltage (Vam), input grid current (Ia), output dc-link voltage (V$_{dc}$), and input pole voltage (Van). (a) Simulation results. (b) Experimental results.

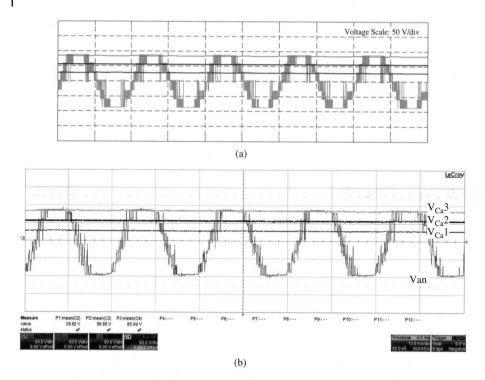

(a)

(b)

Figure 9.13 Results of 4-Cell 7L-RFCR with leading power factor performance based on input pole voltage (Van) and flying capacitor voltages ($V_{Ca}1$, $V_{Ca}2$, and $V_{Ca}3$). (a) Simulation results. (b) Experimental results.

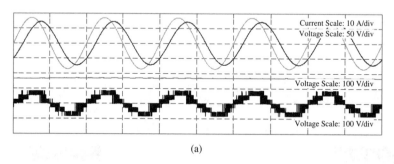

(a)

Figure 9.14 Results of 4-Cell 7L-RFCR with lagging power factor performance based on input grid voltage (Vam), input grid current (Ia), output dc-link voltage (V_{dc}), and input pole voltage (Van). (a) Simulation results. (b) Experimental results.

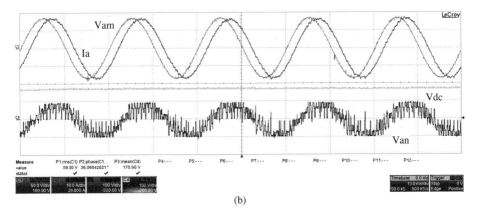

(b)

Figure 9.14 (*Continued*)

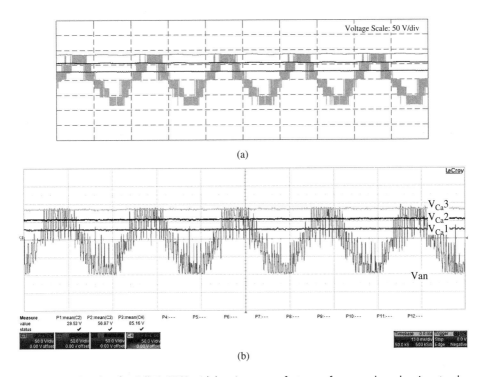

Figure 9.15 Results of 4-Cell 7L-RFCR with lagging power factor performance based on input pole voltage (Van) and flying capacitor voltages ($V_{Ca}1$, $V_{Ca}2$, and $V_{Ca}3$). (a) Simulation results. (b) Experimental results.

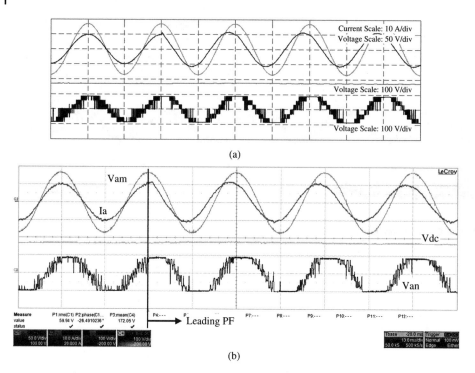

(a)

(b)

Figure 9.16 Results of 4-Cell 7L-RFCR with power factor variation from unity to leading based on input grid voltage (Vam), input grid current (Ia), output dc-link voltage (V_{dc}), and input pole voltage (Van). (a) Simulation results. (b) Experimental results.

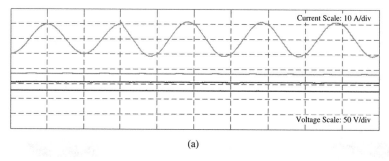

(a)

Figure 9.17 Results of 4-Cell 7L-RFCR with power factor variation from unity to leading based on input grid current (Ia) and flying capacitor voltages ($V_{Ca}1$, $V_{Ca}2$, and $V_{Ca}3$). (a) Simulation results. (b) Experimental results.

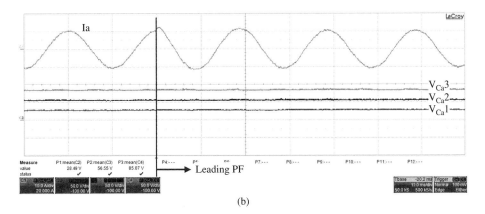

(b)

Figure 9.17 (*Continued*)

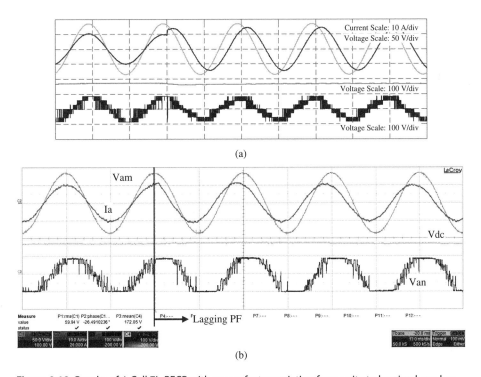

Figure 9.18 Results of 4-Cell 7L-RFCR with power factor variation from unity to lagging based on input grid voltage (Vam), input grid current (Ia), output dc-link voltage (V_{dc}), and input pole voltage (Van). (a) Simulation results. (b) Experimental results.

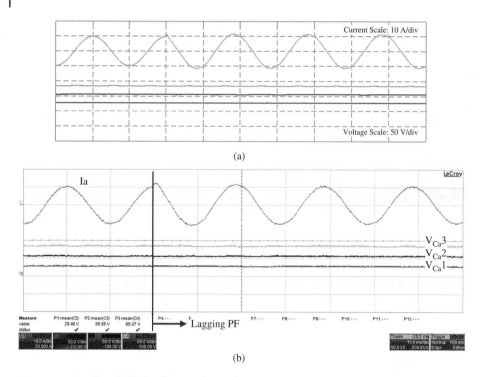

(a)

(b)

Figure 9.19 Results of 4-Cell 7L-RFCR with power factor varies from unity to lagging based on input grid current (Ia) and flying capacitor voltages ($V_{Ca}1$, $V_{Ca}2$, and $V_{Ca}3$). (a) Simulation results. (b) Experimental results.

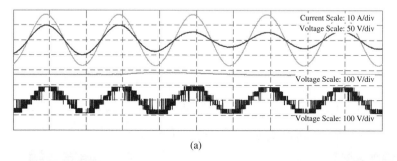

(a)

Figure 9.20 Results of 4-Cell 7L-RFCR under load change from 1 kW to 0.5 kW based on input grid voltage (Vam), input grid current (Ia), output dc-link voltage (V_{dc}), and input pole voltage (Van). (a) Simulation results. (b) Experimental results.

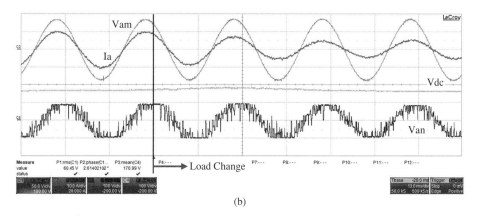

(b)

Figure 9.20 (*Continued*)

to be synthesized differently when compared with those waveforms shown in Figs. 9.10 and 9.11.

Apart from that, it should be noted that the flying capacitor voltages in Figs. 9.13 and 9.15 remain balanced at their desired voltage levels, even in the event of power factor and load change condition as seen in Figs. 9.17, 9.19, and 9.21. The experimental results

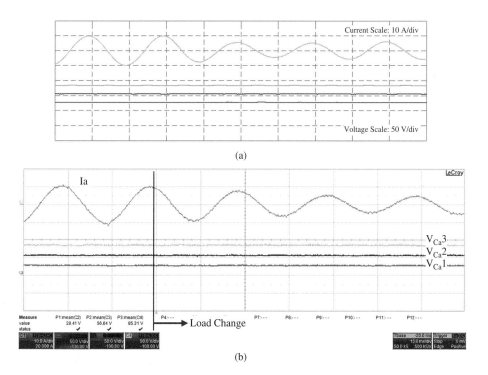

Figure 9.21 Results of 4-Cell 7L-RFCR under load change from 1 kW to 0.5 kW based on input grid current (Ia) and flying capacitor voltages ($V_{Ca}1$, $V_{Ca}2$, and $V_{Ca}3$). (a) Simulation results. (b) Experimental results.

have demonstrated the dynamic response and the stability of the proposed unity power factor operated RFCR using the flying capacitor voltage balancing control.

On matching the experimental results with the simulation result in Fig. 9.9, the practicality of the proposed RFCR in overcoming the challenges in the classical MFCR is confirmed. The proposed RFCR synthesizes higher voltage stepped waveforms with the component counts and costs significantly reduced. Therefore, the four-quadrant operations of RFCR can be attractively applicable for any higher MV drive systems with space constraint as compared with the classical MFCR.

References

1 S. Kouro, M. Malinowski, K. Gopakumar, J. Pou, L. G. Franquelo, W. Bin *et al.*, "Recent advances and industrial applications of multilevel converters," *Industrial Electronics, IEEE Transactions on*, vol. 57, pp. 2553–2580, 2010.

2 H. Abu-Rub, J. Holtz, J. Rodriguez, and B. Ge, "Medium-voltage multilevel converters – State of the art, challenges, and requirements in industrial applications," *Industrial Electronics, IEEE Transactions on*, vol. 57, pp. 2581–2596, 2010.

3 S. Daher, J. Schmid, and F. L. M. Antunes, "Multilevel inverter topologies for stand-alone PV systems," *Industrial Electronics, IEEE Transactions on*, vol. 55, pp. 2703–2712, 2008.

4 F. Blaabjerg, M. Liserre, and M. Ke, "Power electronics converters for wind turbine systems," *Industry Applications, IEEE Transactions on*, vol. 48, pp. 708–719, 2012.

5 X. Lie and V. G. Agelidis, "VSC transmission system using flying capacitor multilevel converters and hybrid PWM control," *Power Delivery, IEEE Transactions on*, vol. 22, pp. 693–702, 2007.

6 M. Perez, R. Lizana, D. Arancibia, J. Espinoza, and J. Rodriguez, "Decoupled currents model and control of modular multilevel converters," *Industrial Electronics, IEEE Transactions on*, vol. PP, pp. 1–1, 2015.

7 Z. Zedong, W. Kui, X. Lie, and L. Yongdong, "A hybrid cascaded multilevel converter for battery energy management applied in electric vehicles," *Power Electronics, IEEE Transactions on*, vol. 29, pp. 3537–3546, 2014.

8 V. Biagini, P. Zanchetta, M. Odavic, M. Sumner, and M. Degano, "Control and modulation of a multilevel active filtering solution for variable-speed constant-frequency more-electric aircraft grids," *Industrial Informatics, IEEE Transactions on*, vol. 9, pp. 600–608, 2013.

9 K. K. Gupta and S. Jain, "A novel multilevel inverter based on switched DC sources," *Industrial Electronics, IEEE Transactions on*, vol. 61, pp. 3269–3278, 2014.

10 V. Dargahi, A. K. Sadigh, M. Abarzadeh, M. R. A. Pahlavani, and A. Shoulaie, "Flying capacitors reduction in an improved double flying capacitor multicell converter controlled by a modified modulation method," *Power Electronics, IEEE Transactions on*, vol. 27, pp. 3875–3887, 2012.

11 D. A. Ruiz-Caballero, R. M. Ramos-Astudillo, S. A. Mussa, and M. L. Heldwein, "Symmetrical hybrid multilevel DC-AC converters with reduced number of insulated dc supplies," *Industrial Electronics, IEEE Transactions on*, vol. 57, pp. 2307–2314, 2010.

12 Y. Ounejjar, K. Al-Haddad, Gre, x, and L. A. goire, "Packed U cells multilevel converter topology: Theoretical study and experimental validation," *Industrial Electronics, IEEE Transactions on*, vol. 58, pp. 1294–1306, 2011.

13 S. H. Hosseini, A. Khoshkbar Sadigh, and M. Sabahi, "New configuration of stacked multicell converter with reduced number of dc voltage sources," in *Power Electronics, Machines and Drives (PEMD 2010), 5th IET International Conference on*, pp. 1–6, 2010.

14 F. Forest, T. A. Meynard, S. Faucher, F. Richardeau, J. J. Huselstein, and C. Joubert, "Using the multilevel imbricated cells topologies in the design of low-power power-factor-corrector converters," *Industrial Electronics, IEEE Transactions on*, vol. 52, pp. 151–161, 2005.

15 M. L. Heldwein, S. A. Mussa, and I. Barbi, "Three-phase multilevel PWM rectifiers based on conventional bidirectional converters," *Power Electronics, IEEE Transactions on*, vol. 25, pp. 545–549, 2010.

16 K. A. Corzine and J. R. Baker, "Reduced-parts-count multilevel rectifiers," *Industrial Electronics, IEEE Transactions on*, vol. 49, pp. 766–774, 2002.

17 J. I. Itoh, Y. Noge, and T. Adachi, "A novel five-level three-phase PWM rectifier with reduced switch count," *Power Electronics, IEEE Transactions on*, vol. 26, pp. 2221–2228, 2011.

18 T. A. Meynard and H. Foch, "Multi-level conversion: high voltage choppers and voltage-source inverters," in *Power Electronics Specialists Conference, 1992. PESC '92 Record., 23rd Annual IEEE*, pp. 397–403, vol. 1, 1992.

19 T. A. Meynard, H. Foch, P. Thomas, J. Courault, R. Jakob, and M. Nahrstaedt, "Multicell converters: basic concepts and industry applications," *Industrial Electronics, IEEE Transactions on*, vol. 49, pp. 955–964, 2002.

20 A. I. Maswood and L. Fangrui, "A unity-power-factor converter using the synchronous-reference-frame-based hysteresis current control," *Industry Applications, IEEE Transactions on*, vol. 43, pp. 593–599, 2007.

Part III

Advanced Multilevel Inverters and their Control Strategies

10

Transformerless Five-Level/Multiple-Pole Multilevel Inverters with Single DC Bus Configuration
Gabriel H. P. Ooi

10.1 Introduction

In this chapter, a new concept of five-level (5L) inverter topology is presented. Numerous topologies on higher-level voltage source inverters have been detailed in Chapters 4 and 7. Reducing the number of components utilized has been quite a trend among researchers and industrial product developers around the globe [1]. In order to maintain high operational efficiency, it is necessary to develop multilevel converters with lower part count while achieving primary objectives such as low total harmonic distortion (THD) and lower Electromagnetic interference (EMI). The importance of achieving reduction in part count has been discussed in Chapter 1. The prices of power electronic equipment are decreasing due to the tremendous advances made in fabrication technologies and logistics. Only potential energy savings in power converters would motivate customers and manufacturers to search for innovative converter designs. The key factors for achieving such optimization are low overall cost, high conversion efficiency, and, most importantly, reduced utilization of semiconductor devices.

The focus of this chapter is to design a transformerless, low-loss, and lightweight converter with lower device ratings. The proposed converter achieves significant reductions in components count and cost. The new multilevel inverter concept is based on the single-pole inverter (see Fig. 10.1) of a multilevel diode-clamped inverter (MDCI) topology. The topologies are developed based on the concept of multiple-pole hierarchy as shown in Fig. 10.2. This type of hierarchy utilizes a lesser number of switching elements, resulting in better efficiency while operating at a lower switching frequency for higher-level incremental voltage stepped waveform.

As mentioned in Chapter 7, the dc-link capacitor voltage of a five-level diode-clamped inverter is unbalanced. The total average current at each node of the dc-link capacitors connected in series is not equal to zero. Thus, a balancing circuit is incorporated to balance the capacitor voltage in order to achieve a balanced output voltage. The balancing circuit also regulates equal voltage among the dc-link capacitors at the input of the 5L inverter.

The balancing circuit for the input distribution voltage level of the five-level inverter circuit was first proposed by Newton [2], and this type of configuration has been successfully installed in motor drive applications [3–5]. The following sections will present

Advanced Multilevel Converters and Applications in Grid Integration, First Edition.
Edited by Ali I. Maswood and Hossein Dehghani Tafti.
© 2019 John Wiley & Sons Ltd. Published 2019 by John Wiley & Sons Ltd.

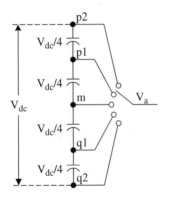

Figure 10.1 Circuit diagram of per-phase leg single-pole existing 5L-MDCI topology with switching position.

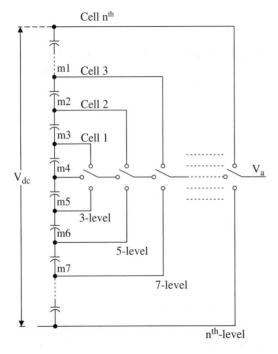

Figure 10.2 Circuit diagram of per-phase leg multiple-pole nL-MDCI topology with switching position.

the proposed five-level inverter topologies with balancing circuits. A comparative study between the proposed and classical inverter is presented in the following subsection.

10.2 Five-Level Multiple-Pole Concept

The basic idea of this multiple-pole is based on a single-pole hierarchy by separating the pole into a parallel structure. Each pole is connected to its respective desired dc voltage value, and each pole is operated at three-level incremental voltage stepped. Higher voltage levels are achieved when the number of poles increase, as shown in Fig. 10.2.

10.3   Circuit Configuration and Operation Principles

The operating principle of the proposed 5L-inverters is presented in this section. The proposed topologies are classified as *multiple-pole multilevel diode-clamped inverters* (M²DCI), *multiple-pole multilevel t-type-clamped inverters* (M²T²CI), and *multiple-pole multilevel single-switch-clamped inverters* (M²S²CI). Their topological design aspects, output characteristics, and power quality are proven to be on par with conventional MDCI configuration. The losses of the discrete components will be investigated later in this chapter, and the performance of gate control for the respective inverters will be verified through the laboratory prototype.

10.3.1   Five-Level/Multilevel Diode-Clamped Inverter (5L-MDCI)

The basic operation of the 5L-MDCI topology with the switching scheme on LS-PWM is presented in Fig. 10.3. The expressions of both output voltage and current flow through the devices are obtained based on the following assumptions: (1) balanced dc-link capacitors; (2) zero voltage ripple in each dc capacitor; and (3) neglected parasitic passive elements. Based on the given assumption, the output voltage is expressed as

$$V_{sm}(t) = \frac{V_{dc}(t)}{4}[T_{s1}(t) + T_{s2}(t) + T_{s3}(t) + T_{s4}(t) - 2] \tag{10.1}$$

The instantaneous current through active switches with their respective operating angles for different modulation depths are listed in Table 10.1. The operating angle is determined by the commutation period for one cycle. Since a topology requires a

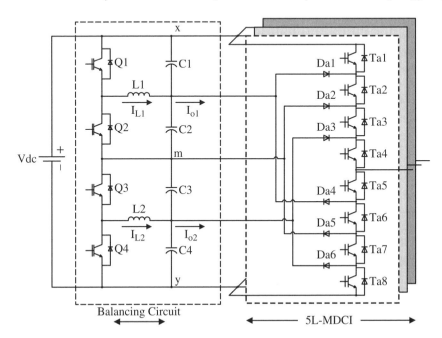

Figure 10.3 Circuit diagram of the classical 5L-MDCI topology associated with the 5L-DC/DC balancing circuit.

Table 10.1 Instantaneous Phase "a" Current Expression of 5L-MDCI with the Respective Angle of the Operating System and the Range of Modulation Depth.

Current	Equation	Operating Angle	Modulation
$I_{Da1}(t)$	$I_a(t)\,T_{a2}(t)[1-T_{a1}(t)]$	$0 \le \omega t < \pi$	$M_a \ge \dfrac{1}{2M_a}$
$I_{Da2}(t)$	$I_a(t)\,T_{a3}(t)[1-T_{a1}(t)]$	$0 \le \omega t < \pi$	
$I_{Da3}(t)$	$I_a(t)\,T_{a4}(t)[1-T_{a3}(t)]$	$0 \le \omega t < \pi$	
$I_{Ta1}(t)$	$I_a(t)\,T_{a1}(t)$	$\sin^{-1}\dfrac{1}{2M_a} \le \omega t < \pi - \sin^{-1}\dfrac{1}{2M_a}$	
$I_{Ta2}(t)$	$I_a(t)\,T_{a2}(t)$	$0 \le \omega t < \pi$	
$I_{Ta3}(t)$	$I_a(t)\,T_{a3}(t)$	$0 \le \omega t < \pi$	
$I_{Ta4}(t)$	$I_a(t)\,T_{a4}(t)$	$0 \le \omega t < \pi$	
$I_{Da1}(t)$	$I_a(t)\,T_{a2}(t)$	$0 \le \omega t < \pi$	$M_a < \dfrac{1}{2}$
$I_{Da2}(t)$	$I_a(t)\,T_{a3}(t)[1-T_{a2}(t)]$	$0 \le \omega t < \pi$	
$I_{Da3}(t)$	0	$0 \le \omega t < \pi$	
$I_{Ta1}(t)$	0	$0 \le \omega t < \pi$	
$I_{Ta2}(t)$	$I_a(t)\,T_{a2}(t)$	$0 \le \omega t < \pi$	
$I_{Ta3}(t)$	$I_a(t)\,T_{a3}(t)$	$0 \le \omega t < \pi$	
$I_{Ta4}(t)$	$I_a(t)$	$0 \le \omega t < \pi$	

Note: Ia(t) is the phase "a" output current of the inverter, which consists of power factor angle between the load and pole voltage. As the switches in the upper and lower legs have similar operating principles, the instantaneous current expression for currents through complementary switches is the same.

complementary switch to synthesize the voltage stepped waveform, the active components in the upper phase leg conduct only during the positive half-cycle, whereas the lower phase leg devices conduct only during the negative half-cycle. Therefore, both the upper phase and lower phase leg devices have the same current expression. The instantaneous current expression for the switches in the upper half of the leg is listed in Table 10.1.

10.3.2   Five-Level/Multiple-Pole Multilevel Diode-Clamped Inverter (5L-M²DCI)

Fig. 10.4 shows the circuit diagram of the proposed 5L-M²DCI topology. In order to operate a stable five-level output pole voltage waveform (i.e. $V_{am} = -V_{dc}/2, -V_{dc}/4, 0, +V_{dc}/4, +V_{dc}/2$), two cells/poles are connected in each phase. Each of the cells is configured based on the classical NPC/MDCI topology, and a five-level/level-shifted PWM technique is used for generating the required gating signals. The switching state of the corresponding output voltage level is shown in Table 10.2.

The inner cell switches ($T_{a1}–T_{a4}$) conduct only when the voltage level is operated at $\pm V_{dc}/4$ and 0. By obtaining this switching condition, the middle two carrier-based comparisons are selected for the switching commutation. The output five-level voltage stepped waveform is then finalized by the outer cell switches, where the first and last carrier comparison is selected for the outer cell switches. The switching scheme and switching state selection for the 5L-M²DCI topology are shown in Fig. 10.5.

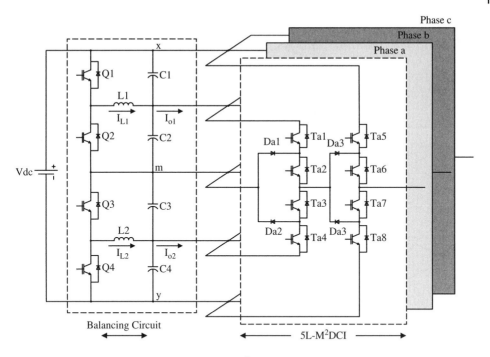

Figure 10.4 Circuit diagram of the proposed 5L-M^{2}DCI topology associated with the 5L-DC/DC balancing circuit.

Table 10.2 5L-M^{2}DCI voltage levels and the corresponding switching states.

States	Switching States								Per-Phase Leg Voltage		
	T_{s1}	T_{s2}	T_{s3}	T_{s4}	T_{s5}	T_{s6}	T_{s7}	T_{s8}	V_{sx}	V_{sy}	V_{sm}
1	1	1	0	0	1	1	0	0	0	V_{dc}	$V_{dc}/2$
2	1	1	0	0	0	1	1	0	$-V_{dc}/4$	$3V_{dc}/4$	$V_{dc}/4$
3	0	1	1	0	0	1	1	0	$-V_{dc}/2$	$V_{dc}/2$	0
4	0	0	1	1	0	1	1	0	$-3V_{dc}/4$	$V_{dc}/4$	$-V_{dc}/4$
5	0	0	1	1	0	0	1	1	$-V_{dc}$	0	$-V_{dc}/2$

Note: Here, "s" represents phase a, b, and c; S_{s1} to S_{s8} are the IGBT switching devices for the individual legs; and logic 1 represents turn-on and logic 0 the turn-off for T_{s1}, T_{s2}, T_{s3}, and T_{s4}.

According to the level-shifted PWM technique, the output pole-voltage expression for amplitude modulation greater than 0.5, under the balance capacitor voltage condition, is written as:

$$V_{sm}(t) = \frac{V_{dc}(t)}{4}[T_{s1}(t) + T_{s2}(t) + T_{s3}(t) + T_{s4}(t) - 2] \qquad (10.2)$$

For amplitude modulation less than 0.5, the modulation wave falls within two center carrier waves. The output pole voltage is expressed as:

$$V_{sm}(t) = \frac{V_{dc}(t)}{4}[T_{s1}(t) + T_{s2}(t) - 1] \qquad (10.3)$$

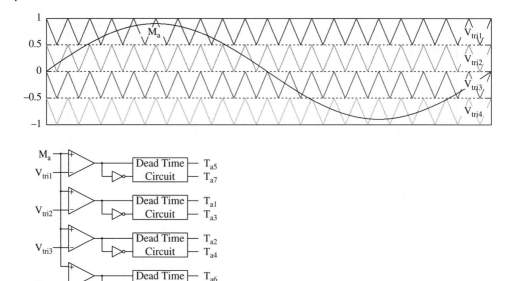

Figure 10.5 Switching scheme based on LS-PWM strategy with the corresponding gating signal of phase "a."

From equations (10.1) and (10.3), it is noted that the modulation indices determine the synthesis of output pole-voltage levels and the selection of particular switching states. The expression of the output pole-voltage for the five-level voltage source inverter with LS-PWM is given by equation (10.1), which is similar to equation (7.10).

The current flow through the switches is determined by the power factor angle between the output pole voltage and current of the inverter, with the dependent modulation depth of the switching state selection at positive half-cycle. Expressions of the current flow through the switches are given in Table 10.3.

10.3.3 Five-Level/Multiple-Pole Multilevel T-Type-Clamped Inverter (5L-M²T²CI)

The schematic of M²T²CI is shown in Fig. 10.6. The T-type-clamped concept is derived according to the direction of the current flow of the diode-clamped inverter, as shown in Fig. 10.7, with the corresponding switching states commutation. According to the switching states in Table 10.2, the current through the center switch (Ta2–Ta3 and Ta6–Ta7) is bidirectional, and is illustrated in Fig. 10.6. The bidirectional switching scheme is observed by the transition of switching states (1 1) ↔ (0 1) ↔ (0 0) of the upper two switches. The transition involves the cumulating of NPC topology and T-clamped inverter into a multiple-pole approach to form a five-level output voltage stepped waveform operation.

The switching states of five-level output voltage stepped waveform are shown in Table 10.4 with the corresponding switching commutation. In order to achieve the desired output voltage level with the corresponding switching states as listed in Table 10.4, simple logic gates circuitry is developed to control two series switches

Table 10.3 Instantaneous Phase "a" Current Expression of 5L-M²DCI with the Respective Angle of the Operating System and the Range of Modulation Depth.

Current	Equation	Operating Angle	Modulation
$I_{Da1}(t)$	$I_a(t)\,T_{a2}(t)[1 - T_{a1}(t)][1 - T_{a5}(t)]$	$0 \le \omega t < \pi$	$M_a \ge \dfrac{1}{2}$
$I_{Da3}(t)$	$I_a(t)[1 - T_{a5}(t)]$	$0 \le \omega t < \pi$	
$I_{Ta1}(t)$	$I_a(t)\,T_{a1}(t)[1 - T_{a5}(t)]$	$0 \le \omega t < \pi$	
$I_{Ta2}(t)$	$I_a(t)\,T_{a2}(t)[1 - T_{a5}(t)]$	$0 \le \omega t < \pi$	
$I_{Ta5}(t)$	$I_a(t)\,T_{a5}(t)$	$\sin^{-1}\dfrac{1}{2M_a} \le \omega t < \pi - \sin^{-1}\dfrac{1}{2M_a}$	
$I_{Ta6}(t)$	$I_a(t)$	$0 \le \omega t < \pi$	
$I_{Da1}(t)$	$I_a(t)[1 - T_{a1}(t)]$	$0 \le \omega t < \pi$	$M_a < \dfrac{1}{2}$
$I_{Da3}(t)$	$I_a(t)$	$0 \le \omega t < \pi$	
$I_{Ta1}(t)$	$I_a(t)\,T_{a1}(t)$	$0 \le \omega t < \pi$	
$I_{Ta2}(t)$	$I_a(t)$	$0 \le \omega t < \pi$	
$I_{Ta5}(t)$	0	$0 \le \omega t < \pi$	
$I_{Ta6}(t)$	$I_a(t)$	$0 \le \omega t < \pi$	

Note: Ia(t) is the phase "a" output current of the inverter, which consists of power factor angle between the load and pole voltages. The principle of operating instantaneous current expression for each complementary switch is similar to the upper phase leg switches.

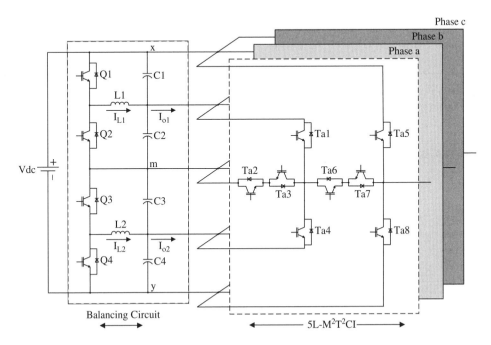

Figure 10.6 Circuit diagram of the proposed 5L-M²T²CI topology associated with the 5L-DC/DC balancing circuit.

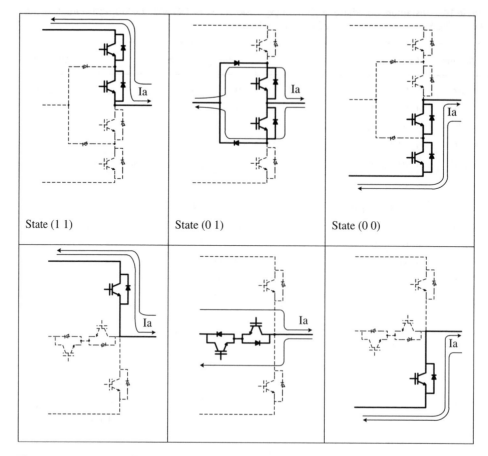

Figure 10.7 Derivation of bidirectional current flow of the t-type-clamped switches based on the switching states transition of the NPC topology.

Table 10.4 5L-M^{2}T^{2}CI voltage levels and the corresponding switching states.

States	Switching States								Per-Phase Leg Voltage		
	T_{s1}	T_{s2}	T_{s3}	T_{s4}	T_{s5}	T_{s6}	T_{s7}	T_{s8}	V_{sx}	V_{sy}	V_{sm}
1	1	0	0	0	1	0	0	0	0	V_{dc}	$V_{dc}/2$
2	1	0	0	0	0	1	1	0	$-V_{dc}/4$	$3V_{dc}/4$	$V_{dc}/4$
3	0	1	1	0	0	1	1	0	$-V_{dc}/2$	$V_{dc}/2$	0
4	0	0	0	1	0	1	1	0	$-3V_{dc}/4$	$V_{dc}/4$	$-V_{dc}/4$
5	0	0	0	1	0	0	0	1	$-V_{dc}$	0	$-V_{dc}/2$

Here, "s" represents phase a, b, and c; S_{s1}–S_{s8} are presented as the IGBT switching devices for the individual legs; and logic 1 represents the turn-on and logic 0 the turn-off for T_{s1}, T_{s2}, T_{s3}, and T_{s4}

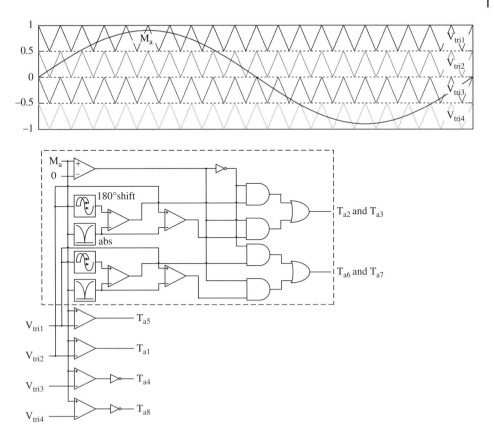

Figure 10.8 Proposed six-carriers switching scheme based on LS-PWM strategy with the corresponding gating signal of phase "a."

connected in between the nodes of each phase and the center point of the dc-link. The control gate circuitry for the 5L-M^2T^2CI topology is shown in Fig. 10.8.

In Fig. 10.8, the control signals of two switches connected in series is obtained from indirect carrier-based comparison. The bidirectional switches for (Ta2, Ta3) and (Ta6, Ta7) are generated by comparing the absolute function of the modulation reference signals with four upper carriers wave (Vtri1, Vtri2, 180° phase shifted of Vtri1 and Vtri2) to achieve the desired switching states.

However, the development of this control unit, as indicated in the dotted box in Fig. 10.8, is more complex and costly for hardware development, and also requires higher computation time for digital control. Thus, Fig. 10.9 shows the modification of the control gating circuit to achieve the same switching states pattern as in Table 10.4.

According to the switching states selection based on the LS-PWM technique, the output pole voltage of 5L-M^2T^2CI under high modulation for more than half the amplitude is expressed as:

$$V_{sm}(t) = \frac{V_{dc}(t)}{4}[T_{s1}(t) + T_{s4}(t) + T_{s5}(t) + T_{s8}(t)] \tag{10.4}$$

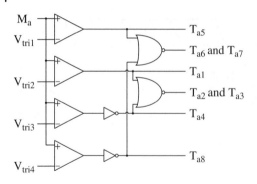

Figure 10.9 Proposed modified four-carriers switching scheme based on LS-PWM strategy with the corresponding gating signal of phase "a."

For amplitude modulation lower than 0.5, the output pole voltage is written as:

$$V_{sm}(t) = \frac{V_{dc}(t)}{4}[T_{s1}(t) - T_{s4}(t)] \tag{10.5}$$

Similarly, the current expressions for switches are determined by the switching function with the range of the modulation depth. The expression of current through the switch is listed in Table 10.5 based on one cycle of the commutation period.

According to Table 10.5, voltage quality of 5L-M²T²CI is poor at low modulation index for M < 0.5. Poor output voltage quality will lead to three-level output pole voltage operation, and this increases the common-mode voltage level when a motor is loaded. Three-level output pole voltage stepped waveform due to low modulation is operated at the inner cell inverter, and the clamping power device of the output terminal of the inner cell inverter is operated in continuous mode. However, operation with low modulation indices is rare in most of the applications, and a momentary period of high modulation indices will also occur during motor start-up. Therefore, low modulation indices will not be the primary focus for this work.

Table 10.5 Instantaneous Phase "a" Current Expression of 5L-M²T²CI with the Respective Angle of the Operating System and the Range of Modulation Depth.

Current	Equation	Operating Angle	Modulation
$I_{Ta1}(t)$	$I_a(t)[T_{a1}(t) - T_{a5}(t)]$	$0 \leq \omega t < \pi$	$M_a \geq \frac{1}{2}$
$I_{Ta2}(t)$	$I_a(t)[1 - T_{a1}(t) - T_{a4}(t)]$	$0 \leq \omega t < 2\pi$	
$I_{Ta4}(t)$	$I_a(t)[T_{a4}(t) - T_{a8}(t)]$	$\pi \leq \omega t < 2\pi$	
$I_{Ta5}(t)$	$I_a(t) T_{a5}(t)$	$\sin^{-1}\frac{1}{2M_a} \leq \omega t < \pi - \sin^{-1}\frac{1}{2M_a}$	
$I_{Ta6}(t)$	$I_a(t)[1 - T_{a5}(t) - T_{a8}(t)]$	$0 \leq \omega t < 2\pi$	
$I_{Ta8}(t)$	$I_a(t) T_{a8}(t)$	$\pi + \sin^{-1}\frac{1}{2M_a} \leq \omega t < 2\pi - \sin^{-1}\frac{1}{2M_a}$	
$I_{Ta1}(t)$	$I_a(t) T_{a1}(t)$	$0 \leq \omega t < \pi$	$M_a < \frac{1}{2}$
$I_{Ta2}(t)$	$I_a(t)[1 - T_{a1}(t) - T_{a4}(t)]$	$0 \leq \omega t < 2\pi$	
$I_{Ta4}(t)$	$I_a(t)[T_{a4}(t) - T_{a8}(t)]$	$\pi \leq \omega t < 2\pi$	
$I_{Ta5}(t)$	0	$0 \leq \omega t < 2\pi$	
$I_{Ta6}(t)$	$I_a(t)$	$0 \leq \omega t < 2\pi$	
$I_{Ta8}(t)$	0	$0 \leq \omega t < 2\pi$	

10.3.4 Five-Level/Multiple-Pole Multilevel Single-Switch-Clamped Inverter (5L-M²S²CI)

Fig. 10.10 shows the five-level/multiple-pole multilevel single-switch-clamped inverter (5L-M²S²CI) topology. The structure of this topology is configured based on a bidirectional switch, which allows the bidirectional flow of current. 5L-M²T²CI shares the same gating signal for two switches connected in series to form a bidirectional current flow. Then, a simple structure of single-switch bidirectional current flow is implemented by incorporating the diode bridge circuit, as shown in Fig. 10.11. Similarly, the same gating logic circuit as 5L-M²T²CI is used in the proposed circuit topology, as shown in Fig. 10.12.

The output pole-voltage of this topology under high modulation depth is expressed as:

$$V_{sm}(t) = \frac{V_{dc}(t)}{4}[T_{s1}(t) + T_{s3}(t) + T_{s4}(t) + T_{s6}(t)] \tag{10.6}$$

Here, the output pole voltage of the inverter for low modulation depth is written as:

$$V_{sm}(t) = \frac{V_{dc}(t)}{4}[T_{s1}(t) - T_{s3}(t)] \tag{10.7}$$

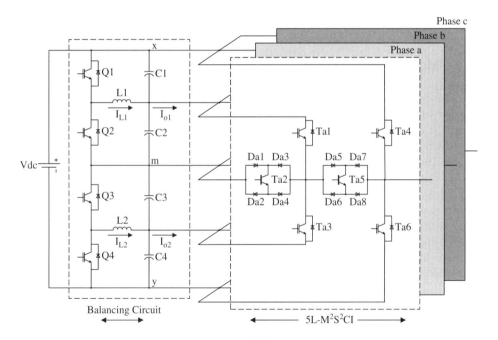

Figure 10.10 Circuit diagram of the proposed 5L-M²S²CI topology associated with the 5L-DC/DC balancing circuit.

Figure 10.11 Alternative bidirectional switch.

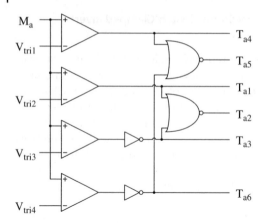

Figure 10.12 Switching scheme based on LS-PWM strategy with the corresponding gating signal of phase "a."

Similarly, a low modulation operation is not recommended to operate, as discussed in subsection 10.2.3. This is due to the design requirement for preventing oversizing, and improves the output quality of the inverter of both proposed five-level inverter topologies with the bidirectional switch configuration. The instantaneous current expression is shown in Table 10.6 with the range of modulation depth.

10.4 Modulation Scheme

The basic derivation of the switching function is detailed in Chapter 5. The 5L-MDCI topology with the use of the LS-PWM technique has been derived and explained in Chapter 7, and is shown in Table 10.7.

The switching function expression for the proposed 5L-inverter topologies is listed in Table 10.8. The switching function is selected according to the required switching state as stated in the instantaneous current table (e.g. Table 10.1 for 5L-MDCI) for the proposed topologies. The switching function expression for global current stress analysis is discussed in the following sections. Power loss derivations for each of the switches are also determined by the global stress analysis with the given switching function expression.

If the voltage expression of the output characteristic is derived, then the product of each local modulation from each switch can be written as $\sum M_{ai}(t) = Ma(t)$.

10.5 Design Consideration

The following stress analysis is used to calculate the conduction and switching losses of power devices. Average and RMS current stress approximation is derived with the switching functions given in Tables 10.7 and 10.8. The power devices for 5L-inverter topologies are selected according to the maximum voltage and current stress levels. The chosen power device ratings are calculated based on the approximation stress analysis as presented in [6–8]. The voltage and current stress expressions are determined based on the following assumptions:

Table 10.6 Instantaneous Phase "a" Current Expression of 5L-M²S²CI with the Respective Angles of the Operating System and the Range of Modulation Depth.

Current	Equation	Operating Angle	Modulation
$I_{Da1}(t)$	$I_a(t)[1 - T_{a1}(t) - T_{a3}(t)]$	$0 \leq \omega t < \pi$	$M_a \geq \dfrac{1}{2}$
$I_{Da3}(t)$	$I_a(t)[1 - T_{a1}(t) - T_{a3}(t)]$	$\pi \leq \omega t < 2\pi$	
$I_{Da5}(t)$	$I_a(t)[1 - T_{a4}(t) - T_{a6}(t)]$	$0 \leq \omega t < \pi$	
$I_{Da7}(t)$	$I_a(t)[1 - T_{a4}(t) - T_{a6}(t)]$	$\pi \leq \omega t < 2\pi$	
$I_{Ta1}(t)$	$I_a(t)[1 - T_{a1}(t) - T_{a4}(t)]$	$0 \leq \omega t < \pi$	
$I_{Ta2}(t)$	$I_a(t)[1 - T_{a1}(t) - T_{a3}(t)]$	$0 \leq \omega t < 2\pi$	
$I_{Ta3}(t)$	$I_a(t)[T_{a3}(t) - T_{a6}(t)]$	$\pi \leq \omega t < 2\pi$	
$I_{Ta4}(t)$	$I_a(t)T_{a4}(t)$	$\sin^{-1}\dfrac{1}{2M_a} \leq \omega t < \pi - \sin^{-1}\dfrac{1}{2M_a}$	
$I_{Ta5}(t)$	$I_a(t)[1 - T_{a4}(t) - T_{a6}(t)]$	$0 \leq \omega t < 2\pi$	
$I_{Ta6}(t)$	$I_a(t)T_{a6}(t)$	$\pi \leq \omega t < 2\pi$	
$I_{Da1}(t)$	$I_a(t)[1 - T_{a1}(t) - T_{a3}(t)]$	$0 \leq \omega t < \pi$	$M_a < \dfrac{1}{2}$
$I_{Da3}(t)$	$I_a(t)[1 - T_{a1}(t) - T_{a3}(t)]$	$\pi \leq \omega t < 2\pi$	
$I_{Da5}(t)$	$I_a(t)$	$0 \leq \omega t < \pi$	
$I_{Da7}(t)$	$I_a(t)$	$\pi \leq \omega t < 2\pi$	
$I_{Ta1}(t)$	$I_a(t)T_{a1}(t)$	$0 \leq \omega t < \pi$	
$I_{Ta2}(t)$	$I_a(t)[1 - T_{a1}(t) - T_{a3}(t)]$	$0 \leq \omega t < 2\pi$	
$I_{Ta3}(t)$	$I_a(t)T_{a3}(t)$	$\pi \leq \omega t < 2\pi$	
$I_{Ta4}(t)$	0	$0 \leq \omega t < \pi$	
$I_{Ta5}(t)$	$I_a(t)$	$0 \leq \omega t < 2\pi$	
$I_{Ta6}(t)$	0	$\pi \leq \omega t < 2\pi$	

Table 10.7 Switching Function of Classical Five-Level Inverter Topologies.

5L-Inverter Topologies	Switches	Switching Function Expression
MDCI	$T_{a1}(t)$	$2M_a(t) - 1$
	$T_{a2}(t)$	$2M_a(t)$
	$T_{a3}(t)$	$2M_a(t) + 1$
	$T_{a4}(t)$	$2M_a(t) + 2$

Here, $M_a(t)$ is the phase "a" modulation signal; $M_a(t) = m_a \sin\omega t$.

1) Current and voltage waveforms are ripple-free from the load and dc-link.
2) Balanced dc-link capacitor voltage.
3) Zero voltage spike caused by the parasitic element due to low switching frequency range.
4) Pure sinusoidal current due to highly inductive load.

Table 10.8 Switching Function of Proposed Five-Level Inverter Topologies.

5L-Inverter Topologies	Switches	Switching Function Expression
M^{2}DCI	$T_{a1}(t)$	$2M_a(t)$
	$T_{a2}(t)$	$2M_a(t)+1$
	$T_{a5}(t)$	$2M_a(t)-1$
	$T_{a6}(t)$	$2M_a(t)+2$
M^{2}T^{2}CI	$T_{a1}(t)$	$2M_a(t)$
	$T_{a4}(t)$	$-2M_a(t)$
	$T_{a5}(t)$	$2M_a(t)-1$
	$T_{a8}(t)$	$-2M_a(t)-1$
M^{2}S^{2}CI	$T_{a1}(t)$	$2M_a(t)$
	$T_{a3}(t)$	$-2M_a(t)$
	$T_{a4}(t)$	$2M_a(t)-1$
	$T_{a6}(t)$	$-2M_a(t)-1$

Note: For global stress analysis, the modulation function for the respective switches is considered as $M_{ai}(t) = M_a(t)$ for all i value (i.e., i = 1, 2, 3,...).

10.5.1 Device Voltage Stress

Voltage stress expressions for each of the power devices in the upper phase leg are listed in Tables 10.9 and 10.10. The rest of the power devices (those not listed in Tables 10.9 and 10.10) are those of the lower phase leg, which are also known as the *symmetrical components* of the upper phase leg devices.

For symmetrical components, the maximum voltage stress levels are the same as those of the upper phase leg power devices. With this expression, the final value of the maximum voltage stress level is calculated by substituting the switching states into the mathematical expression to obtain the maximum voltage stress level, as presented in Tables 10.11 and 10.12.

10.5.2 Devices Current Stress

This section demonstrates the calculation of the current stress of the power devices based on the global stress analysis with the local averaged values over the fundamental period of the switching period, as detailed in [6]. The accuracy of the analysis is found to be commendable as the analytical results are on par with simulation results. In addition, the loss modeling is supported by using the approximation of current stress based on the derived switching function. This approach has been adopted by Infineon Technologies, as described in the application note [9].

The adopted average and RMS current stress analysis is analyzed over one fundamental period of the load current and harmonics are not considered. The load current is assumed to be purely sinusoidal, and is dependent on the operating power factor angle, δ. The summation of the pulse current during the switching commutation of the power device can be simplified by using the integral of local values over one fundamental period of the switching states. For simplification, the general expression of the average

Table 10.9 Voltage Stress Expression for 5L-MDCI and 5L-M^{2}DCI Topologies.

5L-Inverter Topologies	Switches	Voltage Stress Expressions
MDCI	T_{a1}	$\dfrac{V_{dc}(t)}{4}[1 - T_{a1}(t)]$
	T_{a2}	$\dfrac{V_{dc}(t)}{4}[1 - T_{a2}(t)]$
	T_{a3}	$\dfrac{V_{dc}(t)}{4}[1 - T_{a3}(t)]$
	T_{a4}	$\dfrac{V_{dc}(t)}{4}[1 - T_{a4}(t)]$
	D_{a1}	$\dfrac{V_{dc}(t)}{4}[1 - T_{a1}(t)]$
	D_{a2}	$\dfrac{V_{dc}(t)}{4}[T_{a1}(t) + T_{a2}(t)]$
	D_{a3}	$\dfrac{V_{dc}(t)}{4}[T_{a1}(t) + T_{a2}(t) + T_{a3}(t)]$
M^{2}DCI	T_{a1}	$\dfrac{V_{dc}(t)}{4}[1 - T_{a2}(t)]$
	T_{a2}	$\dfrac{V_{dc}(t)}{4}[1 - T_{a2}(t)]$
	T_{a5}	$\dfrac{V_{dc}(t)}{4}[1 - T_{a5}(t)]\left[1 - \dfrac{1}{2}[T_{a1}(t) + T_{a2}(t) - T_{a6}(t)]\right]$
	T_{a6}	$\dfrac{V_{dc}(t)}{2}[1 - T_{a6}(t)]$
	D_{a1}	$\dfrac{V_{dc}(t)}{4}T_{a1}(t)$
	D_{a3}	$\dfrac{V_{dc}(t)}{4}[T_{a1}(t) + T_{a2}(t) - 1]\cdot\left[T_{a5}(t) - \dfrac{1}{2} + \dfrac{T_{a6}(t)}{2}\right]$

and RMS current stress with respect to switching function is approximated (as stated in Tables 10.7 and 10.8) and expressed in the following equation:

$$I_{Tai\langle avg\rangle} = \frac{1}{2\pi}\int_{\alpha}^{\beta}\left[\frac{1}{t_p}\sum_{k=t_{on}/t_p}^{S_{Tai}} I_a(kt_p, \omega t)\right] d\omega t \approx \frac{1}{2\pi}\int_{\alpha}^{\beta}\left[\frac{1}{t_p}\int_{0}^{S_{tai}t_p} I_a(\omega t)\ dt_\mu\right] d\omega t$$

$$I_{Dai\langle avg\rangle} = \frac{1}{2\pi}\int_{\alpha}^{\beta}\left[\frac{1}{t_p}\sum_{k=t_{on}/t_p}^{S_{Tai}} I_a(kt_p, \omega t)\right] d\omega t \approx \frac{1}{2\pi}\int_{\alpha}^{\beta}\left[\frac{1}{t_p}\int_{0}^{S_{Dai}t_p} I_a(\omega t)\ dt_\mu\right] d\omega t$$

(10.8)

$$I_{Tai\langle rms\rangle} = \sqrt{\frac{1}{2\pi}\int_{\alpha}^{\beta}\left[\frac{1}{t_p}\sum_{k=t_{on}/t_p}^{S_{Tai}} I_a^2(kt_p, \omega t)\right] d\omega t} = \sqrt{\frac{1}{2\pi}\int_{\alpha}^{\beta}\left[\frac{1}{t_p}\int_{0}^{S_{tai}t_p} I_a^2(\omega t)\ dt_\mu\right] d\omega t}$$

$$I_{Dai\langle rms\rangle} = \sqrt{\frac{1}{2\pi}\int_{\alpha}^{\beta}\left[\frac{1}{t_p}\sum_{k=t_{on}/t_p}^{S_{Tai}} I_a^2(kt_p, \omega t)\right] d\omega t} = \sqrt{\frac{1}{2\pi}\int_{\alpha}^{\beta}\left[\frac{1}{t_p}\int_{0}^{S_{Dai}t_p} I_a^2(\omega t)\ dt_\mu\right] d\omega t}$$

(10.9)

Here, S_{Tai} is the switching function of the switching scheme as presented in Tables 10.7 and 10.8; and $i = 1,2,3,\ldots$ of the respective power devices; and t_p is the time interval of

Table 10.10 Voltage Stress Expressions for 5L-M^{2}T^{2}CI and 5L-M^{2}S^{2}CI Topologies.

5L-Inverter Topologies	Switches	Voltage Stress Expressions
M^{2}T^{2}CI	T_{a1}	$\dfrac{V_{dc}(t)}{4}[1 - T_{a1}(t) + T_{a4}(t)]$
	T_{a2}	$\dfrac{V_{dc}(t)}{4}T_{a4}(t)$
	T_{a4}	$\dfrac{V_{dc}(t)}{4}[1 - T_{a4}(t) + T_{a1}(t)]$
	T_{a5}	$\dfrac{V_{dc}(t)}{4}[2 - T_{a1}(t) + T_{a4}(t) - T_{a5}(t) + T_{a8}(t)]$
	T_{a6}	$\dfrac{V_{dc}(t)}{4}T_{a8}(t)$
	T_{a8}	$\dfrac{V_{dc}(t)}{4}[2 - T_{a1}(t) + T_{a4}(t) + T_{a5}(t) - T_{a8}(t)]$
M^{2}S^{2}CI	T_{a1}	$\dfrac{V_{dc}(t)}{4}[1 - T_{a1}(t) + T_{a3}(t)]$
	T_{a2}	$\dfrac{V_{dc}(t)}{4}[1 - T_{a1}(t) + T_{a3}(t)]$
	T_{a3}	$\dfrac{V_{dc}(t)}{4}[1 - T_{a1}(t) + T_{a3}(t)]$
	T_{a4}	$\dfrac{V_{dc}(t)}{4}[2 - T_{a1}(t) + T_{a3}(t) - T_{a4}(t) + T_{a6}(t)]$
	T_{a5}	$\dfrac{V_{dc}(t)}{4}[T_{a4}(t) + T_{a6}(t)]$
	T_{a6}	$\dfrac{V_{dc}(t)}{4}[2 + T_{a1}(t) - T_{a3}(t) + T_{a4}(t) - T_{a6}(t)]$
	D_{a1}	$\dfrac{V_{dc}(t)}{4}T_{a1}(t)$
	D_{a3}	$\dfrac{V_{dc}(t)}{4}T_{a3}(t)$
	D_{a5}	$\dfrac{V_{dc}(t)}{4}T_{a4}(t)$
	D_{a7}	$\dfrac{V_{dc}(t)}{4}T_{a6}(t)$

the pulses generated within one period of the switching scheme. α and β are the ranges of the occurring switching pulse during one cycle. Based on equations (10.8) and (10.9), the final general expression of the current stress analysis is as follows:

$$I_{Tai\langle avg\rangle} = \frac{1}{2\pi}\int_{\alpha}^{\beta} I_a(\omega t)\cdot S_{Tai}(\omega t)\ d\omega t = \frac{1}{2\pi}\int_{\alpha}^{\beta} I_{Tai}(\omega t)\ d\omega t$$

$$I_{Dai\langle avg\rangle} = \frac{1}{2\pi}\int_{\alpha}^{\beta} I_a(\omega t)\cdot S_{Dai}(\omega t)\ d\omega t = \frac{1}{2\pi}\int_{\alpha}^{\beta} I_{Dai}(\omega t)\ d\omega t \qquad (10.10)$$

$$I_{Tai\langle rms\rangle} = \sqrt{\frac{1}{2\pi}\int_{\alpha}^{\beta} I_a^2(\omega t)\cdot S_{Tai}(\omega t)\ d\omega t}$$

$$I_{Dai\langle rms\rangle} = \sqrt{\frac{1}{2\pi}\int_{\alpha}^{\beta} I_a^2(\omega t)\cdot S_{Dai}(\omega t)\ d\omega t} \qquad (10.11)$$

$I_{Tai}(\omega t)$ and $I_{Dai}(\omega t)$ are the local currents through the power device with respect to the switching function of the respective device. The local current is expressed as a function

Table 10.11 Maximum Voltage Stress Levels for 5L-MDCI and 5L-M^{2}DCI Topologies.

5L-Inverter Topologies	Maximum Voltage Stress Levels (Based on Table 10.9)
MDCI	$$\begin{cases} \begin{aligned} V_{Ta1} &= V_{Ta2} = V_{Ta3} = V_{Ta4} = V_{Ta5} \\ &= V_{Ta6} = V_{Ta7} = V_{Ta8} = V_{Da6} = \frac{1}{4}V_{dc} \\ V_{Da2} &= V_{Da5} = \frac{1}{2}V_{dc} \\ V_{Da3} &= V_{Da4} = \frac{3}{4}V_{dc} \end{aligned} \end{cases}$$
M^{2}DCI	$$\begin{cases} \begin{aligned} V_{Ta1} &= V_{Ta2} = V_{Ta3} = V_{Ta4} = V_{Ta5} \\ &= V_{Ta6} = V_{Ta7} = V_{Ta8} = V_{Da1} = V_{Da6} = \frac{1}{4}V_{dc} \\ V_{Da2} &= V_{Da5} = \frac{1}{2}V_{dc} \\ V_{Da3} &= V_{Da4} = \frac{3}{4}V_{dc} \end{aligned} \end{cases}$$ $$\begin{cases} \begin{aligned} V_{Ta1} &= V_{Ta2} = V_{Ta3} = V_{Ta4} = V_{Da1} = V_{Da2} = V_{Da3} = V_{Da4} = \frac{1}{4}V_{dc} \\ V_{Ta5} &= V_{Ta8} = \frac{3}{4}V_{dc} \\ V_{Ta6} &= V_{Ta7} = \frac{1}{2}V_{dc} \end{aligned} \end{cases}$$

Table 10.12 Maximum Voltage Stress Levels for 5L-M^{2}T^{2}CI and 5L-M^{2}S^{2}CI Topologies.

5L-Inverter Topologies	Maximum Voltage Stress Levels (Based on Table 10.9)
M^{2}T^{2}CI	$$\begin{cases} V_{Ta1} = V_{Ta4} = \dfrac{V_{dc}}{2} \\ V_{Ta2} = V_{Ta3} = V_{Ta6} = V_{Ta7} = \dfrac{V_{dc}}{4} \\ V_{Ta5} = V_{Ta8} = V_{dc} \end{cases}$$
M^{2}S^{2}CI	$$\begin{cases} \begin{aligned} V_{Ta1} &= V_{Ta3} = \frac{V_{dc}}{2} \\ V_{Ta2} &= V_{Ta5} = V_{Da1} = V_{Da2} = V_{Da3} \\ &= V_{Da4} = V_{Da5} = V_{Da6} = V_{Da7} = V_{Da8} = \frac{V_{dc}}{4} \\ V_{Ta4} &= V_{Ta6} = V_{dc} \end{aligned} \end{cases}$$

of load current and switching function, as shown in Tables 10.1, 10.3, 10.5, and 10.6 for the respective 5L-inverter topologies.

Substitute the boundary limit in the integral form of equations (10.10) and (10.11) to have a better visualization of the current stress expression in terms of amplitude current, power factor angle (operating load angle, δ), and amplitude modulation. A simple graphical approach of the current commutation period with the range of the modulation

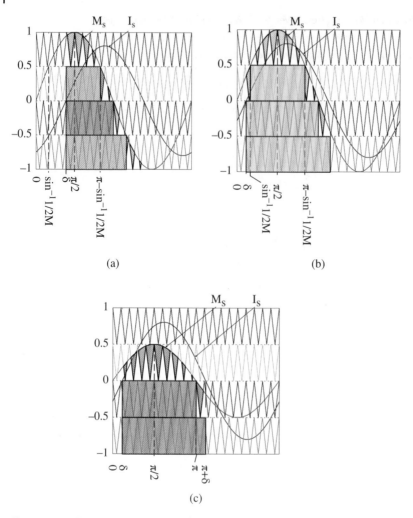

Figure 10.13 Current commutation in the respective switches as shaded in the carrier region for different operating conditions of a 5L-MDCI operation. (a) $\sin^{-1}1/2M_a \leq \delta$ for $M_a \geq 0.5$; (b) $\sin^{-1}1/2M_a > \delta$ for $M_a \geq 0.5$; and (c) $M_a < 0.5$.

depth is shown in Fig. 10.13. These figures illustrate the current through the switches during the switching commutation period of a 5L-MDCI topology with the LS-PWM techniques under various operating conditions. The shaded area represents the current commutation in the power devices of the 5L-MDCI topology.

5L-M²DCI Topology

A similar mathematical approach on the current stress analysis to 5L-MDCI topology, is applied for the proposed 5L-inverter topologies. The current commutation of the power devices within one full period of the switching cycle is shown in Fig. 10.14. With this graphical approach, the boundary limit of the local current for the average and RMS current calculation is analyzed for the lower and upper limits of the integration.

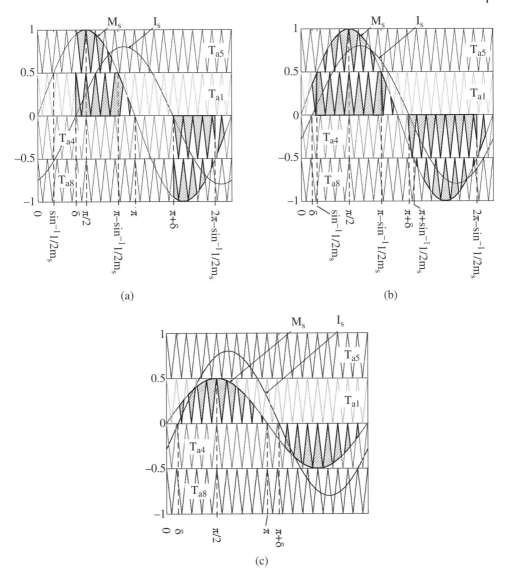

Figure 10.14 Current commutation mode of the respective power switches as shown in the shaded area under the different operating conditions of a 5L-M^{2}T^{2}CI operation. (a) $\sin^{-1}1/2M_a \leq \delta$ for $M_a \geq 0.5$; (b) $\sin^{-1}1/2M_a > \delta$ for $M_a \geq 0.5$; and (c) $M_a < 0.5$.

Based on Fig. 10.14, a zero current switching is achieved for particular switching states (states 1 and 5 of Table 10.2). The zero current switching occurs in T_{a1}–T_{a4} during switching states 1 and 5 as clearly shown in Fig. 10.14, and it is not shaded in the middle two operating regions of the LS-PWM, as shown in the figure.

5L-M^{2}T^{2}CI and 5L-M^{2}S^{2}CI Topologies

The graphical approach of the boundary limit for the integration calculus of the current stress analysis of both proposed 5L-M^{2}T^{2}CI and 5L-M^{2}S^{2}CI topologies are shown in

Fig. 10.14. Zero current switching is also achieved in the power device of the inner-cell switches for both topologies, as illustrated in the two switching states 1 and 5 in Table 10.4.

According to Fig. 10.14, the boundary limits of the current pulses of the bidirectional switches are commutated within the range of $0-2\pi$ of the sinusoidal load current, depending on the power angle and switching state selection. The current stress rating of each discrete diode element of the bidirectional switches in 5L-M^2S^2CI is half that of the active switch of the bidirectional switches in 5L-M^2T^2CI. The remaining devices of 5L-M^2S^2CI have the same current rating as compared to the 5L-M^2T^2CI topology. A list of current stress rating of each of the power devices in the 5L-M^2T^2CI topology is shown in Tables 10.13–10.16.

10.6 Accuracy of the Current Stress Calculation

It is important to know the deviation margin of experimental results with respect to simulation results. Approximation of the stress level of the power device may not be the same as compared to the simulation results, but it is sufficient to prove the concept of current commutation and prevent any oversizing of the converter. The main reason of considering an error in calculations is to validate the previous assumptions. Therefore, the accuracy of the current stress analysis in the previous section is limited based on the assumptions made.

The deviation in the calculated current stress for active switches is partly due to the reverse recovery current through the anti-parallel diode. The current commutation for every active switch and anti-parallel diode in every pulse during the fundamental period is calculated individually using the analytical solution. In order to determine the individual losses incurred in the active switch and anti-parallel diode, the analysis is carried out separately. The on-state resistances for the active switch and anti-parallel diode are listed at different values and characteristics in the manufacturer's datasheet, while the simulation results of the reverse recovery current commutation for the active switch cannot be determined through simulation results.

To verify the current stress expression as stated in section 10.4.2, the data for comparative analysis is presented in Tables 10.13–10.16 for the respective 5L-inverters. The simulated and analytical results for the analytical comparison are obtained from simulation tools such as PSIM and MATLAB R2010b. The current stress on the discrete components is determined by modeling and simulating devices in PSIM. Analytical results are validated using MATLAB Script by setting boundary limits for integration.

Based on the data listed in Tables 10.13–10.16 with the supported simulation toolbox, the results for one of the power devices of the neutral-point-clamped in the presented 5L-inverter topologies is not similar to the simulation results. This is due to the neutral-point clamped diode or bidirectional switches in the inner cell being dependent on the accuracy of the current stress of the series-connected switch of the bridge leg of the inner cell.

However, one of the power devices may deviate to some error value based on the global stress approximation with the local average switching current integration method. Nevertheless, it is sufficient to analyze the overall loss distribution of the power device, where the concept of zero current switching occurs in the multiple-pole

Table 10.13 Comparison Results of the Current Stress on the Power Device Between the Analytical Results and Simulation Results for the Proposed 5L-M²T²Cl Topology.

| Devices | $\sin^{-1}1/2m_a \leq \delta$ and $M_a \geq 0.5$ | | | | $\sin^{-1}1/2m_a > \delta$ and $M_a \geq 0.5$ | | | | $M_a < 0.5$ | | | |
| | Average Values (A) | | RMS Values (A) | | Average Values (A) | | RMS Values (A) | | Average Values (A) | | RMS Values (A) | |
	Analytical	Simulated	Analytical	Simulated	Analytical	Simulated	Analytical	Simulated	Analytical	Simulated	Analytical	Simulated
T_{a1}	0.7633	0.5511	1.5234	1.5949	0.6426	0.6196	1.3898	1.4341	0.2073	0.1888	0.4932	0.4731
$T_{a2} = T_{a3}$	0	0.0003	0.4484	1.2516	0	0.0013	0.4171	0.7084	0	0	0.6848	0.7118
T_{a4}	0.5929	0.5316	1.4095	1.5604	0.6426	0.6112	1.3898	1.4122	0.2073	0.1932	0.4932	0.4780
T_{a5}	0.6390	0.5302	1.6644	1.3984	0.6289	0.5971	1.5623	1.4964	0	0	0	0
$T_{a6} = T_{a7}$	0	0.0192	2.3733	2.5583	0	0.0071	2.2094	2.1336	0	0	0.9775	0.9792
T_{a8}	0.6390	0.4979	1.6644	1.3430	0.6289	0.5695	1.5623	1.4611	0	0	0	0
D_{Ta1}	0.0730	NA	0.3029	NA	0	NA	0	NA	0	NA	0	NA
D_{Ta4}	0.0730	NA	0.3029	NA	0	NA	0	NA	0	NA	0	NA
D_{Ta5}	0.0365	NA	0.2142	NA	0	NA	0	NA	0	NA	0	NA
D_{Ta8}	0.0365	NA	0.2142	NA	0	NA	0	NA	0	NA	0	NA
$D_{Ta2} = D_{Ta3} = D_{Ta6} = D_{Ta7}$	0	NA	0	NA	0	NA	0	NA	0	NA	0	NA

Table 10.14 Comparison Results of the Current Stress on the Power Device Between the Analytical Results and Simulation Results for the Proposed 5L-M^{2}DCI Topology.

| Devices | $\sin^{-1}1/2m_a \leq \delta$ and $M_a \geq 0.5$ | | | | $\sin^{-1}1/2m_a > \delta$ and $M_a \geq 0.5$ | | | | $M_a < 0.5$ | | | |
| | Average Values (A) | | RMS Values (A) | | Average Values (A) | | RMS Values (A) | | Average Values (A) | | RMS Values (A) | |
	Analytical	Simulated	Analytical	Simulated	Analytical	Simulated	Analytical	Simulated	Analytical	Simulated	Analytical	Simulated
$T_{a1} = T_{a4}$	0.7622	0.5579	1.5212	1.6045	0.6392	0.6214	1.6388	1.4356	0.2072	0.1887	0.4938	0.4729
$T_{a2} = T_{a3}$	0.8402	0.8173	1.5538	1.8379	0.7463	0.7514	1.7707	1.5192	0.4396	0.4334	0.6906	0.6896
$T_{a5} = T_{a8}$	0.6382	0.5359	1.6621	1.4117	0.6256	0.5989	1.6212	1.5004	0	0	0	0
$T_{a6} = T_{a7}$	1.4785	1.4835	2.3592	2.3173	1.3373	1.3642	2.3703	2.1347	0.4396	0.4354	0.6906	0.6896
$D_{a1} = D_{a4}$	0.0781	0.2593	0.3166	0.8964	0.1071	0.1300	0.6708	0.4971	0.2325	0.2447	0.4838	0.5019
$D_{a3} = D_{a5}$	0.8404	0.9477	1.6744	1.8376	0.7117	0.7652	1.7291	1.5184	0.4396	0.4354	0.6906	0.6896
D_{Ta1} to D_{Ta4}	0	NA	0	NA	0	NA	0	NA	0	NA	0	NA
D_{Ta5} to D_{Ta8}	0.0365	NA	0.2138	NA	0	NA	0	NA	0	NA	0	NA

Table 10.15 Comparison Results of the Current Stress on the Power Device Between the Analytical Results and Simulation Results for the Proposed 5L-M²T²Cl Topology.

Devices	$\sin^{-1}1/2m_a \le \delta$ and $M_a \ge 0.5$				$\sin^{-1}1/2m_a > \delta$ and $M_a \ge 0.5$				$M_a < 0.5$			
	Average Values (A)		RMS Values (A)		Average Values (A)		RMS Values (A)		Average Values (A)		RMS Values (A)	
	Analytical	Simulated	Analytical	Simulated	Analytical	Simulated	Analytical	Simulated	Analytical	Simulated	Analytical	Simulated
T_{a1}	0.7633	0.5511	1.5234	1.5949	0.6426	0.6196	1.3898	1.4341	0.2073	0.1888	0.4932	0.4731
$T_{a2} = T_{a3}$	0	0.0003	0.4484	1.2516	0	0.0013	0.4171	0.7084	0	0	0.6848	0.7118
T_{a4}	0.5929	0.5316	1.4095	1.5604	0.6426	0.6112	1.3898	1.4122	0.2073	0.1932	0.4932	0.4780
T_{a5}	0.6390	0.5302	1.6644	1.3984	0.6289	0.5971	1.5623	1.4964	0	0	0	0
$T_{a6} = T_{a7}$	0	0.0192	2.3733	2.5583	0	0.0071	2.2094	2.1336	0	0	0.9775	0.9792
T_{a8}	0.6390	0.4979	1.6644	1.3430	0.6289	0.5695	1.5623	1.4611	0	0	0	0
D_{Ta1}	0.0730	NA	0.3029	NA	0	NA	0	NA	0	NA	NA	NA
D_{Ta4}	0.0730	NA	0.3029	NA	0	NA	0	NA	0	NA	NA	NA
D_{Ta5}	0.0365	NA	0.2142	NA	0	NA	0	NA	0	NA	NA	NA
D_{Ta8}	0.0365	NA	0.2142	NA	0	NA	0	NA	0	NA	NA	NA
$D_{Ta2} = D_{Ta3} = D_{Ta6} = D_{Ta7}$	0	NA	0	NA	0	NA	0	NA	0	NA	NA	NA

Table 10.16 Comparison Results of the Current Stress on the Power Device Between the Analytical Results and Simulation Results for the Proposed 5L-M^{2}S^{2}Cl Topology.

| Devices | $\sin^{-1}1/2m_a \leq \delta$ and $M_a \geq 0.5$ | | | | $\sin^{-1}1/2m_a > \delta$ and $M_a \geq 0.5$ | | | | $M_a < 0.5$ | | | |
| | Average Values (A) | | RMS Values (A) | | Average Values (A) | | RMS Values (A) | | Average Values (A) | | RMS Values (A) | |
	Analytical	Simulated	Analytical	Simulated	Analytical	Simulated	Analytical	Simulated	Analytical	Simulated	Analytical	Simulated
T_{a1}	0.7621	0.5578	1.5210	1.6042	0.6392	0.6214	1.3824	1.4356	0.2071	0.1886	0.4928	0.4728
T_{a2}	0.1561	0.5140	0.0443	1.2485	0.1448	0.2608	0.4149	0.7038	0.4649	0.4913	0.3389	0.7112
T_{a3}	0.5919	0.5238	1.4073	1.5441	0.6392	0.6041	1.3824	1.3957	0.2071	0.1929	0.4928	0.4771
T_{a4}	0.6381	0.5357	1.6619	1.4113	0.6256	0.5989	1.5540	1.5001	0	0	0	0
T_{a5}	1.0696	1.8564	3.1068	2.5527	0.9846	1.5140	2.1977	2.1222	0.8791	0.8767	0.9766	0.9782
T_{a6}	0.6381	0.4931	1.6619	1.3300	0.6256	0.5622	1.5540	1.4424	0	0	0	0
D_{a1} to D_{a4}	0.0781	0.2594	0.0222	0.8965	0.0724	0.1301	0.2075	0.4972	0.2325	0.2447	0.1695	0.5018
D_{a5} to D_{a8}	0.5348	0.9476	1.5534	1.8374	0.4923	0.7654	1.0989	1.5184	0.4396	0.4352	0.4883	0.6895
D_{Ta1}	0.0729	NA	0.3023	NA	0	NA	0	NA	0	NA	0	NA
D_{Ta3}	0.0729	NA	0.3023	NA	0	NA	0	NA	0	NA	0	NA
D_{Ta4}	0.0365	NA	0.2138	NA	0	NA	0	NA	0	NA	0	NA
D_{Ta6}	0.0365	NA	0.2138	NA	0	NA	0	NA	0	NA	0	NA
$D_{Ta2} = D_{Ta5}$	0	NA	0	NA	0	NA	0	NA	0	NA	0	NA

hierarchy. The loss modeling of the power device will be discussed in the following sections. The efficiency of the proposed topology is to be significantly high due to reductions in the number of active components.

10.7 Losses in Power Devices

A loss modeling of power devices is discussed in this section. An important aspect of this modeling is to observe the loss distribution across each power device. This allows us to design more economical 5L-inverter topologies with an optimum cost-to-performance ratio.

The correlation between the voltage stress and current stress are related to the losses occurring in every pulse during one cycle. The stress is applied for the conduction and switching losses calculation. The summation of these two losses is the total loss distribution in each power device of the 5L-inverter topologies. The device losses depend on various factors such as temperature, switching frequency, gate voltage level, and power level of the load. For a proper development of the application, each of the devices is selected based on the range of the switching frequency and power level of the application, as shown in Fig. 10.15.

10.7.1 Conduction Loss in Power Devices

The approximation of the conduction loss of the power devices in inverter operation mode is mainly based on multiple pulse calculation. This is also known as the *current stress commutation* of the power device during every pulse in one fundamental period, as discussed in the previous subsection. The conduction losses are also calculated based on the characteristic curve of the individual power semiconductor device.

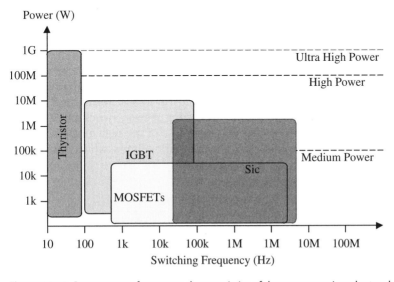

Figure 10.15 Power versus frequency characteristics of the power semiconductor devices (data obtained from the application notes of the manufacturer Infineon Technologies [10]).

Since a low switching frequency range is used for 5L-inverter topologies, two types of semiconductor devices, such as power MOSFET and IGBT devices, are selected for the possible implementation.

In general, a simple approximation of the conduction loss in a power MOSFET device is done by using the on-state resistance for temperature dependency [11, 12]. For the conduction loss in an IGBT device, a linear operation curve of the voltage drop that implies the gate voltage and typical junction temperature of the device is selected for loss approximation [9, 13, 14]. The linear operation curve can be selected from any manufacturer datasheet that provides the output characteristic of collector current versus collector–emitter voltage. Similarly, the conduction loss for the power diode can also be calculated based on the linear expression of the forward voltage drop. The conduction loss expressions for the respective power device are as follows.

MOSFET:

$$P_{Tai\langle avg \rangle} = \frac{1}{2\pi} \int_\alpha^\beta I_a^2(\omega t) \cdot \alpha_{Tai} \cdot R_{DS\langle on \rangle} \ d\omega t = I_{Tai\langle rms \rangle} \cdot R_{DS\langle on \rangle} \tag{10.12}$$

IGBT:

$$P_{Tai\langle avg \rangle} = \frac{1}{2\pi} \int_\alpha^\beta I_a(\omega t) \cdot \alpha_{Tai} \cdot \left[\frac{\Delta V_{CE}}{\Delta I_{Tai}} I_a(\omega t) + V_{CEO} \right] \ d\omega t$$

$$= I_{Tai\langle rms \rangle} \cdot \frac{\Delta V_{CE}}{\Delta I_{Tai}} + I_{Tai\langle avg \rangle} \cdot V_{CEO} \tag{10.13}$$

Diode device:

$$P_{Dai\langle avg \rangle} = \frac{1}{2\pi} \int_\alpha^\beta I_a^2(\omega t) \cdot \alpha_{Dai} \cdot R_{d\langle on \rangle} \cdot R_{d\langle on \rangle} d\omega t = I_{Dai\langle rms \rangle} \cdot R_{d\langle on \rangle} \tag{10.14}$$

$I_{Tai<rms>}$ and $I_{Dai<rms>}$ are the root-mean-square (RMS) pulse current of the active switch and diode, respectively. $R_{DS<on>}$ and $R_{d<on>}$ are the on-state resistance of the power MOSFET and diode, respectively, where V_{CEO} is the on-state zero current collector–emitter voltage and $\Delta V_{CE}/\Delta I_C$ is the slope of the on-state resistance of the IGBT device.

10.7.2 Switching Loss in Power Devices

The theoretical calculation of the switching energy loss in MOSFET and IGBT devices is based on the summation of the turn-on and turn-off energy losses. The average switching loss is the summation of each switching loss occurring at every switching pulse during one cycle period. The summation of the discrete pulse of the switching energy loss at switching frequency, F_s, can be replaced by the integration limits, which is given by:

$$P_{Tai\langle avg \rangle} = \frac{1}{2\pi} \sum_{\lambda=\alpha}^\beta [E_{on}(\lambda, \omega t) + E_{off}(\lambda, \omega t)] \ F_s$$

$$\approx \frac{1}{2\pi} \int_\alpha^\beta [E_{on}(\omega t) + E_{off}(\omega t)] \ F_s d\omega t \tag{10.15}$$

The derivation of each turn-on energy loss, $E_{on}(\omega t)$, and turn-off energy loss, $E_{off}(\omega t)$, has been detailed in [13]. However, the PSIM thermal module does not consider switching periods versus collector current for the simulation result. Hence, a simple approximation of the switching energy loss can be expressed in terms of the voltage stress ratio as a polynomial function of collector current [14]. For theoretical analysis, the tail current of the IGBT device during the turn-off period is assumed to be of very small value, and the small amplitude of overvoltage pulse is also neglected for low switching frequency operation. Based on equation (10.15) and the available information given in the datasheet of the manufacturer (Infineon Technologies), the final average switching loss expression is given by:

$$P_{sw\_Tai\langle avg \rangle} \approx \frac{V_{Tai} F_s}{2\pi V_{nom}} \int_\alpha^\beta [a + b \cdot i_a \sin(\omega t - \delta) + c \cdot i_a^2 \sin^2(\omega t - \delta)] \ d\omega t \qquad (10.16)$$

Here, V_{Tai} is the voltage stress across the power device during turn-off; V_{nom} is the nominal voltage of the semiconductor device; and "a," "b," and "c" are the slope factors of the polynomial function of the energy loss as presented in the datasheet with the respective current rating during operation. As a result, the expression of the total switching energy loss of the device is formulated as a linear function.

10.7.3 Distribution of Power Loss in Devices

This section presents a comparative study of the theoretical and simulated results on the discrete component loss in 5L-inverter topologies. Theoretical and simulated results of the average conduction and switching losses in each power device for 5L-inverter topologies are shown in Figs 10.16–10.19.

The obtained results show that the theoretical calculation is almost the same as compared to the simulated results. The error between the analytical and simulated results is due to the deviation in current stress approximation, and some of the information is not provided in the datasheet. However, the theoretical approximation of the discrete component loss is acceptable to prove the overall efficiency. On top of that, the analytical approximation can prevent the oversizing of thermal management and reduce the cost of production.

The simulated results with PSIM Thermal Module is based on the available components in the laboratory, and all the parameters/specifications of the components used in the simulation tools are listed in Table 10.17.

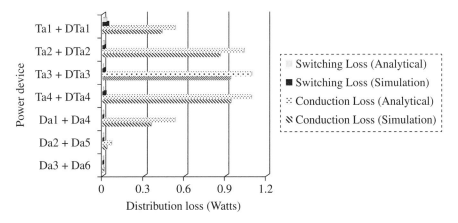

Figure 10.16 Simulation and analytical results of the switching and conduction losses of each power device in 5L-MDCI topology obtained from PSIM Thermal Module toolbox.

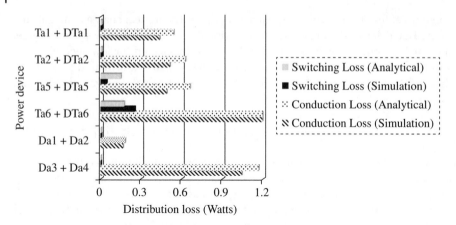

Figure 10.17 Simulation and analytical results of the switching and conduction losses of each power device in 5L-M^{2}DCI topology obtained from PSIM Thermal Module toolbox.

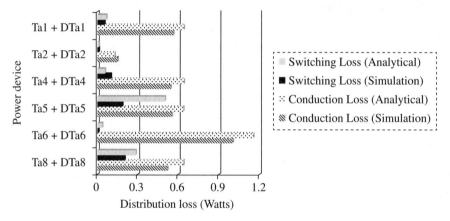

Figure 10.18 Simulation and analytical results of the switching and conduction losses of each power device in 5L-M^{2}T^{2}CI topology obtained from PSIM Thermal Module toolbox.

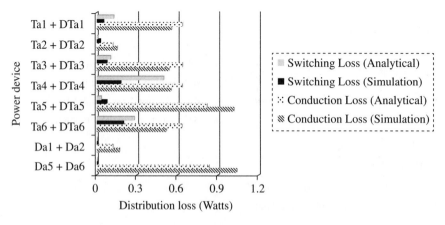

Figure 10.19 Simulation and analytical results of the switching and conduction losses of each power device in 5L-M^{2}S^{2}CI topology obtained from PSIM Thermal Module toolbox.

Table 10.17 Specifications of the Selected Power Devices for each Multilevel Inverter Topologies Test.

Topologies	Switches	Manufacturer	Device Rating	
			Voltage (V)	Current (A)
5L-MDCI	$T_{a1} = T_{a2} = T_{a3} = T_{a4} = T_{a5} = T_{a6} = T_{a7} = T_{a8}$	Infineon Technologies. IGBT model: IKW30N60T	600	30
5L-M²DCI	$T_{a1} = T_{a2} = T_{a3} = T_{a4}$	Infineon Technologies. IGBT model: IKW30N60T	600	30
	$T_{a5} = T_{a8}$	Infineon Technologies. IGBT model: IKW25T120	1200	25
	$T_{a6} = T_{a7}$	Infineon Technologies. IGBT model: IHW30N90T	900	30
5L-M²T²CI	$T_{a1} = T_{a4} = T_{a5} = T_{a8}$	Infineon Technologies. IGBT model: IKW25T120	1200	25
	$T_{a2} = T_{a3} = T_{a6} = T_{a7}$	Infineon Technologies. IGBT model: IKW30N60T	600	30
5L-M²S²CI	$T_{a1} = T_{a3} = T_{a4} = T_{a6}$	Infineon Technologies. IGBT model: IKW25T120	1200	25
	$T_{a2} = T_{a5}$	Infineon Technologies IGBT model: IKW30N60T	600	30

Note: Manufacturer specification of the power diodes on IXYS DSP45-12A is selected for the entire multilevel inverters test.

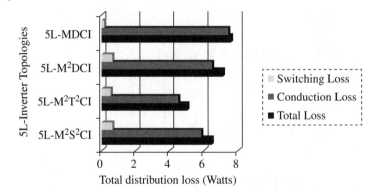

Figure 10.20 Simulation results of the per-phase leg total loss, switching loss, and conduction loss of multilevel inverter topologies obtained from PSIM Thermal Module toolbox.

The simulation results of power losses in the presented topologies are illustrated in Fig. 10.20. According to the simulated results, the efficiency of a multiple-pole hierarchy has been improved even when higher rating devices are selected for the outer cell switch. The total loss reduction in multiple-pole hierarchy is due to the zero current switching that occurs in the inner cell switches, as discussed in the preceding text.

For low and medium motor drive application with better output voltage quality and energy efficient converters, two types of transformerless and filterless multiple-pole multilevel inverters (5L-M^2T^2CI and 5L-M^2S^2CI) are suitable. The device ratings of the switches connected on the outer cell switches of the arm are larger than those of the inner cell switch and bidirectional switch. Hence, high conduction and switching losses are incurred due to high on-state resistance and high voltage stress across the semiconductor switches, respectively. Therefore, both topologies are suitable for low-voltage and medium-power applications by using the available power semiconductors in the market.

On the other hand, the 5L-M^2DCI topology is applicable for medium-voltage and high-power applications. The total losses are significantly higher than the other two proposed topologies (5L-M^2T^2CI and 5L-M^2S^2CI). This topology has several advantages as compared to the 5L-M^2T^2CI and 5L-M^2S^2CI topologies. The advantages that 5L-M^2DCI offers are: lower voltage stress across the outer cell switches; lower device rating; and no shoot-through in the dc-link capacitors during one switch fault. The concept of proposed multiple-pole hierarchy has reduced the total power consumption by a semiconductor device, resulting in higher efficiency and a compact converter as compared to the classical MDCI, particularly for utility applications.

10.7.4 Cost Overview

In this subsection, a cost analysis of the 5L-inverter topologies is presented in order to illustrate the benefits of energy efficient converters that have a lesser number of semiconductor devices with higher power ratings. Total cost calculation considers the costs of power devices, passive elements, and complete analog circuit boards for a comparative study between the classical and proposed 5L inverter topologies.

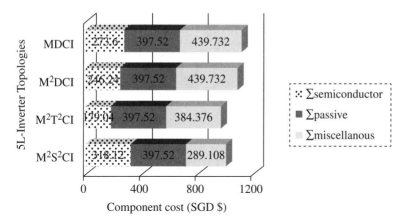

Figure 10.21 Components cost structure for 5L inverter topologies.

The component cost of 5L-inverter topologies considers factors such as:

1) Random cost factors, including country, negotiation, shipment tax, and order quantity.
2) Different materials used in the design and fabrication processes of components.
3) Time dependence on global economy, including market strategy and raw material price.

However, the cost analysis for each component used in the laboratory prototype is based on the components available in the Singapore market. Besides, the approximation of the development cost also depends on the hardware circuit design and type of material selected. With the partial cost calculation shown in Fig. 10.21, the 5L-M^2T^2CI and 5L-M^2S^2CI topologies offer lesser development costs with higher energy efficiency for low- and medium-power applications. For high-power applications, the 5L-M^2DCI topology still offers higher energy efficiency and lesser production costs as compared to the classical 5L-MDCI topology, as shown in Fig. 10.21.

10.7.5 Measured Results

In order to verify the output performance of the proposed and classical 5L-inverter topologies, $500V_{dc}$ with two star-connected output three-phase 50Hz load (R1 = 100Ω, L1 = 122mH; and R2 = 100Ω, L2 = 104mH) associated with the RC filter (Rf = 10Ω and Cf = 18μF) for the enhancement of the balancing features are simulated in PSIM with SimCoupler MATLAB dynamic tools. The control signals of the LS-PWM technique is set at modulation index M = 0.9 and switching frequency Fs = 1kHz.

The simulation results in Fig. 10.22 show the output pole and line-to-line voltages of the respective 5L-inverter topologies. Output voltage THD performance of the respective 5L-inverters topologies is 34% for pole voltage and 17% for line-to-line voltage, as shown in Fig. 10.23. As a result, the proposed 5L-inverter topologies have the similar output characteristics as compared to the classical 5L-MDCI topology.

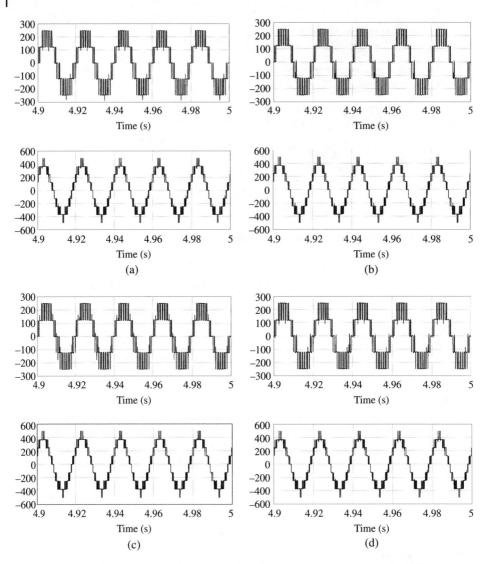

Figure 10.22 Simulation results for the output pole voltage waveform (upper trace) and line-to-line voltage waveform (lower trace) for various 5L-inverter topologies. (a) 5L-MDCI; (b) 5L-M^{2}DCI; (c) 5L-M^{2}T^{2}CI; and (d) 5L-M^{2}S^{2}CI.

For the sake of simplicity, the experimental results are shown in Fig. 10.24 only for the 5L-M^{2}DCI topology with 400V supply. However, the results of the output performance of other proposed 5L-inverter topologies are not listed in this report since their output features are similar to those of 5L-MDCI, as proven in the simulation results.

In order to validate the performance of the zero current switching of the multiple-pole hierarchy, experimental results of the inner cell switches (Ta1 and Ta2) of phase "a" is measured (Fig. 10.25). In general, the built-up voltage and current of the active switch for the LS-PWM technique in multilevel inverters have the same characteristics (e.g. when the switch is turning off, voltage will rise to the desired value and current falls to zero value, similar to when turning on the switch). However, in multiple-pole hierarchy, the

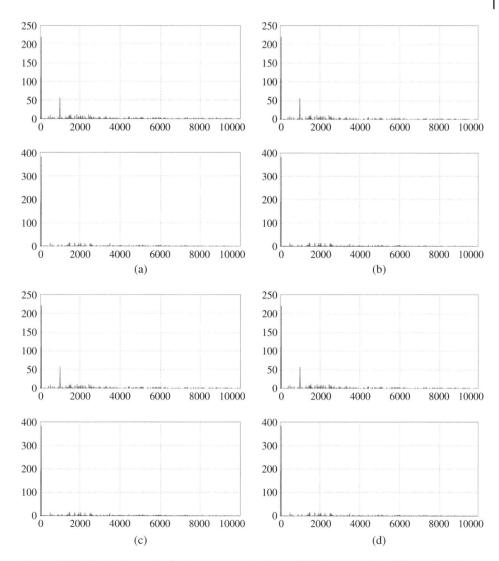

Figure 10.23 Simulation results for the output pole voltage THD (upper trace) and line-to-line voltage THD (lower trace) for different 5L-inverter topologies. (a) 5L-MDCI; (b) 5L-M^2DCI; (c) 5L-M^2T^2CI; and (d) 5L-M^2S^2CI.

current and voltage fall to zero value when the switch is turning on during the particular switching states condition, as discussed the previous subsection. Fig. 10.26 shows the experimentally obtained results of the voltage across the switch and current through the switch during one cycle period.

10.8 Discussion

The proposed inverter structure can develop into different types of multiple-pole multilevel inverter topologies such as M^2DCI, M^2T^2CI, and M^2S^2CI. Among all these topologies, M^2T^2CI and M^2S^2CI have the lowest components cost as compared to the

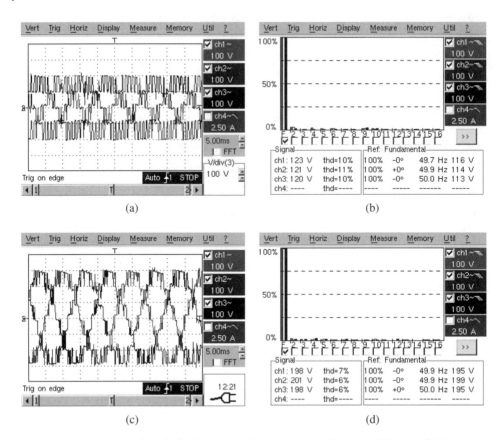

Figure 10.24 Experimental results for the output characteristics performance of the 5L-M²DCI topology. (a) Output pole voltage waveform; (b) THD of the output pole voltage; (c) output line-to-line voltage waveform; and (d) THD of the output line-to-line voltage.

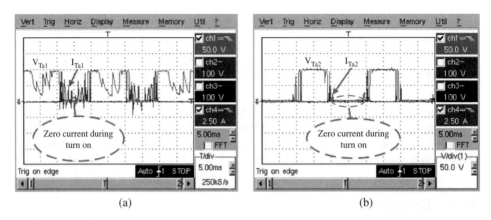

Figure 10.25 Experimental results of the zero current switching of the inner cell switch of 5L-M²DCI during switching state Ta1 = Ta2 = Ta3 = Ta4 = 1 or 0 condition. (a) Ta1 and (b) Ta2.

Table 10.18 Performance Parameter Comparison of Five-Level Inverters.

Five-Level Inverter Topologies	Maximum Voltage Stress Levels		Cost (S$)	Total Loss (Watts)
	IGBT	Diode		
MDCI	$V_{dc}/4$	$3V_{dc}/4$	1074.852	7.5856
M^2DCI	$3V_{dc}/4$	$V_{dc}/4$	1083.492	7.1203
M^2T^2CI	V_{dc}	0	960.936	5.0376
M^2S^2CI	V_{dc}	$V_{dc}/4$	1004.748	6.4557

MDCI and M^2DCI topologies (Fig. 10.22). However, the power rating of the converter is limited to low and medium power, depending on the device rating of the manufacturer.

Mathematical analysis is provided for the device rating selection based on the voltage and current stresses. On top of that, voltage and current stress expressions can be applied to the loss analysis, which proves that the conduction loss is reduced due to the zero current switching of the inner cell switch. The overall performances of five-level inverters are compared in Table 10.18. The multiple-pole approach, higher level and energy-efficient converter with optimum cost, and components count can be investigated for future development associated with the proper balancing circuit on the dc side.

References

1 S. Kouro, M. Malinowski, K. Gopakumar, J. Pou, L. G. Franquelo, W. Bin *et al.*, "Recent advances and industrial applications of multilevel converters," *Industrial Electronics, IEEE Transactions on*, vol. 57, pp. 2553–2580, 2010.

2 C. Newton and M. Sumner, "Novel technique for maintaining balanced internal DC link voltages in diode clamped five-level inverters," *Electric Power Applications, IEE Proceedings*, vol. 146, pp. 341–349, 1999.

3 N. Hatti, K. Hasegawa, and H. Akagi, "A 6.6-kV transformerless motor drive using a five-level diode-clamped pwm inverter for energy savings of pumps and blowers," *Power Electronics, IEEE Transactions on*, vol. 24, pp. 796–803, 2009.

4 K. A. Corzine, J. Yuen, and J. R. Baker, "Analysis of a four-level DC/DC buck converter," *Industrial Electronics, IEEE Transactions on*, vol. 49, pp. 746–751, 2002.

5 N. Hatti, Y. Kondo, and H. Akagi, "Five-level diode-clamped PWM converters connected back-to-back for motor drives," *Industry Applications, IEEE Transactions on*, vol. 44, pp. 1268–1276, 2008.

6 J. W. Kolar, H. Ertl, and F. C. Zach, "A comprehensive design approach for a three-phase high-frequency single-switch discontinuous-mode boost power factor corrector based on analytically derived normalized converter component ratings," *Industry Applications, IEEE Transactions on*, vol. 31, pp. 569–582, 1995.

7 T. Nussbaumer, M. Baumann, and J. W. Kolar, "Comprehensive design of a three-phase three-switch buck-type PWM rectifier," *Power Electronics, IEEE Transactions on*, vol. 22, pp. 551–562, 2007.

8 J. W. Kolar, U. Drofenik, and F. C. Zach, "Current handling capability of the neutral point of a three-phase/switch/level boost-type PWM (VIENNA) rectifier," in *Power Electronics Specialists Conference, 1996. PESC '96 Record., 27th Annual IEEE*, vol. 2, pp. 1329–1336, 1996.

9 D. Graovac and M. Purschel, "IGBT power losses calculation using the data-sheet parameters," ed. Infineon: Application Note, 2009.

10 L. Meng, "Power semiconductor for hybrid and electric vehicles: state of the art, challenges and future roadmap," China Powertrain Segment Marketing, Infineon, July 2012.

11 J. W. Kolar, H. Ertl, and F. C. Zach, "How to include the dependency of the $R_{DS(on)}$ of power MOSFETs on the instantaneous value of the drain current into the calculation of the conduction losses of high-frequency three-phase PWM inverters," *Industrial Electronics, IEEE Transactions on*, vol. 45, pp. 369–375, 1998.

12 F. Gebhardt, H. Vach, and F. W. Fuchs, "Analytical derivation of power semiconductor losses in MOSFET multilevel inverters," in *Power Electronics and Motion Control Conference (EPE/PEMC), 2012 15th International*, pp. DS1b.18-1–DS1b.18-6, 2012.

13 F. Casanellas, "Losses in PWM inverters using IGBTs," *Electric Power Applications, IEE Proceedings*, vol. 141, pp. 235–239, 1994.

14 B. Backlund, R. Schnell, U. Schlapbach, and R. Fischer, "Applying IGBTs," Application Note ed. ABB Switzerland Ltd Semiconductor, 2012.

11

Transformerless Seven-Level/Multiple-Pole Multilevel Inverters with Single-Input Multiple-Output (SIMO) Balancing Circuit

Hossein Dehghani Tafti and Gabriel H. P. Ooi

11.1   Introduction

Solid state devices of high current and voltage rating are utilized in high-voltage and high-power applications in order to tolerate high voltage/current stress. To reduce the amount of voltage or current stress on semiconductor devices, a multilevel approach with more than one switch connected in series for upper phase leg is chosen. Low output voltage total harmonic distortion (THD) with low switching frequency operation can be further improved by implementing a multilevel inverter topology with more than five-level output voltage.

Due to its unbalanced capacitor voltage in the dc-link for seven-level inverter topologies, these topologies have not been practically used in many applications. However, several balancing methods have been reported in the literature, such as balancing algorithm in generic n-level inverter [1] and isolated multi-winding transformer [2]. The first method requires more effort and time for engineers to design the control algorithm owing to the complexity involved, and this method has not yet been verified practically. The second method is not well suited for applications where space and weight are major concerns. Although the transformer in this type provides galvanic isolation between the dc and ac sides of the converter, the capital cost and losses incurred are significantly high. Therefore, an active balancing circuit for seven-level inverter topologies is presented in this chapter. The seven-level balancing circuit is known as the *single-input multiple-output* (SIMO) balancing circuit, which maintains six identical voltage levels in the dc bus.

In order to verify the performance of the SIMO balancing circuit, a new seven-level active-clamped multiple-pole multilevel diode-clamped inverter (7L-AM^{2}DCI) topology and a multiple-pole multilevel diode-clamped inverter (7L-M^{2}DCI) topology are chosen.

11.2   Circuit Configuration and Operating Principles

In this section, the basic operating principle of the new seven-level inverter topologies and dc/dc balancing circuit is discussed. The derived concept of a multiple-pole hierarchy is detailed in Chapter 8 with the complete mathematical analysis. The proposed seven-level inverter topologies achieve zero current switching for a particular

Advanced Multilevel Converters and Applications in Grid Integration, First Edition.
Edited by Ali I. Maswood and Hossein Dehghani Tafti.
© 2019 John Wiley & Sons Ltd. Published 2019 by John Wiley & Sons Ltd.

switching state in the LS-PWM technique. The realization of zero-current switching is similar to 5L multiple-pole multilevel inverter topologies. Hence, the following subsection will briefly explain the switching operation of the converters.

11.2.1 Seven-Level/Multiple-Pole Multilevel Diode-Clamped Inverter (7L-M²DCI)

Fig. 11.1 shows a complete schematic diagram of the per-phase three-poles half-bridge 7L-M²DCI topology. The mathematical analysis for the output phase voltage of phase "a" n-level M²DCI topology is presented here. A generalized equation for the output phase voltage is given as follows:

$$V_{am}(t) = \frac{V_{dc}(t)}{n-1} \left\{ \sum \left[\frac{\text{upper switching devices}}{\text{per – phase leg}} \right] - \frac{n-1}{2} \right\} \qquad (11.1)$$

Here, n represents the number of incremental output voltage steps for a single-phase n-level multiple-pole multilevel diode-clamped inverter (nL-M²DCI). From equation (11.1), a seven-level output phase voltage as a function of switching states is written as:

$$V_{am}(t) = \frac{V_{dc}(t)}{6} \{ T_{a1}(t) + T_{a2}(t) + T_{a5} + T_{a6} + T_{a9} + T_{a10} - 3 \} \qquad (11.2)$$

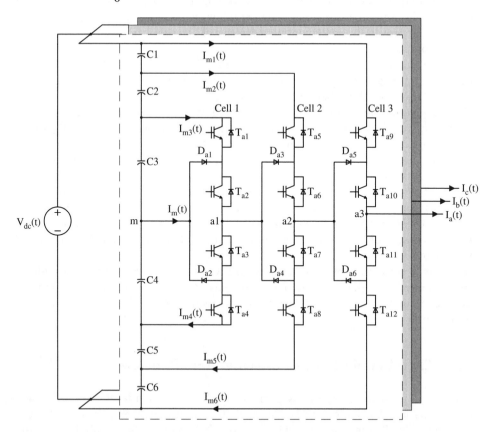

Figure 11.1 Configuration of a three-phase seven-level/multiple-pole multilevel diode-clamped inverter (7L-M²DCI).

Table 11.1 7L-M²DCI output voltage level and the corresponding switching states.

	Switching States						Pole Voltage
States	T_{s1}	T_{s2}	T_{s5}	T_{s6}	T_{s9}	T_{s10}	V_{sm}
1	1	1	1	1	1	1	$V_{dc}/2$
2	1	1	1	1	0	1	$V_{dc}/3$
3	1	1	0	1	0	1	$V_{dc}/6$
4	0	1	0	1	0	1	0
5	0	0	0	1	0	1	$-V_{dc}/6$
6	0	0	0	0	0	1	$-V_{dc}/3$
7	0	0	0	0	0	0	$-V_{dc}/2$

From equation (11.2), we can infer that an accurate switching sequence is required to synthesize a good seven-level output. The switching states and sequence along with the corresponding output phase voltage is shown in Table 11.1.

The output voltage levels for 7L-M²DCI topology are obtained by comparing the modulation control signal and each carrier frequency zone (see Fig. 11.2. Each of the modulated signals with the respective carrier zone is fed to the respective switching device, as shown in Fig. 11.2. The output pole voltage waveform of each cell in the 7L-M²DCI topology shown in Fig. 11.3 is obtained with high modulation index values (e.g. reference control signals are above the amplitude of the carrier frequency of the upper-most triangle wave in Fig. 11.2. A case of low modulation index (e.g. reference control signals below the amplitude of the second top and second bottom carrier wave in Fig. 11.2 will lead to a five-level output voltage waveform. Hence, operating at a low modulation index is not recommend in this topology.

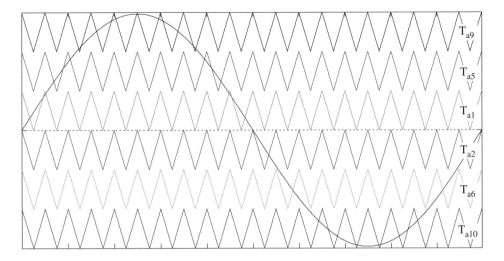

Figure 11.2 Level-shifted PWM technique for the 7L-M²DCI topology.

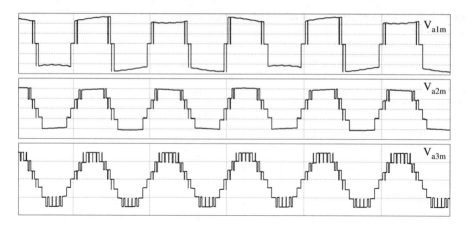

Figure 11.3 Output pole voltage waveform of each inverter cell of the 7L-M²DCI topology, across node a1 to node m of cell 1, node a2 to node m of cell 2, and node a3 to node m of cell 3 (see Fig. 11.1), under balanced dc-link voltage conditions.

11.2.2 Seven-Level/Active-Clamped Multiple-Pole Multilevel Diode-Clamped Inverter (7L-AM²DCI)

While extending the 7L-M²DCI topology concept to a 7L-AM²DCI configuration (as shown in Fig. 11.4 with the same incremental output voltage level, it is important to use switches of a lower rating so as to limit the conduction losses to a minimum. The load terminal of this proposed topology is connected to cell 2 with a seven-level switching characteristic. Fulfilling the above-mentioned design considerations, two series-connected switching devices are directly connected to dc capacitors of C1 and C6 through a small inductance (L_{ai1} and L_{ai2}). The purpose of connecting an inductor in between the active switch clamp of C1 and C6 and the cell 2 NPC inverter is to prevent circulating current, which causes high voltage stress across each IGBT device.

Similar switching strategy is applied as shown in Fig. 11.5 and used in Fig. 11.4. This generates the multiple gate signals required to synthesize the seven-level switching characteristics. The PWM signals are generated by the comparison of the carrier frequency with three sinusoidal waveforms shifted in phase by 120° with each other. The PWM signals are connected to the switching devices T_{a1}–T_{a12}. Note that the upper-most and bottom-most gating signals are connected to the switching devices of T_{a1} and T_{a7}, respectively (similarly for phases "b" and "c").

Based on the switching sequence shown in Fig. 11.5, the seven-level output voltage steps with the corresponding gating signals per-phase are listed in Table 11.2, and output pole voltage waveform of each cell in 7L-AM²DCI topology is shown in Fig. 11.6.

Referring to Table 11.2, the final output distinct voltage levels with the corresponding switching function of a per-phase configuration can be written as:

$$V_{am}(t) = \frac{V_{dc}(t)}{6}\{T_{a1}(t) + T_{a3}(t) + T_{a4} + T_{a7} + T_{a9} + T_{a10} - 3\} \qquad (11.3)$$

11.3 SIMO Voltage Balancing Circuit

The proposed voltage balancing circuit is shown in Fig. 11.7. The voltage balancing circuit is comprised of three dc/dc converters. Two identical balancing converters are

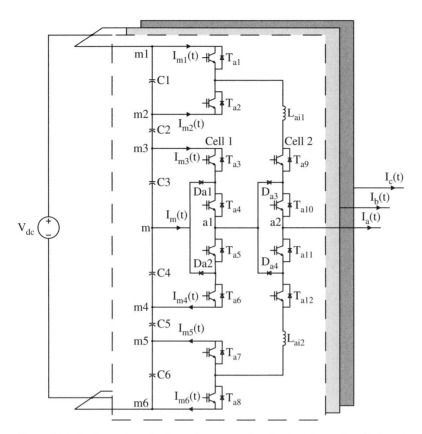

Figure 11.4 Configuration of a three-phase seven-level/active-clamped multiple-pole multilevel diode-clamped inverter (7L-AM^{2}DCI).

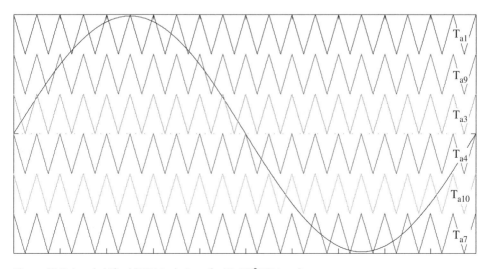

Figure 11.5 Level-shifted PWM technique for 7L-AM^{2}DCI topology.

Table 11.2 7L-AM^{2}DCI output voltage level and the corresponding switching states.

	Switching States						Pole Voltage
States	T_{s1}	T_{s3}	T_{s4}	T_{s7}	T_{s9}	T_{s10}	V_{sm}
1	1	1	1	1	1	1	$V_{dc}/2$
2	0	1	1	1	1	1	$V_{dc}/3$
3	0	1	1	1	0	1	$V_{dc}/6$
4	0	0	1	1	0	1	0
5	0	0	0	1	0	1	$-V_{dc}/6$
6	0	0	0	1	0	0	$-V_{dc}/3$
7	0	0	0	0	0	0	$-V_{dc}/2$

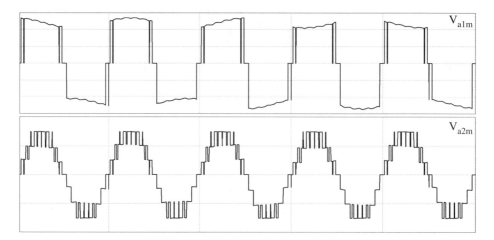

Figure 11.6 Output pole voltage waveform of each inverter cell of a 7L-AM^{2}DCI topology, across node a1 to node m of cell 1, and node a2 to node m of cell 2 under (see Fig. 11.4) balanced dc-link voltage condition.

connected to the upper two dc capacitors (C1 and C2) and lower two dc capacitors (C5 and C6). These two balancing converters are operated at half the duty cycle in order to distribute equal dc voltage across C1 and C2 for the upper converter and C5 and C6 for the lower converter. The dc capacitors connected to the neutral point are connected to a dual-buck/boost converter to obtain a balance dc voltage across C3 and C4. Accurate switching periods are required for the switching devices for obtaining equalized voltage across C3 and C4. These capacitors operate with step-down voltage functionality. These switches are operated at one-third the duty cycle of the inverter switches. Q_{inner1} and Q_{inner4} are connected to switching devices Q_{inner2} and Q_{inner3} for the step-down voltage operating mode. The duty cycle for switching devices Q_{inner1} and Q_{inner4} can be expressed as:

$$\frac{V_{C3}}{V_{Cs1}} = \frac{V_{C4}}{V_{Cs2}} = d_{inner} \approx \frac{1}{3} \tag{11.4}$$

Here, d_{inner} is the duty cycle of the dual-buck/boost balancing converter. As can be seen from equation (11.4), the voltage across the dc capacitor (C3–C4) is one-third the main dc-link voltage. The voltage transfer gain (V_{C3}/V_{dc}) is equal to the duty ratio of the inner dual-buck/boost balancing converter.

The voltage across the capacitors (C3 and C4) connected to the neutral point m in Fig. 11.7 is one-third of half the dc-link voltage. Therefore, the voltage across the nodes (m1, m3) and (m4, m6) is also two-third of half the dc-link voltage. This is due to the transfer of energy from higher potential (C1, C2) to lower potential (C3, C4). From Fig. 11.7, the voltage transfer gain for the upper converter and lower converter can be expressed as:

$$
\begin{cases}
\dfrac{V_{C1}}{V_{Cs3}} = \dfrac{V_{C2}}{V_{Cs3}} = d_{top} \approx \dfrac{1}{2} \\[4mm]
\dfrac{V_{C5}}{V_{Cs4}} = \dfrac{V_{C6}}{V_{Cs4}} = d_{bot} \approx \dfrac{1}{2}
\end{cases}
\tag{11.5}
$$

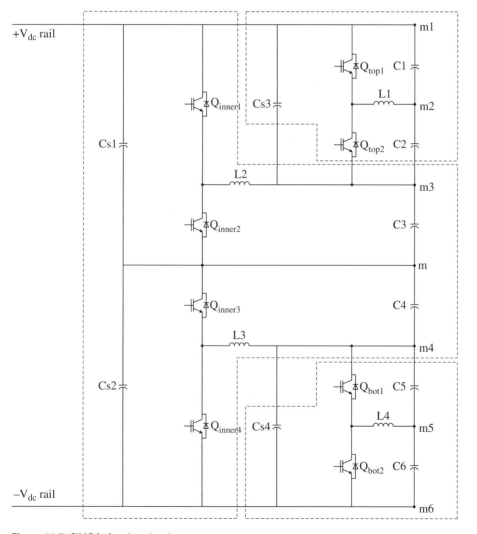

Figure 11.7 SIMO balancing circuit.

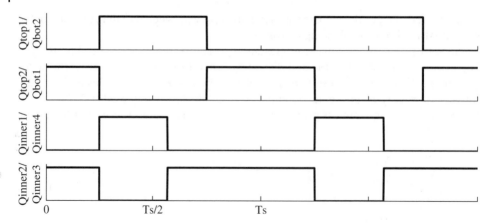

Figure 11.8 SIMO balancing circuit.

Here, d_{top} and d_{bot} are the duty ratios of the upper converter of C1 and C2, and lower converter of C5 and C6, respectively. The switching sequence for the voltage balancing circuit is shown in Fig. 11.8.

11.4 Design Considerations

In order to provide a general guideline regarding component selection for the minimum safety operation of the 7L-converter design, the stresses of the active components of the converters are calculated analytically with dependence on the switching states operation. The analytical approximation of the current stress for each active component is not provided in this chapter. However, the analytical derivation of the current stresses can be done by using the global and local current stresses approximation, which is detailed in Chapter 8.

11.4.1 Voltage Stress on Seven-Level/Multiple-Pole Multilevel Diode-Clamped Inverter

The voltage stress on inverter switches for the 7L-M²DCI topology can be written as:

$$\begin{cases} V_{Ta1}(t) = \dfrac{V_{dc}(t)}{6}[1 - T_{a1}(t)] \\[2ex] V_{Ta2}(t) = \dfrac{V_{dc}(t)}{6}[1 - T_{a2}(t)] \\[2ex] V_{Ta3}(t) = \dfrac{V_{dc}(t)}{6}[1 - T_{a3}(t)] \\[2ex] V_{Ta4}(t) = \dfrac{V_{dc}(t)}{6}[1 - T_{a4}(t)] \end{cases} \qquad (11.6)$$

Similarly, the voltage stress for each of the IGBT devices on cell 2 can be written as:

$$
\begin{cases}
V_{Ta5}(t) = \dfrac{V_{dc}(t)}{3}[1 - T_{a5}(t)] \cdot \left[1 - \dfrac{1}{2}T_{a6}(t)[T_{a1}(t) + T_{a2}(t) - 1]\right] \\[3mm]
V_{Ta6}(t) = \dfrac{V_{dc}(t)}{3}[1 - T_{a6}(t)] \\[3mm]
V_{Ta7}(t) = \dfrac{V_{dc}(t)}{3}[1 - T_{a7}(t)] \\[3mm]
V_{Ta8}(t) = \dfrac{V_{dc}(t)}{3}[1 - T_{a8}(t)] \cdot \left[1 - \dfrac{1}{2}T_{a7}(t)[T_{a3}(t) + T_{a4}(t) - 1]\right]
\end{cases}
\tag{11.7}
$$

And the voltage stress across each IGBT on cell 3 can be written as:

$$
\begin{cases}
V_{Ta9}(t) = \dfrac{V_{dc}(t)}{2}[1 - T_{a9}(t)] \cdot \left[1 - \dfrac{1}{3}T_{a10}(t)[T_{a1}(t) + T_{a2}(t) + T_{a5}(t) + T_{a6}(t) - 2]\right] \\[3mm]
V_{Ta10}(t) = \dfrac{V_{dc}(t)}{2}[1 - T_{a10}(t)] \\[3mm]
V_{Ta11}(t) = \dfrac{V_{dc}(t)}{2}[1 - T_{a11}(t)] \\[3mm]
V_{Ta12}(t) = \dfrac{V_{dc}(t)}{2}[1 - T_{a12}(t)] \cdot \left[1 - \dfrac{1}{3}T_{a11}(t)[T_{a3}(t) + T_{a4}(t) + T_{a7}(t) + T_{a8}(t) - 2]\right]
\end{cases}
\tag{11.8}
$$

From equations (11.6), (11.7), and (11.8), a generalized mathematical expression for voltage stress on each IGBT for cell 2 can be formulated as:

$$
\begin{cases}
V_{Ta(5+4i)}(t) = \dfrac{V_{dc}(t)}{n-1}[1 - T_{a(5+4i)}(t)] \cdot \left\{[i+2] - T_{a(6+4i)}(t)\begin{bmatrix} T_{a1}(t) + T_{a2}(t) - i - 1 \\ + \displaystyle\sum_{m\neq0}^{i}\begin{bmatrix} T_{a(1+4m)}(t) \\ + T_{a(2+4m)}(t) \end{bmatrix} \end{bmatrix}\right\} \\[8mm]
V_{Ta(6+4i)}(t) = \dfrac{V_{dc}(t)}{2}[1 - T_{a(6+4i)}(t)] \\[3mm]
V_{Ta(7+4i)}(t) = \dfrac{V_{dc}(t)}{2}[1 - T_{a(7+4i)}(t)] \\[3mm]
V_{Ta(8+4i)}(t) = \dfrac{V_{dc}(t)}{n-1}[1 - T_{a(8+4i)}(t)] \cdot \left\{[i+2] - T_{a(7+4i)}(t)\begin{bmatrix} T_{a3}(t) + T_{a4}(t) - i - 1 \\ + \displaystyle\sum_{m\neq0}^{i}\begin{bmatrix} T_{a(3+4m)}(t) \\ + T_{a(4+4m)}(t) \end{bmatrix} \end{bmatrix}\right\}
\end{cases}
\tag{11.9}
$$

Here, n and i are the incremental voltage level and the cell number for n^{th}-level M^2DCI converter, respectively. The range of any incremental cell i can be determined as:

$$
\min\langle 0 \rangle \leq i \leq \max\left\langle \dfrac{n-5}{2} \right\rangle
\tag{11.10}
$$

Applying equations (11.6) to (11.9), the maximum voltage stress on each IGBT for 7L-M^2DCI configuration is listed in Table 11.3 with the corresponding switching states. From Table 11.3, the maximum voltage stress level for each of the inverter switches

Table 11.3 7L-M²DCI voltage stress and the corresponding switching states.

Voltage Stress (Volts)	States						
	1	2	3	4	5	6	7
V_{Ta1}	0	0	0	$V_{dc}/6$	$V_{dc}/6$	$V_{dc}/6$	$V_{dc}/6$
V_{Ta2}	0	0	0	0	$V_{dc}/6$	$V_{dc}/6$	$V_{dc}/6$
V_{Ta3}	$V_{dc}/6$	$V_{dc}/6$	$V_{dc}/6$	0	0	0	0
V_{Ta4}	$V_{dc}/6$	$V_{dc}/6$	$V_{dc}/6$	$V_{dc}/6$	0	0	0
V_{Ta5}	0	0	$V_{dc}/6$	$V_{dc}/3$	$V_{dc}/2$	$V_{dc}/3$	$V_{dc}/3$
V_{Ta6}	0	0	0	0	0	$V_{dc}/3$	$V_{dc}/3$
V_{Ta7}	$V_{dc}/3$	$V_{dc}/3$	0	0	0	0	0
V_{Ta8}	$V_{dc}/3$	$V_{dc}/3$	$V_{dc}/2$	$V_{dc}/3$	$V_{dc}/6$	0	0
V_{Ta9}	0	$V_{dc}/6$	$V_{dc}/3$	$V_{dc}/2$	$2V_{dc}/3$	$5V_{dc}/6$	$V_{dc}/2$
V_{Ta10}	0	0	0	0	0	0	$V_{dc}/2$
V_{Ta11}	$V_{dc}/2$	0	0	0	0	0	0
V_{Ta12}	$V_{dc}/2$	$5V_{dc}/6$	$2V_{dc}/3$	$V_{dc}/2$	$V_{dc}/3$	$V_{dc}/6$	0

in 7L-M²DCI topology is described in the following parameters:

$$
\begin{cases}
V_{Ta1}(t) = V_{Ta2}(t) = V_{Ta3}(t) = V_{Ta4}(t) = V_{dc}(t)/6 \\
V_{Ta5}(t) = V_{Ta8}(t) = V_{Ta10}(t) = V_{Ta11}(t) = V_{dc}(t)/2 \\
V_{Ta6}(t) = V_{Ta7}(t) = V_{dc}(t)/3 \\
V_{Ta9}(t) = V_{Ta12}(t) = 5V_{dc}(t)/6
\end{cases}
\tag{11.11}
$$

11.4.2 Voltage Stress on Seven-Level/Active-Clamped Multiple-Pole Multilevel Diode-Clamped Inverter

The voltage stress across each of the switching devices in the clamping circuit and cell 1 NPC inverter can be estimated as:

$$
\begin{cases}
V_{Ta1}(t) = \dfrac{V_{dc}(t)}{6}[1 - T_{a1}(t)] \\[4pt]
V_{Ta2}(t) = \dfrac{V_{dc}(t)}{6}[1 - T_{a2}(t)] \\[4pt]
V_{Ta3}(t) = \dfrac{V_{dc}(t)}{6}[1 - T_{a3}(t)] \\[4pt]
V_{Ta4}(t) = \dfrac{V_{dc}(t)}{6}[1 - T_{a4}(t)] \\[4pt]
V_{Ta5}(t) = \dfrac{V_{dc}(t)}{6}[1 - T_{a5}(t)] \\[4pt]
V_{Ta6}(t) = \dfrac{V_{dc}(t)}{6}[1 - T_{a6}(t)] \\[4pt]
V_{Ta7}(t) = \dfrac{V_{dc}(t)}{6}[1 - T_{a7}(t)] \\[4pt]
V_{Ta8}(t) = \dfrac{V_{dc}(t)}{6}[1 - T_{a8}(t)]
\end{cases}
\tag{11.12}
$$

The voltage stress across each switching device on the cell 2 NPC inverter can be esti‑mated by measuring the voltage across the switching devices (T_{a5}–T_{a8}). Each IGBT in the cell 2 NPC inverter will experience a distinct voltage stress level, depending on the conduction period of each IGBT.

The voltage across the collector and node m ($V_{c5m(Ta5)}$, $V_{c6m(Ta6)}$, $V_{c7m(Ta7)}$, and $V_{c8m(Ta8)}$) of each series‑connected IGBT device in the cell 2 NPC inverter can be calculated as a function of the switching states shown in equation (11.13):

$$
\begin{cases}
V_{c9m(Ta9)}(t) = \dfrac{V_{dc}(t)}{6}[2 + T_{a1}(t)][1 - T_{a9}(t)] \\[2mm]
V_{c10m(Ta10)}(t) = \dfrac{V_{dc}(t)}{6}[T_{a3}(t) + T_{a4}(t) - 1][1 - T_{a10}(t)] \\[2mm]
V_{c11m(Ta11)}(t) = \dfrac{V_{dc}(t)}{6}[2 + T_{a1}(t)][1 - T_{a11}(t)] \\[2mm]
V_{c12m(Ta12)}(t) = \dfrac{V_{dc}(t)}{6}T_{a11}(t)[T_{a3}(t) + T_{a4}(t) - 1][1 - T_{a12}(t)]
\end{cases}
\tag{11.13}
$$

The voltage across the emitter and node m ($V_{e5m(Ta5)}$, $V_{e6m(Ta6)}$, $V_{e7m(Ta7)}$, and $V_{e8m(Ta8)}$) of each series‑connected IGBT device in the cell 2 NPC inverter can be calculated as a function of the switching states shown in equation (11.14):

$$
\begin{cases}
V_{e9m(Ta9)}(t) = \dfrac{V_{dc}(t)}{6}T_{a10}(t)[T_{a3}(t) + T_{a4}(t) - 1][1 - T_{a9}(t)] \\[2mm]
V_{e10m(Ta10)}(t) = \dfrac{V_{dc}(t)}{6}[T_{a7}(t) - 3][1 - T_{a10}(t)] \\[2mm]
V_{e11m(Ta11)}(t) = \dfrac{V_{dc}(t)}{6}[T_{a3}(t) + T_{a4}(t) - 1][1 - T_{a11}(t)] \\[2mm]
V_{e12m(Ta12)}(t) = \dfrac{V_{dc}(t)}{6}[T_{a7}(t) - 3][1 - T_{a12}(t)]
\end{cases}
\tag{11.14}
$$

Equations (11.13) and (11.14) do not consider the voltage drop across the inductors as it is negligible when compared to the dc‑link voltage. A low switching frequency at 1 kHz does not create any voltage spike with a slow rate of voltage change (dv/dt). The voltage stress across each IGBT of the cell 2 NPC inverter is obtained in equation (11.15) by subtracting equations (11.13) and (11.14).

$$
\begin{cases}
V_{Ta9}(t) = \dfrac{V_{dc}(t)}{6}[1 - T_{a9}(t)]\{2 + T_{a1}(t) - T_{a10}(t)[T_{a3}(t) + T_{a4}(t) - 1]\} \\[2mm]
V_{Ta10}(t) = \dfrac{V_{dc}(t)}{6}\{[T_{a3}(t) + T_{a4}(t) - 1][1 - T_{a10}(t)] - [T_{a7}(t) - 3][1 - T_{a10}(t)]\} \\[2mm]
V_{Ta11}(t) = \dfrac{V_{dc}(t)}{6}\{[2 + T_{a1}(t)][1 - T_{a11}(t)] - [T_{a3}(t) + T_{a4}(t) - 1][1 - T_{a11}(t)]\} \\[2mm]
V_{Ta12}(t) = \dfrac{V_{dc}(t)}{6}[1 - T_{a12}(t)]\{3 - T_{a7}(t) + T_{a11}(t)[T_{a3}(t) + T_{a4}(t) - 1]\}
\end{cases}
\tag{11.15}
$$

Based on equations (11.12) and (11.15), the maximum voltage stress level for each of the switches with their respective switching states is listed in Table 11.4. It can be inferred from Table 11.4 that the 7L‑AM^{2}DCI switches have different maximum

Table 11.4 7L-AM^{2}DCI voltage stress and the corresponding switching states.

Voltage Stress (Volts)	States						
	1	2	3	4	5	6	7
V_{Ta1}	0	$V_{dc}/6$	$V_{dc}/6$	$V_{dc}/6$	$V_{dc}/6$	$V_{dc}/6$	$V_{dc}/6$
V_{Ta2}	$V_{dc}/6$	0	0	0	0	0	0
V_{Ta3}	0	0	0	$V_{dc}/6$	$V_{dc}/6$	$V_{dc}/6$	$V_{dc}/6$
V_{Ta4}	0	0	0	0	$V_{dc}/6$	$V_{dc}/6$	$V_{dc}/6$
V_{Ta5}	$V_{dc}/6$	$V_{dc}/6$	$V_{dc}/6$	0	0	0	0
V_{Ta6}	$V_{dc}/6$	$V_{dc}/6$	$V_{dc}/6$	$V_{dc}/6$	0	0	0
V_{Ta7}	0	0	0	0	0	0	$V_{dc}/6$
V_{Ta8}	$V_{dc}/6$	$V_{dc}/6$	$V_{dc}/6$	$V_{dc}/6$	$V_{dc}/6$	$V_{dc}/6$	0
V_{Ta9}	0	0	$V_{dc}/6$	$V_{dc}/3$	$V_{dc}/2$	$V_{dc}/3$	$V_{dc}/3$
V_{Ta10}	0	0	0	0	0	$V_{dc}/6$	$V_{dc}/3$
V_{Ta11}	$V_{dc}/3$	$V_{dc}/6$	0	0	0	0	0
V_{Ta12}	$V_{dc}/3$	$V_{dc}/3$	$V_{dc}/2$	$V_{dc}/3$	$V_{dc}/6$	0	0

voltage stress levels. For optimum efficiency and performance, IGBTs with suitable voltage ratings must be selected.

From Table 11.4, the maximum voltage stress level for each of the inverter switches in the 7L-AM^{2}DCI topology is described in the following parameters:

$$\begin{cases} V_{Ta1}(t) = V_{Ta2}(t) = V_{Ta3}(t) = V_{Ta4}(t) = V_{dc}(t)/6 \\ V_{Ta5}(t) = V_{Ta6}(t) = V_{Ta7}(t) = V_{Ta8}(t) = V_{dc}(t)/6 \\ V_{Ta9}(t) = V_{Ta12}(t) = V_{dc}(t)/2 \\ V_{Ta10}(t) = V_{Ta11}(t) = V_{dc}(t)/3 \end{cases} \tag{11.16}$$

11.4.3 Voltage Stress on the SIMO Voltage Balancing Circuit

The voltage stress of the IGBT in the SIMO balancing circuit is calculated based on voltage ripple of each dc capacitor and zero voltage drops across each buck/boost inductor, and the voltage stress of the switching device, as follows:

$$\begin{cases} V_{Qinner1}(t) = V_{Qinner2}(t) = V_{Qinner3}(t) = V_{Qinner4}(t) = V_{dc}(t)/2 \\ V_{Qtop1}(t) = V_{Qtop2}(t) = V_{Qbot1}(t) = V_{Qbot2}(t) = V_{dc}(t)/3 \end{cases} \tag{11.17}$$

11.5 Experimental Verification

The down-scaled experimental setup, is shown in Fig. 11.9 was used to demonstrate the performance of the voltage balancing capability of the SIMO balancing circuit on

Figure 11.9 Experimental setup on 7L-inverter configuration with the SIMO balancing circuit.

the 7L-inverter. The parameters of the laboratory prototype are as follows:

1. DC supply, $V_{dc} = 100$ V
2. Switching frequency, $F_s = 1$ kHz
3. Modulation frequency, $F = 50$ Hz
4. Switching frequency of balancing circuit, $F_b = 5$ kHz
5. Resistive load, $R = 150$ Ω
6. Inductive load, $L = 122$ mH
7. Balancing inductor, $L1 = L2 = L3 = L4 = 5$ mH
8. DC capacitor, $C1 = C2 = C3 = C4 = C5 = C6 = 2200$ μF

The output resistive and inductive loads of the inverter are connected in series and finally form a star configuration for three-phase system.

Fig. 11.10 shows the output voltage waveform of the SIMO balancing circuit. The dc capacitor voltages are almost equalized for an open loop control. However, there is a significant small error between the actual dc capacitor voltage value and the desired value. This is due to the unregulated duty cycle of the SIMO balancing circuit, which is manually set at a fixed duty cycle by using the function generator. However, in this chapter, a 7L SIMO balancing circuit with the open loop system was used to demonstrate the capability of the proposed circuit diagram in balancing the capacitor voltages.

With the balancing circuit connected at the input terminal of the 7L-M^2DCI topology, the output voltage waveform of 7L-M^2DCI is shown in Fig. 11.11. The output THD value of the line-to-line voltage terminal, shown in Fig. 11.12, is noted to be less distorted owing to the 13-level incremental voltage steps (shown in Fig. 11.12 (b)). The voltage quality of a three-phase 7L-inverter is improved due to low THD as observed in the output voltage waveform with component count reduction.

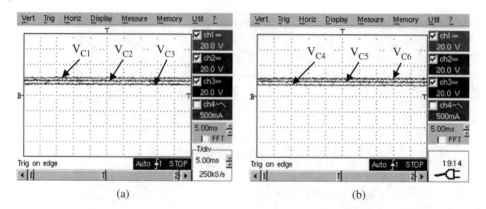

(a) (b)

Figure 11.10 Experimental results of the dc-link voltage waveform for the upper three dc capacitors (V_{c1}, V_{c2}, and V_{c3}) and lower three dc capacitors (V_{c4}, V_{c5}, and V_{c6}) of the output SIMO balancing circuit.

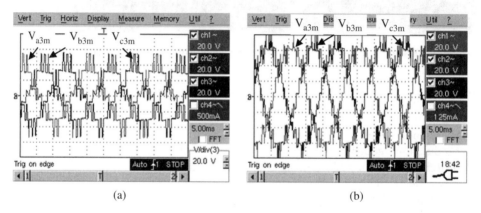

(a) (b)

Figure 11.11 Experimental results of the output voltage waveform of the 7L-M²DCI topology. (a) Pole voltage and (b) line-to-line voltage.

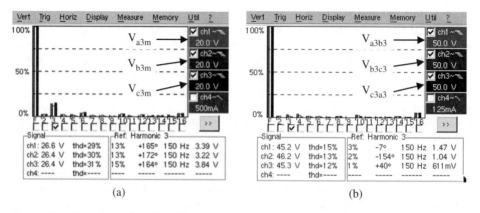

(a) (b)

Figure 11.12 Experimental results of the output voltage THD of the 7L-M²DCI topology. (a) Pole voltage and (b) line-to-line voltage.

11.6 Summary

This chapter presents a new balancing circuit for the seven-level inverter topology with multi-input dc capacitor components connected in series in a single dc-link configuration. The balancing circuit is able to provide six distinct voltage levels for the 7L-inverter topologies.

Besides, two proposed 7L-inverter topologies (M^2DCI and AM^2DCI) are discussed in this chapter. By comparing both proposed inverter topologies, we find that the 7L-AM^2DCI topology is able to limit the circulating current to the minimum and provide low voltage stress across the IGBT devices. However, six discrete inductive components are required to prevent any voltage spike occurring across the switches. Nevertheless, the size of the high-frequency inductors can be reduced by using the mutual inductor design (four terminals with single core). Therefore, significant improvements are needed for future development, such as balancing control algorithms for the SIMO balancing circuit and inductive component design with different materials and configuration.

References

1 L.M. Grzesiak and J.G. Tomasik, "Novel DC link balancing scheme in generic n-level back-to-back converter system," in *Power Electronics, 2007. ICPE '07. 7th International Conference on*, pp. 1044–1049, 2007.

2 S. Daher, J. Schmid, and F.L.M., Antunes, "Multilevel inverter topologies for stand-alone PV systems," *Industrial Electronics, IEEE Transactions on*, vol. 55, 2703–2712, 2008.

12

Three-Phase Seven-Level Three-Cell Lightweight Flying Capacitor Inverter

Ziyou Lim

12.1 Introduction

Despite the proven facts that producing high-resolution waveforms are favorable at all times, there are still many complications involved, such as the increase in component counts, cost, size, as well as design complexity. As a result, the overall reliability and efficiency of the system are also affected. Such implications motivate many researchers to design and develop alternative multilevel topologies.

Several multilevel topologies have been developed to synthesize higher number of m levels mL output voltages with reduced component counts, and have been reported in the recent literature, such as: (I) multilevel inverter based on switched dc sources [1]; (II) improved double flying capacitor multicell converter [2]; (III) symmetric hybrid multilevel inverter [3]; (IV) packed-u cells converter [4]; and (V) stacked multicell inverter with reduced dc voltage sources [5].

Multilevel inverters (I) to (V) share a similarity in their circuit designs with the H-bridge configuration, as shown in Fig. 12.1. It can be observed that one leg is being replaced with the multilevel flying capacitor inverter (MFCI) or active neutral point clamped inverter (ANPC) topology, while the other leg has two low-frequency (LF) power switches. These converters (I) to (V) have the ability to synthesize higher mL output voltages with reduced component counts as compared to the commercially available MFCI and ANPC. High flexibility is also found in these converters, owing to the modular cells concept being utilized in their designs.

Nevertheless, a bulky phase-shifted transformer is required to provide isolated dc sources for (I) and (II) in a single-phase system and (III) to (V) in a three-phase system. Thus, multiple bridge rectifiers are also needed at the secondary side of the transformer. As a result, additional cost and losses are incurred, and more installation space is required [6].

Therefore, two approaches are proposed for developing transformerless modular cell inverters with reduced flying capacitors and switching devices, using novel hybrid carrier-based modulation methods. The two proposed multilevel reduced modular cell converters are the *lightweight flying capacitor inverter* (LFCI), explored in this chapter, and the *reduced flying capacitor inverter* (RFCI), discussed in Chapter 13. Both topologies are designed based on clamping a series of modular cells onto the low operational frequency cell.

Advanced Multilevel Converters and Applications in Grid Integration, First Edition.
Edited by Ali I. Maswood and Hossein Dehghani Tafti.
© 2019 John Wiley & Sons Ltd. Published 2019 by John Wiley & Sons Ltd.

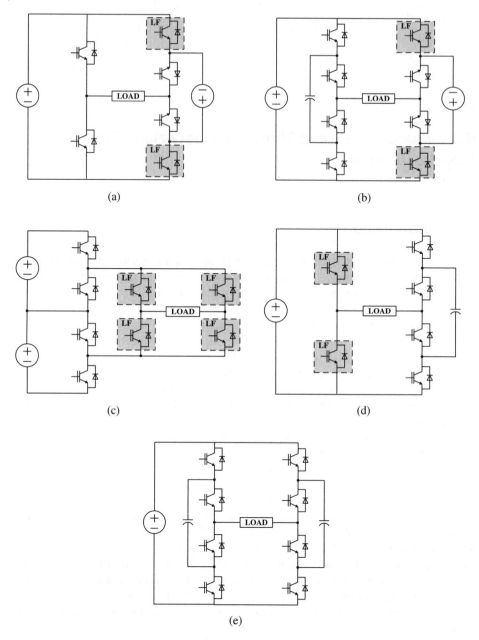

Figure 12.1 Recent single-phase multilevel topologies (I) to (V) with reduced component counts: (a) 5L multilevel inverter with switched dc sources; (b) 11L improved double flying capacitor multicell inverter; (c) 5L symmetric hybrid multilevel inverter; (d) 7L packed-u cells inverter; and (e) 5L stacked multicell inverter with reduced dc sources.

By doing so, both LFCI and RFCI retain the modularity approach, and eliminate the need for a transformer at all times for both single- and three-phase topologies. On top of that, higher mL output voltages can be synthesized, with the component counts significantly reduced. Similarities in converter design is found between the proposed topologies and the classical MFCI.

Some advanced hybrid SPWM schemes are also proposed in this chapter, namely the *phase-shifted carrier phase-opposition-disposition* PWM (PSC-POD-PWM) and the *phase-shifted carrier phase-disposition* PWM (PSC-PD-PWM). In Chapter 3, the harmonic characteristics analysis of both hybrid SPWM schemes are detailed and proven to have successfully overcome the issues of classical modulation LSC-PWM and PSC-PWM, with their respective advantageous properties well attained.

The classical MFCI proposed by Meynard *et al.* [7] has been commercially developed in the industries by ALSTOM Power Conversion [8, 9] for medium voltage (MV) up to 4.16kV drives. Commonly, the output voltages of classical MFCI are synthesized using phase-shifted carrier PWM (PSC-PWM), as explained in Chapter 4. However, the following factors make MFCI unfeasible for higher-voltage applications [6, 9]: (a) high-switching-frequency operation; (b) large inrush dc-link current; and (c) increased flying capacitor energy storage.

Hence, a detailed comparative study made between the proposed LFCI, the proposed RFCI, and the classical MFCI based on seven-level (7L) output voltages is presented to highlight the enhanced performances. The perceived merits of both proposed LFCI and RFCI allow the converter to be attractively suitable for higher-MV drives and also for applications where space constraint is a major concern. The performances of both LFCI and RFCI with hybrid SPWM is evaluated under dynamic conditions to verify their feasibilities.

12.2 LFCI Topology

The problems associated with the classical MFCI based on PSC-PWM modulation, as addressed earlier on, make the topology impractical to realize for higher mL output phase voltage. Moreover, MFCI is not suitable for high-power applications due to its high-switching-frequency PSC-PWM [8].

As higher mL output voltage is always desired to be achieved for high-power applications, those challenges encountered in classical MFCI motivates the designing of a lighter-weight MFCI while achieving higher mL output voltage. Hence, a new approach has been proposed to overcome these limitations.

In order to achieve higher mL output voltage with reduced number of components, the proposed method is designed in such the way that the output voltage levels are achieved based on the binary progression when more n_{cells} are connected. Therefore, the number of output voltage levels for the light-weight multilevel flying capacitor inverter (LFCI) is expressed in equation (12.1).

$$(mL)LFCI = 2^{n_{cells}} - 1 \qquad (12.1)$$

According to equation (12.1), the three-level output phase voltage is synthesized with two-cell LFCI, which is exactly the same as the classical two-cell 3L-MFCI. However, LFCI is perfectly suitable for achieving higher incremental voltage stepped waveform at an exponential rate. This shows that a seven-level output phase voltage is achievable with a three-cell LFCI and that a 15-level output voltage is synthesized by the four-cell LFCI.

Since the natural balancing technique is still required for higher mL-MFCI despite the excessive redundant switching states generated by PSC-PWM, capacitor voltage balancing will no longer be an issue for flying capacitors–based converter topology. Instead of having $2^6 = 64$ states in the six-cell 7L-MFCI, now only $2^3 = 8$ switching states are needed for the three-cell 7L-LFCI. This shows that the minimum number of switching states (n_{states}) needed for the mL-LFCI or mL-MFCI is actually dependent on the n_{cells} implemented, which can be determined with equation (12.2).

$$n_{states} = 2^{n_{cells}} \tag{12.2}$$

The reason for expressing the n_{states} in equation (12.2) based on binary progression is simple – i.e. because of the given condition in equation (3.1), only either "turn on" or "turn off" state is permitted for a normal operation of the semiconductor devices. Thus, in order to minimize the number of components required, the n_{states} must also be reduced when synthesizing the desired mL output voltage.

The performance of the proposed LFCI under the two advanced hybrid sinusoidal PWM schemes (PSC-POD-PWM and PSC-PD-PWM) is analyzed. The feasibility of the LFCI is also validated through both simulation and experimental results to highlight the enhanced performances achieved as compared to classical MFCI.

12.3 Circuit Configuration

The general circuit configuration of the proposed three-phase LFCI is shown in Fig. 12.2. Each modular cell consists a pair of semiconductor switches (S_x and its complementary $S_x{}'$) and a capacitor p clamped in between the cells. Hence, the number of switches ($i_{switches}$) and clamped capacitors ($p_{capacitors}$) can be determined based on the number of modular cells (n_{cells}) used.

The proposed LFCI is constructed using series-connected modular cells (Cell 1 to Cell n) that are based on the binary progression ratio. It is intended to synthesize the desired mL output voltage waveforms (12.1) using the maximized n_{states} generated (12.2) according to the n_{cells} implemented. By doing this, the number of components required is further minimized as well, even though the proposed LFCI share similarities in converter design with the classical MFCI topology.

In order to maximize (12.2), the pair of switches (S_{x1} and $S_{x1}{}'$) in Cell 1 must always operate at a low frequency of 50Hz. However, these two LF switches do not perform as one H-bridge leg, such as those LF switches in Fig. 12.1. Thus, it allows the proposed LFCI to remain transformerless for three-phase systems because of its single dc-link configuration.

12.4 Operational Principles

Having three modular cells (Cell 1, Cell 2, and Cell 3) connected in series for each phase-leg results a single-phase seven-level (7L) LFCI, as presented in Fig. 12.3(a). In this case, a minimum number of six $i_{switches}$ (Sx1 to Sx3 and Sx1' to Sx3') and two $p_{capacitors}$

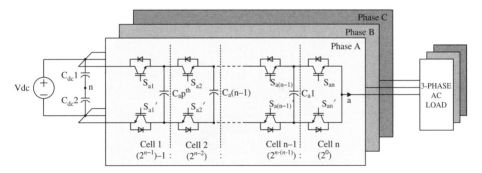

Figure 12.2 Proposed transformerless three-phase *n*-cell LFCI topology.

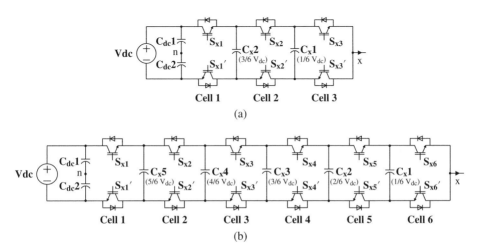

Figure 12.3 (a) Proposed Three-Cell 7L-LFCI. (b) Classical Six-Cell 7L-MFCI.

(Cx1 and Cx2) clamped in between the modular cells are required for each phase-leg, depending on the voltage ratings of the converter design. Thus, the total component counts are greatly reduced when comparing with the classical six-cell 7L-MFCI (12 i_{switches} and five $p_{\text{capacitors}}$ per phase-leg), which can be clearly observed from Fig. 12.3(b). The difference would be greater when designing three-phase systems.

Since three cells of six active switches are used for the proposed three-cell 7L-LFCI, there is a combination of $2^3 = 8$ operating modes from "000" to "111," as summarized in Table 12.1. It should be clearly observed that only two redundant states are being produced for the zero output level. By reducing the excessive switching states, higher mL output voltage waveforms can now be synthesized using fewer components.

Besides that, the pair of switches (S_{x1} and S_{x1}') in Cell 1 is always performing at the fundamental frequency of 50Hz – where all the positive output levels are produced when S_{x1} conducts, and vice versa, i.e. S_{x1}' conducts in all the negative voltage levels. Therefore, the output seven-level phase voltage waveforms are synthesized correspondingly at $\pm V_{\text{dc}}/2, \pm V_{\text{dc}}/3, \pm V_{\text{dc}}/6$, and zero level, based on the given switching states in Table 12.1.

The charging and discharging states of the two flying capacitors (C_x1 and C_x2) for the three-cell 7L-LFCI in Table 12.1 are observed to be the same as those of the three-cell

Table 12.1 Three-Cell 7L-LFCI Voltage Levels and the Corresponding Switching States.

States	Signal Level	Switching States			Charging/ Discharging States of Capacitors		Phase-Leg xn Voltage
		Sx1	Sx2	Sx3	Cx1	Cx2	Vxn
1	2	1	1	1			$+V_{dc}/2$
2	3	1	1	0	+		$+V_{dc}/3$
3	4	1	0	1	−	+	$+V_{dc}/6$
4	5	1	0	0		+	0
5	−5	0	1	1	−		
6	−4	0	1	0	+	−	$-V_{dc}/6$
7	−3	0	0	1	−		$-V_{dc}/3$
8	−2	0	0	0			$-V_{dc}/2$

Note: Here, x represents phase a, b and c; logic "1" and "0" represent the "on" and "off" states, respectively; and "+" and "−" represent the "charging" and "discharging" states of capacitors, while "▋" represents neither states.

4L-MFCI in Table 4.2. Therefore, the operation modes of the three-cell 7L-LFCI can be referred to the same illustrations of current flows with respect to the corresponding switching states shown in Fig. 4.7. The reduction in n_{states} allows the charging time for the C_x1 and C_x2 of the proposed three-cell 7L-LFCI to be shortened, allowing it to reach its steady-state operation faster as compared to the classical six-cell 7L-MFCI.

The operational principles of the proposed three-cell 7L-LFCI are achieved using two proposed advanced hybrid sinusoidal PWM schemes, named *phase-shifted carrier phase-opposition-disposition PWM* (PSC-POD-PWM) and *phase-shifted carrier phase-disposition PWM* (PSC-PD-PWM), as shown in Fig. 12.4. Generally, for all carrier-based PWM, $(mL - 1)$ number of triangular carriers (n_{tri}) would be required for any mL modulation scheme, despite the converter topological design.

In the case of the proposed three-cell 7L-LFCI, n_{tri} are used and divided into two sets equally, with one set level-shifted above zero reference magnitude $(tri_{TOP(1,2,3,...,(ntri/2))})$ and the other set below the zero reference magnitude $(tri_{BOT(1,2,3,...,(ntri/2))})$. The top and bottom carriers are 120° phase-shifted evenly, based on equation (12.3).

$$\Delta_{phase} = \frac{360°}{(n_{carrier}/2)}(\text{deg}) = \frac{2\pi}{(n_{carrier}/2)}(\text{rad}) \qquad \text{for} \qquad n_{carrier} = even > 2$$

(12.3)

The modulation technique for the *n*-cell *mL*-LFCI under the proposed two advanced hybrid sinusoidal PWM schemes (PSC-POD-PWM and PSC-PD-PWM) is achieved based on the analog design shown in Fig. 12.5. The top and bottom set of triangular carriers, 120° phase-shifted apart, are fed into the respective comparators in Stage A of Fig. 12.5. The six switching pulses (Sp-x1, Sp-x2, Sp-x3, Sp-x4, Sp-x5, and Sp-x6) in Fig. 12.6 are modulated after comparing the reference sinusoidal signal (Mx_{sine}) with the respective carriers.

The switching pulses Sp-x1, Sp-x2, and Sp-x3 are generated during the positive half-cycle of Mx_{sine}, which is the only period when the sinusoidal signal will be greater

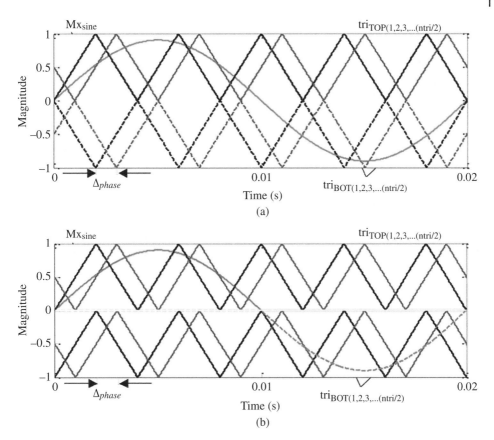

Figure 12.4 Proposed advanced hybrid sinusoidal PWM schemes: (a) PSC-POD-PWM; (b) PSC-PD-PWM.

than the three phase-shifted (120°) carriers above the zero reference magnitude. The other three switching pulses Sp-x4, Sp-x5, and Sp-x6 are generated when Mx_{sine} is greater than the three inverted phase-shifted (120°) triangular carriers below the zero reference magnitude. From Sp-x4, Sp-x5, and Sp-x6 in Fig. 12.6, it can be seen that switching pulses are always in the "on" state during the positive half-cycle of Mx_{sine}, where the conditions described in equation (3.1) are met.

After the six switching pulses (Sp-x1, Sp-x2, Sp-x3, Sp-x4, Sp-x5, and Sp-x6) are modulated in Stage A, these signals are further level-shifted again, respectively, in Stage B, based on Fig. 12.5. During the positive cycle of Mx_{sine}, the first two top switching pulses modulated with respect to tri_{TOP1} and tri_{TOP2} are always level shifted to the +1V above the zero reference, while the rest of the switching pulses modulated with respect to tri_{TOP3} up to $tri_{TOP(ntri/2)}$ are not required to be level-shifted. With respect to the negative cycle of Mx_{sine}, the switching pulses modulated with respect to tri_{BOT1} up to the $tri_{BOT((ntri/2)-2)}$ are level-shifted to −1V below the zero reference, and the remaining last two bottom switching pulses modulated with respect to $tri_{BOT((ntri/2)-1)}$ and $tri_{BOT(ntri/2)}$ are always level-shifted to −2V below the zero reference.

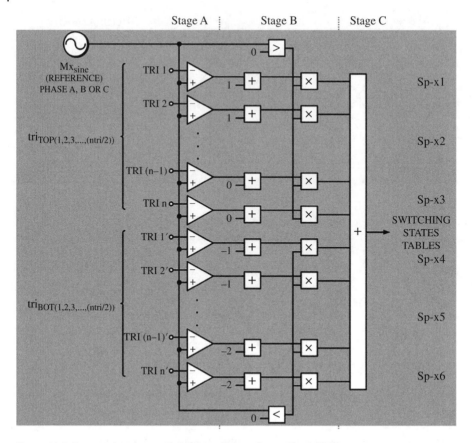

Figure 12.5 Proposed analog modulation technique for *n*-cell *mL*-LFCI.

The results of the six switching pulses modulated after level-shifting in Stage B (LS-x1, LS-x2, LS-x3, LS-x4, LS-x5, and LS-x6) are shown clearly in Fig. 12.7 according to the conditions explained. After level-shifting the six switching pulses in Stage B, LS-x1, LS-x2, LS-x3, LS-x4, LS-x5, and LS-x6 are then summed together in Stage C of Fig. 12.5. The resulting waveform of the summed signal (SSignal) is shown in Fig. 12.8.

The purpose of level-shifting the six switching pulses in Stage B and then summing together the signals in Stage C is to create the SSignalLevel states as shown in Table 12.1. These SSignalLevel states are mapped to the eight switching states conditions designed previously by determining the amplitude level of the SSignal using the look-up table, which is also referred as the "switching state table" in Fig. 12.5, based on Table 12.1. Hence, the significant reduction of excessive switching states from 64 to the minimum required eight switching states can now be practically achieved.

Hence, the final gating signals for the switches S_{x1}, S_{x2}, and S_{x3} of the three-cell 7L-LFCI are achieved from the SSignal according to conditions set in the switching state table. The resulting gate signals can be observed from Fig. 12.9 over two fundamental cycle periods. As for the complementary switches S_{x1}', S_{x2}', and S_{x3}', the gating signals are in fact the inverted signals of the switches S_{x1}, S_{x2}, and S_{x3}, which can be obtained simply with the use of the inverter NOT gates.

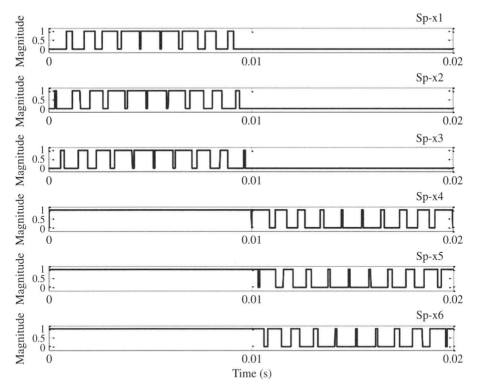

Figure 12.6 Output switching pulses (Sp-x) at Stage A of Fig. 12.5 under the PSC-POD-PWM modulation scheme.

On the other hand, due to the symmetrical geometry in the modulation carriers (combination of in-phase disposition and phase-shifted properties) of the proposed advanced hybrid PSC-PD-PWM scheme, the bottom sets of carriers ($\text{tri}_{BOT(1,2,3,...(ntri/2))}$) in Fig. 12.4 can be totally eliminated, as shown in Fig. 12.10. It can also be alternatively named as *reduced PSC-PD-PWM* (R-PSC-PD-PWM). Instead of having n_{tri} to synthesize the desired mL output phase voltage methods using any SPWM based on equation (3.2), the R-PSC-PD-PWM can successfully synthesize the same mL output phase voltage with 50% less n_{tri}, which is now re-expressed in equation (12.4).

$$n_{tri(RPSCPDPWM)} = \frac{n_{tri(SPWM)}}{2} = \frac{(mL) - 1}{2} \tag{12.4}$$

Despite greatly reducing the n_{tri} (by 50%), the Δ_{phase} between each triangular carrier must still be evenly phase-shifted according to equation (12.3). However, another set of same-reference sinusoidal signal is required to be level-shifted up with respect to the peak of the triangular carriers in order to achieve the same PWM modulation effect, as shown in Fig. 12.11. By rebuilding it this way, the control signal Mx_{sine} will not be affected.

The motivation behind R-PSC-PD-PWM is to avoid the complexity of generating switching pulses with digital signal processors (DSPs) or the designing of analog comparator circuitries, especially when more n_{tri} is required for higher mL output phase voltage. Besides this, the computational effort is greatly reduced with the reduction

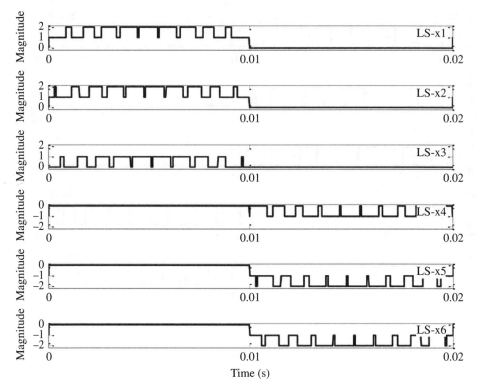

Figure 12.7 Output level-shifted switching pulses (LS-x) at Stage B of Fig. 12.5 under PSC-POD-PWM modulation scheme.

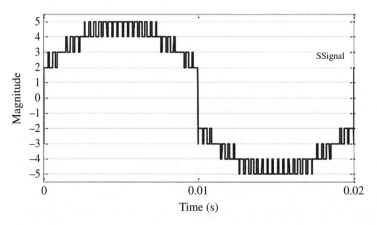

Figure 12.8 Output summed signal (SSignal) at Stage C of Fig. 12.5 under PSC-POD-PWM modulation scheme.

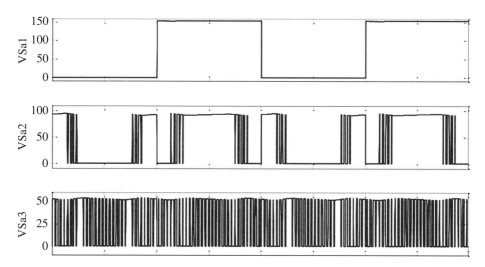

Figure 12.9 Voltage stress across Sa1 (top) to Sa3 (bottom) for three-cell 7L-LFCI based on hybrid PSC-PD-PWM with Mx = 0.9 and V_{dc} = 300V.

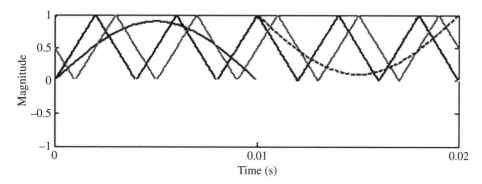

Figure 12.10 Reduced carrier of PSC-POD-PWM.

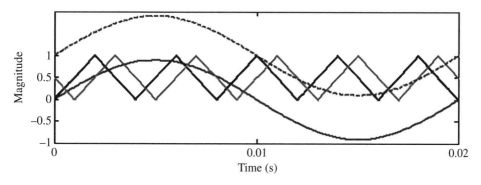

Figure 12.11 Level-shifting of reference sinusoidal signal in R-PSC-PD-PWM.

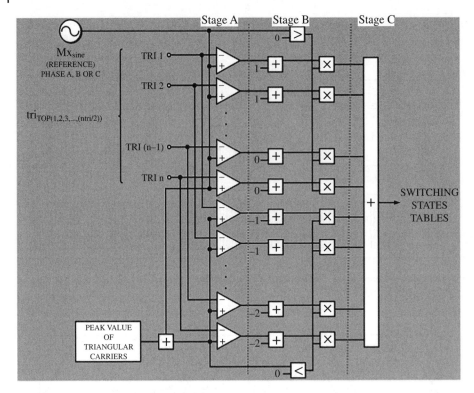

Figure 12.12 Proposed analog modulation technique for *n*-cell *mL*-LFCI.

of n_{tri} as compared to level-shifting another set of reference signal Mx_{sine}. Thus, the sampling time needed for DSP can be decreased to minimize the delays.

In the case of R-PSC-PD-PWM, only Stage A of the modulation technique in Fig. 12.5 has to be modified again to satisfy the described operating condition with the rebuilding of Mx_{sine}, as shown in Fig. 12.12. Hence, the modulation method in Stages B and C of Fig. 12.5 remains the same for the R-PSC-PD-PWM modulation technique, shown in Fig. 12.12.

12.5 Modulation Scheme

The modulation technique to achieve the desired gating signals for LFCI as presented previously is based on the analog approach. The modulation technique presented in both Figs. 12.5 and 12.12 is achievable with the DSPs. But it may become more complicated to achieve the same desired gating signals by developing the hardware circuit due to the level-shifter, multiplier, and summation properties shown in Stages B and C in both Figs. 12.5 and 12.12.

An alternative approach for avoiding these complications during hardware development is to use the digital logic gates instead. The design of alternative digital modulation techniques for the same three-cell 7L-LFCI is presented in Fig. 12.13. This digital modulation technique will replace Stages B and C in Figs. 12.5 and 12.12, including the need

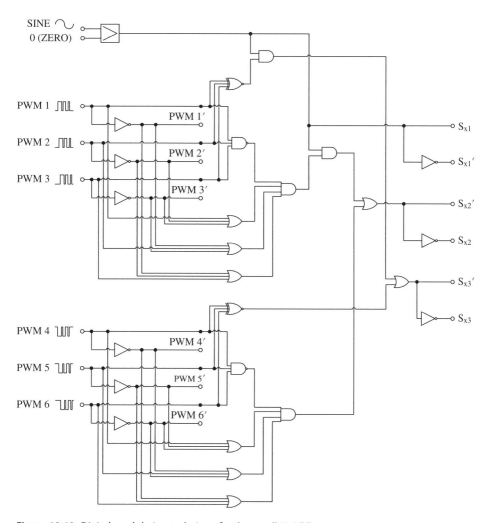

Figure 12.13 Digital modulation technique for three-cell 7L-LFCI.

for determining the switching state conditions in Table 12.1 by using the look-up tables. The digital logic gates proposed in Fig. 12.13 consists of the AND gate, NAND gate, OR gate, XNOR gate, as well as the inverter NOT gates.

PWM1, PWM2, PWM3, PWM4, PWM5, and PWM6 in Fig. 12.13 are referred to as the *switching pulses* (Sp-x), which are modulated right after the comparators in Stage A in Figs. 12.5 and 12 modulation technique presented.12. S_{x1}, S_{x2}, S_{x3}, S_{x3}', S_{x2}', and S_{x1}' at the output in Fig. 12.13 are referred to as the *output gating signals* for the respective switches of three-cell 7L-LFCI.

According to the digital modulation technique shown in Fig. 12.13 using logic gates, the final expression of the gating signal for the switch S_{x1} is given as:

$$S_{x1}(t) = \begin{cases} 1 & \text{if } \sin(\omega t) > 0 \\ 0 & \text{if } \sin(\omega t) < 0 \end{cases} \tag{12.5}$$

The result of the final gating signal for S_{x1} will be a square wave when the reference sinusoidal signal is greater than the zero reference. The gating signal of S_{x1} is achieved with a single analog comparator for each phase-leg. This is the only analog comparator in this case that cannot be replaced with any digital logic gates due to the need for controlling the reference signal Mx_{sine}.

The final expression of the modulated gating signal for the switch S_{x2}' is formulated with equation (12.6) based on the digital logic gates used in Fig. 12.13.

$$\overline{S_{x2}(t)} = \begin{bmatrix} (\overline{PWM1(t) \bullet PWM2(t) \bullet PWM3(t)}) \bullet (PWM1(t) + \overline{PWM2(t)} + \overline{PWM3(t)}) \\ \bullet(\overline{PWM1(t)} + PWM2(t) + \overline{PWM3(t)}) \bullet (\overline{PWM1(t)} + \overline{PWM2(t)} + PWM3(t)) \\ \bullet S_{x1}(t) \end{bmatrix}$$
$$+ \begin{bmatrix} (\overline{PWM4(t) \bullet PWM5(t) \bullet PWM6(t)}) \bullet (PWM4(t) + \overline{PWM5(t)} + \overline{PWM6(t)}) \\ \bullet(\overline{PWM4(t)} + PWM5(t) + \overline{PWM6(t)}) \bullet (\overline{PWM4(t)} + \overline{PWM5(t)} + PWM6(t)) \end{bmatrix}$$

(12.6)

The method applied to achieve the gating signal for the switch S_{x2}' requires several logic gates such as the NAND, AND, and OR gates based on equation (12.6). Besides this, a part of the expression in equation (12.6) also depends on the switching state condition of S_{x1} based on equation (12.5). Hence, the gating signal for the switch S_{x2} is achieved with aid of the NOT gate to obtain its inverted signals from its complementary switch S_{x2}', as shown in Fig. 12.13.

The final expression of the gating signal for switch S_{x3}' is given in equation (12.7).

$$\overline{S_{x3}(t)} = \begin{cases} \left[(PWM1(t) \bullet PWM2(t) \bullet PWM3(t)) + (\overline{PWM1(t)} \bullet \overline{PWM2(t)} \bullet \overline{PWM3(t)}) \bullet S_{x1}(t) \right] \\ + \left[(PWM4(t) \bullet PWM5(t) \bullet PWM6(t)) + (\overline{PWM4(t)} \bullet \overline{PWM5(t)} \bullet \overline{PWM6(t)}) \right] \end{cases}$$

(12.7)

The gating signal for S_{x3}' is achieved simply with two XNOR gates, one AND gate, and one OR gate, shown in Fig. 12.13, based on equation (12.7). The gating signal for S_{x3} can be achieved with the inverter NOT gate as well to obtain its inverted signal from its complementary switch S_{x3}'.

The same gating signals modulated based on the proposed digital modulation technique shown in Fig. 12.13 can achieve the same results as the gating signal from the previous analog modulation technique presented in Figs. 12.5 and 12.12. Even though the proposed alternative approach for modulation technique uses digital logic gates, the control of the reference sinusoidal signal based on Mx will not be affected for practical applications. As a result, this proposed method can help to reduce the complication process in hardware development. Moreover, the computational effort required can also be reduced and simplified in the DSPs to provide faster and dynamic control response in the converters.

12.6 Design Considerations

For the n-cell mL-LFCI, the voltage across the respective capacitor clamped at p^{th} position (C_{xp}^{th}) in Fig. 12.2 can be determined based on equation (12.8).

$$V_{Cx(p^{th}) \atop LFCI} = \frac{2^{(p^{th})} - 1}{mL - 1} \times V_{dc}$$

(12.8)

Table 12.2 Capacitor Voltages for Classical MFCI and Proposed LFCI.

| Capacitors Clamped at p^{th} Position | Phase x Capacitor Voltages VCx,p^{th} | | | |
| | MFCI | | LFCI | |
	Four-Cell (Five-Level)	Six-Cell (Seven-Level)	Three-Cell (Seven-Level)	Four-Cell (15-Level)
1	$V_{dc}/4$	$V_{dc}/6$	$V_{dc}/6$	$V_{dc}/14$
2	$2V_{dc}/4$	$2V_{dc}/6$	$3V_{dc}/6$	$3V_{dc}/14$
3	$3V_{dc}/4$	$3V_{dc}/6$	$\sim$	$7V_{dc}/14$
4	$\sim$	$4V_{dc}/6$	$\sim$	$\sim$
5	$\sim$	$5V_{dc}/6$	$\sim$	$\sim$

Note: Here, x represents phase a, b, and c and p^{th} represents the placing position number of flying capacitors being clamped.

In the case of the proposed three-cell 7L-LFCI, the flying capacitor voltages $V_{Cx}1$ and $V_{Cx}2$ are maintained at $V_{dc}/6$ and $V_{dc}/2$, respectively, under the steady-state operation. Even though the three-cell 7L-LFCI has the same operation modes as the three-cell 4L-MFCI, the voltage across the two flying capacitors are now different, as can be observed from Table 12.2.

It should be noted that not only are the number of flying capacitors being reduced when comparing based on the number of output phase voltage levels (proposed three-cell 7L-LFCI and classical six-cell 7L-MFCI), but also the voltage ratings of these flying capacitors are significantly reduced based on the number of modular cells used (proposed four-cell 15L-LFCI and classical three-cell 5L-MFCI).

Since the voltage across Cx1 and Cx2 are now maintained at $(1/6)V_{dc}$ and $(3/6)V_{dc}$, the voltage stress across the six switches are no longer the same as compared to those in the classical MFCI approach. The voltage stress across the switches is affected by the voltage differences between $V_{Cxp}{}^{th}$ and $V_{Cx}(p^{th}{}_{-1})$ using equation (12.8) and the switching states given in Table 12.1. Hence, the voltage stress across the respective switches is determined using equation (12.9).

$$V_{Sx(cell\ n^{th})\atop LFCI}(t) \approx \begin{cases} -\left(\dfrac{mL - 2^{(p^{th})}}{mL - 1}\right) V_{dc} \times S_{x(cell\ n^{th})}(t) & \text{if } cell\ n^{th} = 1 \\ -\left(\dfrac{2^{(p^{th})} - 2^{(p^{th}-1)}}{mL - 1}\right) V_{dc} \times S_{x(cell\ n^{th})}(t) & \text{if } cell\ n^{th} \geq 2 \end{cases} \tag{12.9}$$

Therefore, the voltage stress across the switches S_{x1}, S_{x2}, and S_{x3} (V_{Sx1}, V_{Sx2}, and V_{Sx3}) in the three-cell 7L-LFCI are approximately at $V_{dc}/2$, $V_{dc}/3$, and $V_{dc}/6$, respectively, as shown in Fig. 12.9, with the dc voltage source of $300V_{dc}$ supplied. According to equation (12.9), it is worth noting that the blocking voltage capability of the two LF switches S_{x1} and S_{x1}' in Cell 1 must be able to withstand $V_{dc}/2$ for n-cell mL-LFCI at all times. The same amount of voltage stress also applies to its respective complementary switches.

The output phase voltages and output line-to-line voltages for LFCI are expressed with equations (12.10) and (12.11), respectively, based on the corresponding switching states

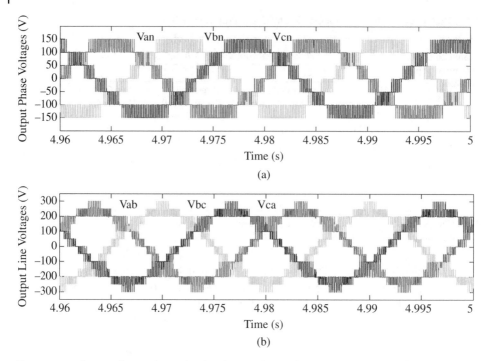

Figure 12.14 Output phase voltages (top) and output line-to-line voltages (bottom) of classical six-cell 7L-MFCI using PSC-PWM with Mx = 0.9 and V_{dc} = 300V.

in Table 12.1, and the flying capacitor voltages are calculated using equation (12.8).

$$Vxn(t) = V_{dc}(t)(S_{x1}(t) - 0.5) + \sum_{m=2}^{n_{cells}} V_{Cx((P_{capacitors}+2)-m)}(t)(S_{x(m)}(t) - S_{x(m-1)}(t))$$

(12.10)

$$\begin{bmatrix} Vab \\ Vbc \\ Vca \end{bmatrix}_{mL} = V_{dc}(t) \begin{bmatrix} (S_{a1}(t) - S_{b1}(t)) \\ (S_{b1}(t) - S_{c1}(t)) \\ (S_{c1}(t) - S_{a1}(t)) \end{bmatrix}$$

$$+ \sum_{m=2}^{n_{cells}} V_{Cx((P_{capacitors}+2)-m)}(t) \begin{bmatrix} (S_{a(m)}(t) - S_{a(m-1)}(t) - S_{b(m)}(t) + S_{b(m-1)}(t)) \\ (S_{b(m)}(t) - S_{b(m-1)}(t) - S_{c(m)}(t) + S_{c(m-1)}(t)) \\ (S_{c(m)}(t) - S_{c(m-1)}(t) - S_{a(m)}(t) + S_{a(m-1)}(t)) \end{bmatrix}_{x \in \{a,b,c\}}$$

(12.11)

Both equations of phase voltages (12.10) and line-to-line voltages (12.11) are also applicable for the classical *n*-cell *mL*-MFCI using PSC-PWM. The output voltages of both the classical six-cell 7L-MFCI with PSC-PWM and the proposed three-cell 7L-LFCI under the two advanced hybrid sinusoidal PWM schemes are shown in Figs. 12.14 and 12.15, respectively.

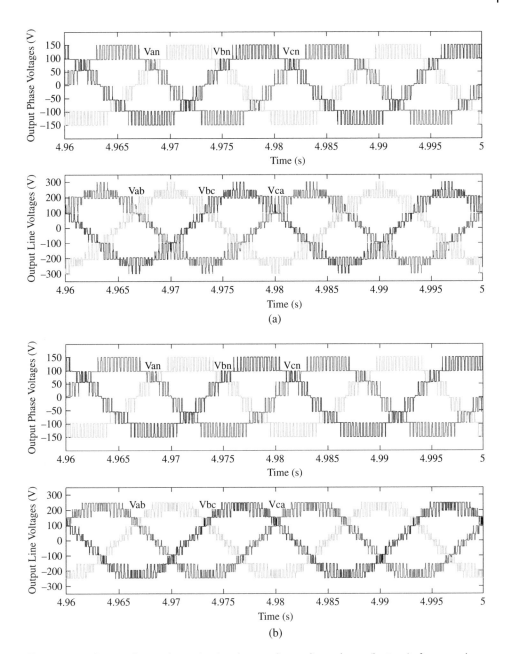

Figure 12.15 Output phase voltages (top) and output line-to-line voltages (bottom) of proposed three-cell 7L-LFCI using (a) PSC-POD-PWM and (b) PSC-PD-PWM with Mx = 0.9 and V$_{dc}$ = 300V.

However, one can observe that the overall switching frequency is greatly reduced for the proposed three-cell 7L-LFCI employing the hybrid PWM schemes, as compared to the classical six-cell 7L-MFCI using PSC-PWM. The harmonic characteristics of both proposed hybrid PSC-POD-PWM and PSC-PD-PWM are detailed and analyzed in subsection 12.1.5 of this chapter.

12.7 Harmonic Characteristics

In this chapter, two hybrid sinusoidal carrier-based PWM schemes are proposed for the transformerless n-cell mL-LFCI, namely PSC-POD-PWM and PSC-PD-PWM, as described in Fig. 12.4. Both proposed modulation schemes greatly reduce the overall switching frequency operation by 50% as compared to the classical PSC-PWM modulation. These will be proven analytically in the following text using the double Fourier series approach.

12.7.1 PSC-POD-PWM

For the first proposed hybrid PSC-POD-PWM, same number of triangular carriers is required to achieve m-level modulation according to equation (3.2). As shown in Fig. 12.4(a), half of the n_{tri} is level-shifted above the reference magnitude zero $(\text{tri}_{TOP(1,2,3,...,(ntri/2))})$, while the other half is inverted (phase-opposition-disposition) and level-shifted below the reference magnitude zero $(\text{tri}_{BOT(1,2,3,...,(ntri/2))})$.

It can be observed that the triangular carriers are phase-shifted away with one another instead of further level-shifting like those in POD-PWM. On top of that, the hybrid PSC-POD-PWM can avoid the complication of phase shifting carriers for higher mL output voltage as compared to PSC-PWM, since only half of the n_{tri} are phase-shifted for both above and below the reference zero. Thus, the Δ_{phase} for both upper and lower triangular carriers is determined based on equation (12.3).

The harmonic analysis of PSC-POD-PWM is simplified using a single-phase reference sinusoidal signal and two fundamental triangular carriers, as shown in Fig. 12.16. The switching function of PSC-POD-PWM expressed in Table 12.3 can be derived based on the three-level modulation representation shown in Fig. 12.16. Thus, there are three conditions $(+V_{dc}, 0V, \text{and} -V_{dc})$ generated according to the condition given in

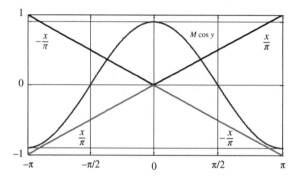

Figure 12.16 Reference sinusoidal signal with two fundamental triangular carriers – PSC-POD-PWM.

Table 12.3 Switching Function of PSC-POD-PWM.

$f(x, y)$	$-\pi < x < 0$	$0 < x < \pi$
$+V_{dc}$	$M \cos y > -\dfrac{x}{\pi}$	$M \cos y > \dfrac{x}{\pi}$
0	$\dfrac{x}{\pi} < M \cos y < -\dfrac{x}{\pi}$	$-\dfrac{x}{\pi} < M \cos y < \dfrac{x}{\pi}$
$-V_{dc}$	$M \cos y < \dfrac{x}{\pi}$	$M \cos y < -\dfrac{x}{\pi}$

equation (12.1). The lower and upper boundaries for the transition conditions stated in Table 12.3 are determined by equations (12.12) and (12.13), respectively.

$$x_{LOWER} = -\pi M \cos y \tag{12.12}$$

$$x_{UPPER} = \pi M \cos y \tag{12.13}$$

With the boundary conditions known, equations (12.12) and (12.13) are substituted into the complex form of double Fourier integrals, as expressed previously in equation (4.6). Thus, the complex form of double Fourier series for the PSC-POD-PWM modulation is expressed in equation (12.14).

$$
C_{mn} = \frac{1}{2\pi^2} \left[\int_{-\frac{\pi}{2}}^{\frac{\pi}{2}} \int_{-\pi M \cos y}^{\pi M \cos y} V_{dc}\, e^{j(mx+ny)}\, dx\, dy - \int_{-\pi}^{-\frac{\pi}{2}} \int_{\pi M \cos y}^{-\pi M \cos y} V_{dc}\, e^{j(mx+ny)}\, dx\, dy \right.
$$
$$
\left. - \int_{\frac{\pi}{2}}^{\pi} \int_{\pi M \cos y}^{-\pi M \cos y} V_{dc}\, e^{j(mx+ny)}\, dx\, dy \right]
$$
$$
= \frac{V_{dc}}{2\pi^2} \left[\int_{-\frac{\pi}{2}}^{\frac{3\pi}{2}} \int_{-\pi M \cos y}^{\pi M \cos y} e^{j(mx+ny)}\, dx\, dy \right] \tag{12.14}
$$

Based on the general spectrum analysis equation (12.5), the dc offset component of the PSC-POD-PWM can be determined using equation (12.15) by letting $m = n = 0$ in equation (12.14).

$$
A_{00} + jB_{00} = \frac{V_{dc}}{2\pi^2} \int_{-\frac{\pi}{2}}^{\frac{3\pi}{2}} \int_{-\pi M \cos y}^{\pi M \cos y} e^{j(0+0)}\, dx\, dy
$$
$$
= \frac{V_{dc}}{2\pi^2} \int_{-\frac{\pi}{2}}^{\frac{3\pi}{2}} [2\pi M \cos y]\, dy
$$
$$
= 0 \tag{12.15}
$$

In order to determine the fundamental baseband component, let $m = 0$ and $n > 0$ for equation (12.14). Thus, the baseband harmonic is determined based on equation (12.16).

$$
A_{0n} + jB_{0n} = \frac{V_{dc}}{2\pi^2} \left[\int_{-\frac{\pi}{2}}^{\frac{3\pi}{2}} \int_{-\pi M \cos y}^{\pi M \cos y} e^{j(ny)}\, dx\, dy \right]
$$
$$
= \frac{M V_{dc}}{\pi} \int_{-\frac{\pi}{2}}^{\frac{3\pi}{2}} [\cos y] e^{j(ny)}\, dy
$$

$$= \frac{MV_{dc}}{2\pi} \int_{-\frac{\pi}{2}}^{\frac{3\pi}{2}} (e^{jy(n+1)} + e^{jy(n-1)}) dy \tag{12.16}$$

In equation (12.16), the integral of the term $e^{j(ny)}$ is equal to zero for any non-zero condition of n. $A_{0n} + jB_{0n} = 0$ for all conditions when $n > 1$; therefore, let $n = 1$ for equation (12.16). Hence, equation (12.16) can be further expressed with equation (12.17).

$$A_{01} + jB_{01} = \frac{MV_{dc}}{2\pi} \int_{-\frac{\pi}{2}}^{\frac{3\pi}{2}} (e^{jy(0)}) dy$$

$$= MV_{dc} \tag{12.17}$$

The carrier harmonic components can be expressed in equation (12.18) by letting $m > 0$ and $n = 0$ for equation (12.14).

$$A_{m0} + jB_{m0} = \frac{V_{dc}}{2\pi^2} \left[\int_{-\frac{\pi}{2}}^{\frac{3\pi}{2}} \int_{-\pi M \cos y}^{\pi M \cos y} e^{j(mx)} dx\, dy \right]$$

$$= \frac{V_{dc}}{m\pi^2} \int_{-\frac{\pi}{2}}^{\frac{3\pi}{2}} 2 \sum_{k=1}^{\infty} \sin k\frac{\pi}{2} J_k(m\pi M) \cos ky\, dy$$

$$= 0 \tag{12.18}$$

Let $m > 0$ and $n \neq 0$ for equation (12.14) to determine the sideband harmonic components, which is derived as shown in equation (12.19).

$$A_{mn} + jB_{mn} = \frac{V_{dc}}{2\pi^2} \left[\int_{-\frac{\pi}{2}}^{\frac{3\pi}{2}} \int_{-\pi M \cos y}^{\pi M \cos y} e^{j(mx+ny)} dx\, dy \right]$$

$$= \frac{V_{dc}}{m\pi^2} \int_{-\frac{\pi}{2}}^{\frac{3\pi}{2}} e^{jny} \left[2 \sum_{k=1}^{\infty} J_k(m\pi M) \sin k\frac{\pi}{2} \cos ky \right] dy$$

$$= \frac{V_{dc}}{m\pi^2} \sum_{k=1}^{\infty} J_k(m\pi M) \sin k\frac{\pi}{2} \int_{-\frac{\pi}{2}}^{\frac{3\pi}{2}} [e^{jy(n+k)} + e^{jy(n-k)}]\, dy$$

$$= \left(\begin{array}{c} \frac{4V_{dc}}{m\pi^2} \sum_{k=1}^{\infty} J_k(m\pi M) \sin k\frac{\pi}{2} \left[\frac{\pi}{2} \Big|_{n=k} \right] \\ + \frac{4V_{dc}}{m\pi^2} \sum_{k=1}^{\infty} J_k(m\pi M) \sin k\frac{\pi}{2} \left[\frac{\sin \frac{\pi}{2}(n+k)}{(n+k)} \Big|_{-n\neq k} + \frac{\sin \frac{\pi}{2}(n-k)}{(n-k)} \Big|_{n\neq k} \right] \end{array} \right)$$

$$= \frac{2V_{dc}}{m\pi} J_n(m\pi M) \cos n\pi \tag{12.19}$$

Substitute equations (12.15), (12.17), (12.18), and (12.19) into equation (12.5) to achieve the spectrum analysis expression for the switching function in Table 12.3, which describes Fig. 12.16. The expression is shown in equation (12.20).

$$f(t) = \frac{MV_{dc}}{2} \cos(\omega_o t + \theta_0)$$

$$+ \frac{V_{dc}}{\pi} \sum_{m=1}^{\infty} \sum_{\substack{n=-\infty \\ (n\neq 0)}}^{\infty} \frac{1}{m} J_n(m\pi M) \cos n\pi \times \cos(m[\omega_c t + \theta_c] + n[\omega_o t + \theta_0]) \tag{12.20}$$

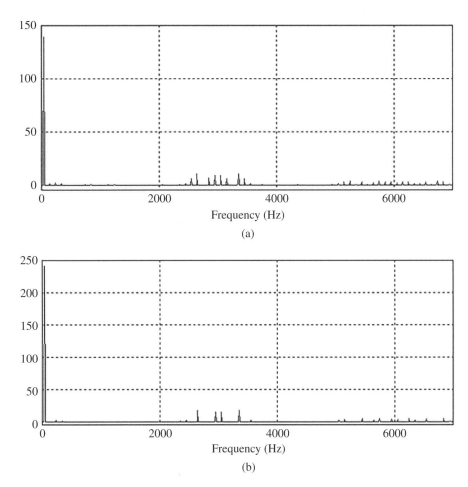

Figure 12.17 Voltage spectrum based on seven-level modulation PSC-POD-PWM: (a) output phase voltage; (b) output line-to-line voltage.

By observing the seven-level PSC-PWM modulation in Fig. 12.15 with the seven-level PSC-POD-PWM modulation in Fig. 12.17, it should be noted that the first group of sideband harmonic is being shifted away to 3kHz from the fundamental baseband harmonic for PSC-POD-PWM instead of 6kHz for the classical PSC-PWM. This is because half the n_{tri} of PSC-POD-PWM is used in upper level-shifted (above the reference magnitude zero) and the other half for lower level-shifted (below the reference magnitude zero).

Therefore, the behavior of sideband harmonic phase-shifted in PSC-POD-PWM can be determined generally based on the frequency (f_{shift}) or multiple of harmonic order[th], which are redefined in equations (12.21) ad (12.22), respectively.

$$f_{shift} = f_c \times \frac{n_{tri}}{2} \tag{12.21}$$

$$Harmonic \ order^{th} = \frac{f_c \times n_{tri}}{2 \times f_o} \tag{12.22}$$

Hence, the final m-level output phase voltage spectrum expression for the PSC-POD-PWM scheme is derived in equation (12.23).

$$Vxn_{PSCPODPWM-xL}(t) = \frac{MV_{dc}}{2} \cos(\omega_o t + \theta_0)$$

$$+ \frac{V_{dc}}{\pi} \sum_{m=1}^{\infty} \sum_{\substack{n=-\infty \\ (n \neq 0)}}^{\infty} \frac{1}{(Nm)} J_{(2n-1)}((Nm)\pi M) \cos[(2n-1)\pi]$$

$$\times \cos((Nm)[\omega_o t + \theta_c] + (2n-1)[\omega_o t + \theta_0]) \tag{12.23}$$

Here, N = half the total number of triangular carriers used in PSC-POD-PWM.

Based on the expression of equation (12.23), there is no existence of dc offset or carrier harmonic components. Thus, only fundamental baseband harmonic and sideband harmonic are created in PSC-POD-PWM. The sideband harmonic carries the term of $\cos[(2n-1)\pi]$, which shows that only odd sidebands occur at multiples of the harmonic order[th], based on equation (12.22). Equation (12.23) describes the characteristic of output phase voltage for seven-level PSC-POD-PWM modulation, shown in Fig. 12.17(a).

The numerical values of equation (12.23) are calculated for analytical results based on the seven-level modulation under the same condition of $Mx_{sine} = 0.9$ and f_c = 1kHz. The analytical results match similarly to the simulation results shown in Fig. 12.18(a). The experimental results are also obtained in Fig. 12.18(b) for the seven-level modulation. Once again, this proves that the theoretical analysis from equation (12.23) describes the characteristics of PSC-POD-PWM for any mL output phase voltage.

The spectrum analysis of the output line-to-line voltage Vab using the PSC-POD-PWM method is determined with equation (12.24).

$$Vab_{PSCPODPWM-xL}(t) = \frac{\sqrt{3}MV_{dc}}{2} \cos\left(\omega_o t + \frac{\pi}{6}\right)$$

$$+ \frac{2V_{dc}}{\pi} \sum_{m=1}^{\infty} \sum_{\substack{n=-\infty \\ (n \neq 0)}}^{\infty} \frac{1}{(Nm)} J_{(2n-1)}((Nm)\pi M)$$

$$\cos[(2n-1)\pi] \sin\left((2n-1)\frac{\pi}{3}\right)$$

$$\times \cos\left((Nm)[\omega_o t] + (2n-1)\left[\omega_o t - \frac{\pi}{3}\right] + \frac{\pi}{3}\right) \tag{12.24}$$

Because of the three-phase power system, the crossover point between the two-phase reference sinusoidal signals at multiples of 60° cancels away all the triplen sideband harmonic only, since there is no carrier harmonic existing in the output phase voltage. Thus, the same $\sin((2n-1)\pi/3)$ term appears in the sideband harmonic content of equation (12.24). The cancellation of the triplen sideband harmonic can be clearly observed in Fig. 12.17 between the output phase voltage and the output line-to-line voltage.

The analytical results of output line-to-line voltage for the seven-level PSC-POD-PWM modulation are obtained with the numerical calculation from equation (12.24).

The comparison between the analytical results and the simulation results is shown in Fig. 12.19(a). Besides this, experimental results are obtained in Fig. 12.19(b) as well to verify the characteristics of PSC-POD-PWM, expressed in equation (12.24) for any mL output line-to-line voltage.

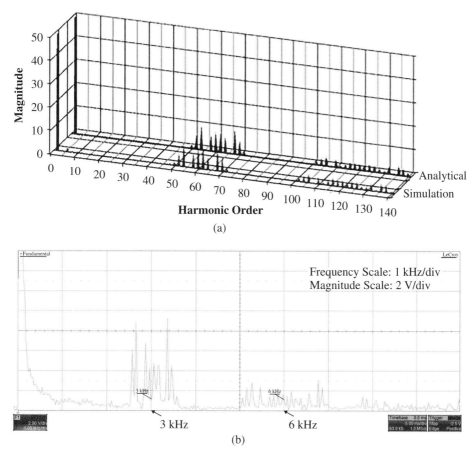

Figure 12.18 Output phase voltage spectrum – seven-level PSC-POD-PWM modulation: (a) simulation and analytical results; (b) experimental results.

12.7.2 PSC Phase-Disposition PWM (PSC-PD-PWM)

The derivation of the second proposed advanced hybrid PSC-PD-PWM (or alternatively equivalent to R-PSC-PD-PWM in Fig. 12.10) is the resulting combination of both LSC-PWM and PSC-PWM. Hence, the harmonic spectrum analysis for hybrid PSC-PD-PWM is simplified using the single-phase reference sinusoidal signal and the two level-shifted fundamental triangular carriers in Fig. 12.20.

The switching function of PSC-PD-PWM is expressed in Table 12.4, which is derived based on Fig. 12.20. Since two fundamental level-shifted carriers are used in Fig. 12.20 for analysis, there are only three conditions generated, which are $+V_{dc}$, 0V, and $-V_{dc}$, following the switching condition expressed in equation (3.1).

However, the boundary conditions for LSC-PWM (referred to as "PD-PWM") in this case are slightly more complex. Thus, the boundary conditions for the transitions between 0V to $+V_{dc}$ and $+V_{dc}$ to 0V are given in equations (12.25) and (12.26).

$$x_{LOWER} = -\pi M \cos y \tag{12.25}$$

$$x_{UPPER} = \pi M \cos y \tag{12.26}$$

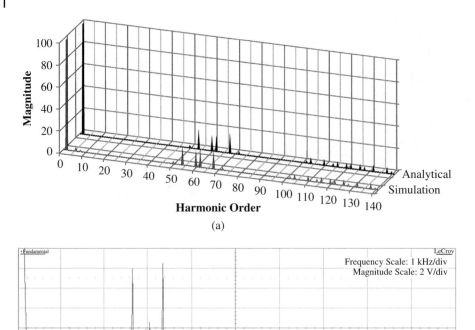

(a)

(b)

Figure 12.19 Output line-to-line voltage spectrum – seven-level PSC-POD-PWM modulation: (a) simulation and analytical results; (b) experimental results.

In order to achieve the $-V_{dc}$ level, the reference sinusoidal signal has to be lesser than both upper and lower triangular carriers. Therefore, the boundary conditions for the transitions between 0V to $-V_{dc}$ and $-V_{dc}$ to 0V are given based on the upper carrier and lower carrier.

The lower and upper boundary conditions for this transition based on the upper carrier are given in equations (12.27) and (12.28), respectively.

$$x_{LOWER} = -\pi \tag{12.27}$$

$$x_{UPPER} = -\pi(1 + M \cos y) \tag{12.28}$$

The lower and upper boundary conditions based on the lower carrier are given in equations (12.29) and (12.30), respectively.

$$x_{LOWER} = \pi(1 + M \cos y) \tag{12.29}$$

$$x_{UPPER} = \pi \tag{12.30}$$

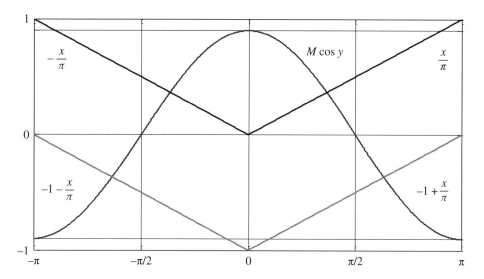

Figure 12.20 Reference sinusoidal signal with two fundamental level-shifted carriers – PSC-PD-PWM.

Table 12.4 Switching Function of PSC-PD-PWM.

$f(x, y)$	$-\pi < x < 0$	$0 < x < \pi$
$+V_{dc}$	$M \cos y > -\dfrac{x}{\pi}$	$M \cos y > \dfrac{x}{\pi}$
0	$-1 - \dfrac{x}{\pi} < M \cos y < -\dfrac{x}{\pi}$	$-1 + \dfrac{x}{\pi} < M \cos y < \dfrac{x}{\pi}$
$-V_{dc}$	$M \cos y < -1 - \dfrac{x}{\pi}$	$M \cos y < -1 + \dfrac{x}{\pi}$

The complex form of double Fourier analysis for PSC-PD-PWM modulation is expressed in equation (12.31) based on the boundary conditions equations applied to equation (12.6).

$$
\begin{aligned}
C_{mn} &= \frac{1}{2\pi^2}
\begin{bmatrix}
\displaystyle\int_{-\frac{\pi}{2}}^{\frac{\pi}{2}} \int_{-\pi M \cos y}^{\pi M \cos y} V_{dc}\, e^{j(mx+ny)}\, dx\, dy \\[2ex]
-\displaystyle\int_{-\pi}^{-\frac{\pi}{2}} \left[\int_{-\pi}^{-\pi(1+M\cos y)} V_{dc}\, e^{j(mx+ny)}\, dx + \int_{\pi(1+M\cos y)}^{\pi} V_{dc}\, e^{j(mx+ny)}\, dx \right] dy \\[2ex]
-\displaystyle\int_{\frac{\pi}{2}}^{\pi} \left[\int_{-\pi}^{-\pi(1+M\cos y)} V_{dc}\, e^{j(mx+ny)}\, dx + \int_{\pi(1+M\cos y)}^{\pi} V_{dc}\, e^{j(mx+ny)}\, dx \right] dy
\end{bmatrix} \\[4ex]
&= \frac{V_{dc}}{2\pi^2} \left[\int_{-\frac{\pi}{2}}^{\frac{\pi}{2}} \int_{-\pi M \cos y}^{\pi M \cos y} e^{j(mx+ny)}\, dx\, dy + \int_{\frac{\pi}{2}}^{\frac{3\pi}{2}} \int_{\pi(1-M\cos y)}^{\pi(1+M\cos y)} e^{j(mx+ny)}\, dx\, dy \right] \quad (12.31)
\end{aligned}
$$

Again, the dc offset component can be determined with equation (12.32) by letting $m = n = 0$ in equation (12.31).

$$
A_{00} + jB_{00} = \frac{V_{dc}}{2\pi^2} \left[\int_{-\frac{\pi}{2}}^{\frac{\pi}{2}} \int_{-\pi M \cos y}^{\pi M \cos y} e^{j(0+0)}\, dx\, dy + \int_{\frac{\pi}{2}}^{\frac{3\pi}{2}} \int_{\pi(1-M\cos y)}^{\pi(1+M\cos y)} e^{j(0+0)}\, dx\, dy \right]
$$

$$= \frac{V_{dc}}{2\pi^2} \int_{-\frac{\pi}{2}}^{\frac{\pi}{2}} [2\pi M \cos y] dy + \frac{V_{dc}}{2\pi^2} \int_{\frac{\pi}{2}}^{\frac{3\pi}{2}} [2\pi M \cos y] dy$$

$$= 0 \tag{12.32}$$

The fundamental baseband component is determined with equation (12.33) based on the condition $m = 0$ and $n > 0$. Equation (12.33) will experience same term $e^{j(ny)}$ during the integral process. The integral of the term $e^{j(ny)}$ will be equal to zero for any non-zero value of n. Therefore, n will be equal to 1 since $A_{0n} + jB_{0n} = 0$ for all conditions when $n > 1$. The expression is further simplified into equation (12.34).

$$
\begin{aligned}
A_{0n} + jB_{0n} &= \frac{V_{dc}}{2\pi^2} \left[\int_{-\frac{\pi}{2}}^{\frac{\pi}{2}} \int_{-\pi M \cos y}^{\pi M \cos y} e^{j(ny)} \, dx \, dy + \int_{\frac{\pi}{2}}^{\frac{3\pi}{2}} \int_{\pi(1-M \cos y)}^{\pi(1+M \cos y)} e^{j(ny)} \, dy \, dy \right] \\
&= \frac{MV_{dc}}{\pi} \left[\int_{-\frac{\pi}{2}}^{\frac{\pi}{2}} [\cos y] e^{j(ny)} \, dy + \int_{\frac{\pi}{2}}^{\frac{3\pi}{2}} [\cos y] e^{j(ny)} \, dy \right] \\
&= \frac{MV_{dc}}{2\pi} \left[\int_{-\frac{\pi}{2}}^{\frac{\pi}{2}} (e^{jy(n+1)} + e^{jy(n-1)}) dy + \int_{\frac{\pi}{2}}^{\frac{3\pi}{2}} (e^{jy(n+1)} + e^{jy(n-1)}) dy \right]
\end{aligned}
\tag{12.33}
$$

$$
\begin{aligned}
A_{01} + jB_{01} &= \frac{MV_{dc}}{2\pi} \left[\int_{-\frac{\pi}{2}}^{\frac{\pi}{2}} (e^{jy(0)}) dy + \int_{\frac{\pi}{2}}^{\frac{3\pi}{2}} (e^{jy(0)}) dy \right] \\
&= MV_{dc}
\end{aligned}
\tag{12.34}
$$

The carrier harmonic components are determined with equation (12.35) based on the condition of $m > 0$ and $n = 0$ for equation (12.31).

$$
\begin{aligned}
A_{m0} + jB_{m0} &= \frac{V_{dc}}{2\pi^2} \left[\int_{-\frac{\pi}{2}}^{\frac{\pi}{2}} \int_{-\pi M \cos y}^{\pi M \cos y} e^{j(mx)} \, dx \, dy + \int_{\frac{\pi}{2}}^{\frac{3\pi}{2}} \int_{\pi(1-M \cos y)}^{\pi(1+M \cos y)} e^{j(mx)} \, dx \, dy \right] \\
&= \frac{V_{dc}}{m\pi^2} 2 \sum_{k=1}^{\infty} \sin k \frac{\pi}{2} J_k(m\pi M) \left[\begin{matrix} 2 \cos \dfrac{k\left(\frac{\pi}{2}-\frac{\pi}{2}\right)}{2} \sin \dfrac{k\left(\frac{\pi}{2}+\frac{\pi}{2}\right)}{2} \\ \dfrac{}{k} \\ +(e^{jm\pi}) \dfrac{2 \cos \dfrac{k\left(\frac{3\pi}{2}+\frac{\pi}{2}\right)}{2} \sin \dfrac{k\left(\frac{3\pi}{2}-\frac{\pi}{2}\right)}{2}}{k} \end{matrix} \right] \\
&= \frac{4V_{dc}}{m\pi^2} \sum_{k=1}^{\infty} J_k(m\pi M) \frac{\left(\sin \frac{k\pi}{2}\right)^2}{k} (1 - e^{jm\pi})
\end{aligned}
\tag{12.35}
$$

The term $(\sin k\pi/2)^2$ in equation (12.35) is equal to a non-zero solution if the value of k is odd. Hence, the term $(1 - e^{jm\pi})$ will be equal to 2 when the value of m is odd as well. Thus, m must be an odd value at all times, and the final expression for carrier harmonic is simplified into equation (12.36).

$$A_{m0} + jB_{m0} = \frac{8V_{dc}}{\pi^2} \frac{1}{2m-1} \sum_{k=1}^{\infty} \frac{J_{2k-1}((2m-1)\pi M)}{2k-1} \tag{12.36}$$

Finally, the sideband harmonic is defined by equation (12.37) with the condition of $m > 0$ and $n \neq 0$ set for equation (12.31).

$$A_{mn} + jB_{mn} = \frac{V_{dc}}{2\pi^2} \left[\int_{-\frac{\pi}{2}}^{\frac{\pi}{2}} \int_{-\pi M \cos y}^{\pi M \cos y} e^{j(mx+ny)} \, dx \, dy + \int_{\frac{\pi}{2}}^{\frac{3\pi}{2}} \int_{\pi(1-M \cos y)}^{\pi(1+M \cos y)} e^{j(mx+ny)} \, dx \, dy \right]$$

$$= \frac{V_{dc}}{m\pi^2} \left[\int_{-\frac{\pi}{2}}^{\frac{\pi}{2}} e^{jny} \sin(m\pi M \cos y) \, dy + e^{jm\pi} \int_{\frac{\pi}{2}}^{\frac{3\pi}{2}} e^{jny} \sin(m\pi M \cos y) \, dy \right]$$

$$= \frac{2V_{dc}}{m\pi^2} \sum_{k=1}^{\infty} J_k(m\pi M) \sin k\frac{\pi}{2} \left[\frac{\pi}{2} \Big|_{n=k} \right]$$

$$+ \frac{2V_{dc}}{m\pi^2} \sum_{k=1}^{\infty} J_k(m\pi M) \sin k\frac{\pi}{2} \left[\frac{\sin \frac{\pi}{2}(n+k)}{(n+k)} \Big|_{-n \neq k} + \frac{\sin \frac{\pi}{2}(n-k)}{(n-k)} \Big|_{n \neq k} \right]$$

$$+ \frac{2V_{dc}}{m\pi^2} \sum_{k=1}^{\infty} J_k(m\pi M) \sin k\frac{\pi}{2}(e^{jm\pi}) \left[\frac{\pi}{2} \Big|_{n=k} \right]$$

$$+ \frac{2V_{dc}}{m\pi^2} \sum_{k=1}^{\infty} J_k(m\pi M) \sin k\frac{\pi}{2}(e^{jm\pi}) \left[-\frac{\sin \frac{\pi}{2}(n+k)}{(n+k)} \Big|_{-n \neq k} - \frac{\sin \frac{\pi}{2}(n-k)}{(n-k)} \Big|_{n \neq k} \right]$$

$$\tag{12.37}$$

Since k must be an odd value due to the term $(\sin k\pi/2)$ in equation (12.37) to determine a non-zero solution, the value of n can only be an even value for the term $[\sin (\pi(n+k))/2]$. Equation (12.37) is further formulated as equation (12.38).

$$A_{mn} + jB_{mn} = \frac{V_{dc}}{m\pi} \sum_{n=1}^{\infty} J_n(m\pi M) \sin n\frac{\pi}{2}(1 + e^{jm\pi})$$

$$- \frac{4V_{dc}}{m\pi^2} \sum_{k=1}^{\infty} J_{2k-1}(m\pi M) \left[\frac{(2k-1) \cos \left(\frac{n\pi}{2} \right)}{(n+(2k-1))(n-(2k-1))} \Big|_{n \neq 2k-1} \right] (1 - e^{jm\pi}) \tag{12.38}$$

The exponential terms of equation (12.38) are further simplified by considering the conditions of equation (12.39):

$$\begin{cases} (1 - e^{jm\pi}) = 2 & \text{if } m = ODD \\ (1 + e^{jm\pi}) = 2 & \text{if } m = EVEN \end{cases} \tag{12.39}$$

The final expression for the sideband harmonic is given in equation (12.40).

$$A_{mn} + jB_{mn} = \frac{2V_{dc}}{\pi} \frac{1}{2m} \sum_{n=1}^{\infty} J_n(2m\pi M) \sin n\frac{\pi}{2}$$

$$- \frac{8V_{dc}}{\pi^2} \frac{1}{2m-1} \sum_{k=1}^{\infty} J_{2k-1}((2m-1)\pi M) \left[\frac{(2k-1) \cos \left(\frac{n\pi}{2} \right)}{(n+(2k-1))(n-(2k-1))} \Big|_{n \neq 2k-1} \right]$$

$$\tag{12.40}$$

The spectrum analysis for the switching function based on the two fundamental level-shifted carriers can be derived with the substitution of equations (12.32), (12.34),

(12.36), and (12.40) back into equation (12.5). Due to the lengthy equation, the final expression after substitution will not be presented.

The phase-shifting of the carriers and sideband harmonic in the PSC-PD-PWM can be determined using the same equations (12.21) and (12.22) that were applied earlier on for PSC-POD-PWM methods.

The final expression for mL output phase voltage spectrum analysis based on the PSC-PD-PWM scheme is given in equation (12.41).

$$
\begin{aligned}
Vxn_{PSCPDPWM-xL}(t) = {} & \frac{MV_{dc}}{2}\cos(\omega_o t + \theta_0) \\
& + \frac{4V_{dc}}{\pi^2}\sum_{m=1}^{\infty}\frac{1}{N(2m-1)}\sum_{k=1}^{\infty}\frac{J_{2k-1}(N(2m-1)\pi M)}{2k-1}\times\cos(N(2m-1)\omega_c t + \theta_c) \\
& + \left(\begin{aligned} & \frac{V_{dc}}{\pi}\sum_{m=1}^{\infty}\frac{1}{N(2m)}\sum_{\substack{n=-\infty \\ (n\neq 0)}}^{\infty}J_{2n-1}(N(2m)\pi M)\sin\left((2n-1)\frac{\pi}{2}\right) \\ & \times\cos(N(2m)[\omega_c t + \theta_c] + (2n-1)[\omega_o t + \theta_0]) \end{aligned}\right) \\
& + \left(\begin{aligned} & \sum_{\substack{n=-\infty \\ (n\neq 0)}}^{\infty}\sum_{k=1}^{\infty}J_{2k-1}(N(2m-1)\pi M)\left[\begin{aligned} & \frac{4V_{dc}}{\pi^2}\sum_{m=1}^{\infty}\frac{1}{N(2m-1)} \\ & \frac{(2k-1)\cos\left(\frac{(2n)\pi}{2}\right)}{((2n)+(2k-1))((2n)-(2k-1))}\bigg|_{n\neq 2k-1} \end{aligned}\right] \\ & \times\cos(N(2m-1)[\omega_c t + \theta_c] + (2n)[\omega_o t + \theta_0]) \end{aligned}\right)
\end{aligned}
\tag{12.41}
$$

Here, N = the total number of triangular carriers used in PSC-PD-PWM.

Equation (12.41) describes that the output phase voltage based on PSC-PD-PWM carries only the fundamental baseband harmonic, carrier harmonic, and sideband harmonic. Equation (12.41) shows that the carrier harmonic will occur only at odd multiples of harmonic order[th] of equation (12.22). On top of that, even sideband harmonic occurs around the odd multiples of harmonic order[th], and odd sideband harmonic occurs around the even multiples of the harmonic order[th]. The characteristics of output phase voltage for seven-level PSC-PD-PWM modulation in Fig. 12.21 is described similarly by equation (12.10).

The analytical results are obtained based on the numerical values calculated from equation (12.41) for the seven-level PSC-PD-PWM modulation with the same conditions of $Mx_{sine} = 0.9$ and $f_c = 1$kHz. The analytical results shown in Fig. 12.22(a) based on the theoretical description earlier on are matched with the simulation results shown in Fig. 12.21(a). On top of that, the experimental results have also verified the exact characteristic of carrier and sideband harmonics for PSC-PD-PWM derived from the analytical equations.

The output line-to-line voltage Vab spectrum analysis for PSC-PD-PWM is expressed in equation (12.42).

$$
Vab_{PSCPDPWM-xL}(t) = \frac{\sqrt{3}MV_{dc}}{2}\cos\left(\omega_o t + \frac{\pi}{6}\right)
$$

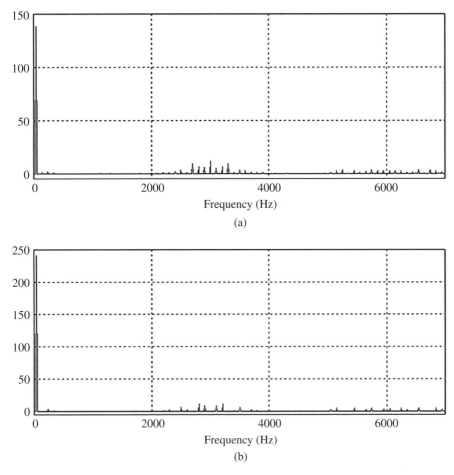

Figure 12.21 Voltage spectrum based on seven-level modulation PSC-PD-PWM: (a) output phase voltage; (b) output line-to-line voltage.

$$
\begin{aligned}
&+\left(\begin{array}{l}
\dfrac{2V_{dc}}{\pi}\displaystyle\sum_{m=1}^{\infty}\dfrac{1}{N(2m)}\\[2mm]
\displaystyle\sum_{\substack{n=-\infty\\(n\neq 0)}}^{\infty} J_{2n-1}(N(2m)\pi M)\sin\left((2n-1)\dfrac{\pi}{2}\right)\sin\left((2n-1)\dfrac{\pi}{3}\right)\\[2mm]
\times\cos\left(N(2m)[\omega_c t]+(2n-1)\left[\omega_o t-\dfrac{\pi}{3}\right]+\dfrac{\pi}{2}\right)
\end{array}\right)\\[4mm]
&+\left(\begin{array}{l}
\dfrac{8V_{dc}}{\pi^2}\displaystyle\sum_{m=1}^{\infty}\dfrac{1}{N(2m-1)}\\[2mm]
\displaystyle\sum_{\substack{n=-\infty\\(n\neq 0)}}^{\infty}\sum_{k=1}^{\infty} J_{2k-1}(N(2m-1)\pi M)\left[\left.\dfrac{(2k-1)\cos\left(\frac{(2n)\pi}{2}\right)}{((2n)+(2k-1))((2n)-(2k-1))}\right|_{n\neq 2k-1}\right]\sin\left((2n)\dfrac{\pi}{3}\right)\\[2mm]
\times\cos\left(N(2m-1)[\omega_c t]+(2n)\left[\omega_o t-\dfrac{\pi}{3}\right]+\dfrac{\pi}{2}\right)
\end{array}\right)
\end{aligned}
$$

$$(12.42)$$

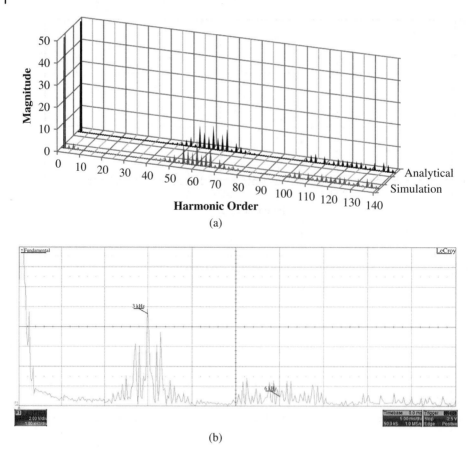

(a)

(b)

Figure 12.22 Output phase voltage spectrum – seven-level PSC-PD-PWM modulation: (a) simulation and analytical results; (b) experimental results.

The crossover point between the two-phase reference sinusoidal signals at multiples of 60° will also cancel away the carrier harmonic and triplen sideband harmonic by its nature. As a result, the terms $\sin((2n-1)\ \pi/3)$ and $\sin((2n)\ \pi/3)$ are added in equation (12.42) for the sideband harmonic components. Thus, the cancellation of both carrier and sideband harmonics can be clearly observed from the output phase voltage in Fig. 12.21(a) and output line-to-line voltage in Fig. 12.21(b).

The analytical results for the seven-level PSC-PD-PWM modulation based on the output line-to-line voltage are calculated with equation (12.42). Once again, the analytical results are compared with the simulation results of Fig. 12.21(b), as shown in Fig. 12.23(a). Finally, the experimental results in Fig. 12.23(b) have validated both the analytical and simulation results. Thus, equation (12.42) has exactly described the characteristics of PSC-PD-PWM for any mL output line-to-line voltage.

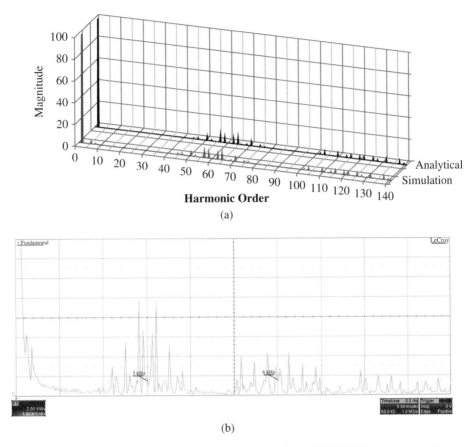

Figure 12.23 Output line-to-line voltage spectrum – seven-level PSC-PD-PWM modulation: (a) simulation and analytical results; (b) experimental results.

12.8   Experimental Verification

The studies of both classical six-cell 7L-MFCI using PSC-PWM and the proposed four-cell 7L-RFCI under PSC-PD-PWM are co-simulated using PSIM v9.0.4 Simcouple and MATLAB R2010b Simulink®. The hardware prototype has been developed with the aid of dSPACE RT1103 Real-Time Workshop (RTW) to verify the results discussed previously. All results presented are based on the detailed settings given in Table 12.5.

The experimental harmonic spectrums of the two proposed hybrid PSC-POD-PWM and PSC-PD-PWM for the output phase voltages (Figs. 12.18 and 12.22) as well as the line-to-line voltages (Figs. 12.19 and 12.23) are closely matched with the mathematical harmonic characteristics analysis. Therefore, one can conclude that reduced overall switching frequency operation is achievable with the proposed modulation method for the modular cell converter.

Table 12.5 Configuration Settings for Simulation and Hardware Prototype of Three-Cell 7L-LFCI.

System Parameters	Settings
Reference sinusoidal frequency (f_o)	50 Hz
Triangular carrier frequency (f_c)	1 kHz
Modulation amplitude (M_a)	0.9
dc voltage source	300 V_{dc}
Flying capacitors (C_{xp}^{th})	2200 μF
Load – resistors / inductors	37.5 Ω/40 mH
Filter – resistors / capacitors	4 Ω/25 μF

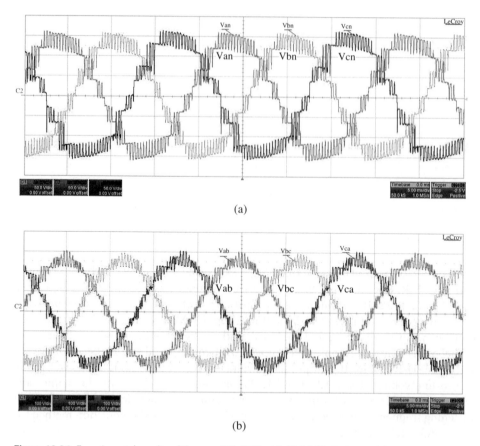

(a)

(b)

Figure 12.24 Experimental results of three-cell 7L-LFCI with PSC-POD-PWM modulation: (a) output phase voltages; (b) output line-to-line voltages.

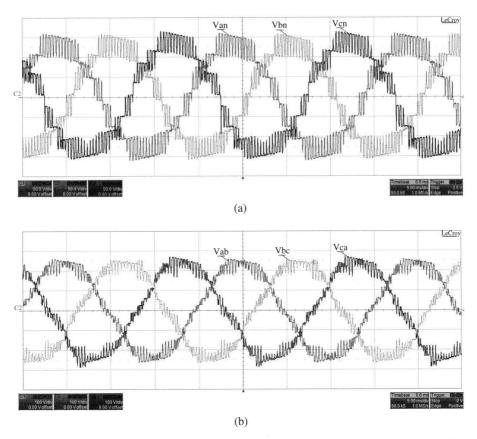

(a)

(b)

Figure 12.25 Experimental results of three-cell 7L-LFCI with PSC-PD-PWM modulation: (a) output phase voltages; (b) output line-to-line voltages.

Besides this, the inherent natural balancing of LFCI using PSC-POD-PWM and PSC-PD-PWM is well maintained with the aid of the RC balance booster filter. The synthesized output seven-level phase voltage stepped waveforms can be clearly observed from Figs. 12.24(a) and 12.25(b) based on the hybrid PSC-POD-PWM and PSC-PD-PWM, respectively, while the experimental output line-to-line voltages are shown in Figs. 12.24(b) and 12.25(b). The experimental results of the three-cell 7L-LFCI matched with the simulation results was presented previously in Fig. 12.15. Hence, it can be clearly observed that there is a slight difference in the magnitude of the synthesized output line-to-line voltages between the PSC-POD-PWM and PSC-PD-PWM.

The agreement between simulation and experimental results has validated the feasibility of the proposed LFCI using the two advanced hybrid sinusoidal PWM modulation schemes. In addition to this, the aforementioned issues such as the high switching frequency operation of PSC-PWM and the increased flying capacitor energy storage that have restricted classical MFCI for high/medium voltage drive applications are successfully overcome. Because of these advantageous features, the proposed LFCI can now be more attractively suitable for higher-voltage applications where space is a major concern.

References

1 K. K. Gupta and S. Jain, "A novel multilevel inverter based on switched DC sources," *Industrial Electronics, IEEE Transactions on*, vol. 61, pp. 3269–3278, 2014.

2 V. Dargahi, A. K. Sadigh, M. Abarzadeh, M. R. A. Pahlavani, and A. Shoulaie, "Flying capacitors reduction in an improved double flying capacitor multicell converter controlled by a modified modulation method," *Power Electronics, IEEE Transactions on*, vol. 27, pp. 3875–3887, 2012.

3 D. A. Ruiz-Caballero, R. M. Ramos-Astudillo, S. A. Mussa, and M. L. Heldwein, "Symmetrical hybrid multilevel DC–AC converters with reduced number of insulated dc supplies," *Industrial Electronics, IEEE Transactions on*, vol. 57, pp. 2307–2314, 2010.

4 Y. Ounejjar, K. Al-Haddad, and L. A. Gregoire, "Packed U cells multilevel converter topology: Theoretical study and experimental validation," *Industrial Electronics, IEEE Transactions on*, vol. 58, pp. 1294–1306, 2011.

5 S. H. Hosseini, A. Khoshkbar Sadigh, and M. Sabahi, "New configuration of stacked multicell converter with reduced number of dc voltage sources," in Power Electronics, Machines and Drives (PEMD 2010), 5th IET International Conference on, pp. 1–6, 2010.

6 S. Daher, J. Schmid, and F. L. M. Antunes, "Multilevel inverter topologies for stand-alone PV systems," *Industrial Electronics, IEEE Transactions on*, vol. 55, pp. 2703–2712, 2008.

7 T. A. Meynard and H. Foch, "Multi-level conversion: high voltage choppers and voltage-source inverters," in Power Electronics Specialists Conference, 1992. PESC '92 Record., 23rd Annual IEEE, vol. 1, pp. 397–403, 1992.

8 S. Kouro, M. Malinowski, K. Gopakumar, J. Pou, L. G. Franquelo, W. Bin *et al.*, "Recent advances and industrial applications of multilevel converters," *Industrial Electronics, IEEE Transactions on*, vol. 57, pp. 2553–2580, 2010.

9 H. Abu-Rub, J. Holtz, J. Rodriguez, and B. Ge, "Medium-voltage multilevel converters – State of the art, challenges, and requirements in industrial applications," *Industrial Electronics, IEEE Transactions on*, vol. 57, pp. 2581–2596, 2010.

13

Three-Phase Seven-Level Four-Cell Reduced Flying Capacitor Inverter

Ziyou Lim

13.1 Introduction

The proposed reduced flying capacitor inverter (RFCI) is designed to synthesize the output voltage levels doubly according to the number of series-connected modular cells (n_{cells}) as compared to the classical multilevel flying capacitor inverter (MFCI). Therefore, the number of output voltage levels for the RFCI can be determined with the following equation (13.1).

$$(mL)RFCI = (2 \times n_{cells}) - 1 \tag{13.1}$$

It is clearly observed that the proposed approach (two-cell 3L-RFCI) also synthesizes the three-level output phase voltage using two modular cells, which is exactly the same as the classical two-cell 3L-MFCI. However, the differences between the classical MFCI and the proposed RFCI would be obvious when producing higher mL output stepped voltage waveforms with every incremental modular cell implemented.

The key objectives of designing the proposed RFCI is to overcome the aforementioned issues faced in the classical MFCI, such as (a) high switching frequency operation, (b) large inrush dc-link current, and (c) increased flying capacitor energy storage. On top of that, the RFCI also greatly enhanced the performances of LFCI as presented previously. The performance of the proposed RFCI using both hybrid PSC-POD-PWM and PSC-PD-PWM schemes is evaluated under dynamic conditions to verify its feasibility.

13.2 Circuit Configuration

The general circuit configuration of transformerless three-phase modular cell inverter with RFCI is shown in Fig. 13.1. Each modular cell consists a pair of semiconductor switches S_x, including its complementary S_x' and a capacitor p clamped in between the cells. Hence, the number of switches ($i_{switches}$) and clamped capacitors ($p_{capacitors}$) can be determined based on the number of modular cells (n_{cells}) used. The first pair of power switches S_{x1} and S_{x1}' in Cell 1 nearest to the dc-link is operating at a low frequency of 50 Hz based on the modulating frequency (f_o), while the rest of the switches in Cell 2 to Cell n are operating at a constant frequency, depending on the carrier frequency (f_c).

Advanced Multilevel Converters and Applications in Grid Integration, First Edition.
Edited by Ali I. Maswood and Hossein Dehghani Tafti.
© 2019 John Wiley & Sons Ltd. Published 2019 by John Wiley & Sons Ltd.

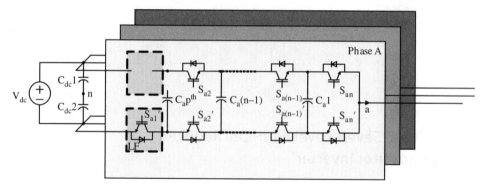

Figure 13.1 Proposed transformerless three-phase *n*-cell reduced modular cell inverter topology.

Instead of having two LF switches served as an H-bridge leg, such as those in Fig. 5.1, the proposed RFCI clamps the series-connected modular cells (Cell 2 to Cell n) onto a low operational frequency cell (Cell 1). High modularity and flexibility are retained in the proposed RFCI converter. Thus, the proposed RFCI is allowed to synthesize higher *mL* output voltages easily by extending the n_{cells}.

Even though similarity is found between the proposed RFCI and the classical MFCI, the number of components, cost and size of the converter are significantly reduced with the proposed approach of clamping the modular cells onto a low frequency operated cell. The output voltage levels doubly increase with every incremental modular cell without the aid of any phase-shifted transformer. Moreover, lower and better output voltage harmonic distortion is also achieved.

13.3 Operation Principles

Clamping three modular cells onto the low operational frequency Cell 1 results in a single-phase seven-level (7L) transformerless RFCI, as presented in Fig. 13.2(a). Each phase-leg requires a minimum of eight $i_{switches}$ (S_{x1} to S_{x4} and S_{x1}' to S_{x4}') and three $p_{capacitors}$ (C_x1 to C_x3) clamped in between the modular cells, depending on the device voltage ratings considered during the design of the converter. It can be clearly observed that the total component counts are greatly reduced as compared to the classical six-cell 7L-MFCI (12 $i_{switches}$ and five $p_{capacitors}$ per phase-leg as shown in Fig. 13.2(b)). The difference would be even more obvious when the converter is designed for three-phase systems.

Since four pairs of active switches are used in the proposed four-cell 7L-RFCI, there is a combination of $2^4 = 16$ valid operating modes from '0000' to '1111', as summarized in Table 13.1. Two redundant states are produced for zero output level, and three redundant states for each $\pm V_{dc}/3$ and $\pm V_{dc}/6$ output level. It should be noted that all positive output levels are achieved when S_{x1} conducts, while all negative output levels are obtained when S_{x1}' conducts. Therefore, the two switches S_{x1} and S_{x1}' in Cell 1 are always performing at a fundamental frequency of 50 Hz. Hence, the seven-level output phase voltage stepped waveforms are synthesized accordingly at $\pm V_{dc}/2, \pm V_{dc}/3, \pm V_{dc}/6$ and zero level, based on the given corresponding switching states in Table 13.1.

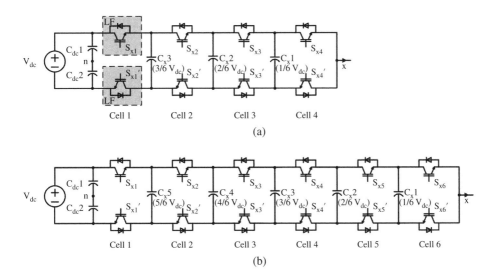

Figure 13.2 (a) Proposed Four-Cell 7L-RFCI. (b) Classical Six-Cell 7L-MFCI.

Table 13.1 Four-Cell 7L-RFCI Voltage Levels and Corresponding Switching States.

States	Switching States				Charging/Discharging States of Capacitors			Phase Leg xn Voltage
	S_{x1}	S_{x2}	S_{x3}	S_{x4}	C_x1	C_x2	C_x3	V_{xn}
1	1	1	1	1				$+V_{dc}/2$
2	1	1	1	0	+			$+V_{dc}/3$
3	1	1	0	1	−	+		
4	1	1	0	0		+		$+V_{dc}/6$
5	1	0	1	1	−	+		$+V_{dc}/3$
6	1	0	1	0	+	−	+	$+V_{dc}/6$
7	1	0	0	1	−		+	
8	1	0	0	0			+	0
9	0	1	1	1			−	
10	0	1	1	0	+		−	$-V_{dc}/6$
11	0	1	0	1	−	+	−	
12	0	1	0	0		+	−	$-V_{dc}/3$
13	0	0	1	1	−			$-V_{dc}/6$
14	0	0	1	0	+	−		$-V_{dc}/3$
15	0	0	0	1	−			
16	0	0	0	0				$-V_{dc}/2$

Here, 'x' represents phase a, b and c; '1' and '0' represent the 'on' and 'off' states, respectively; and '+' and '−' represent the 'charging' and 'discharging' states of capacitors, respectively, while '■' represents neither states.

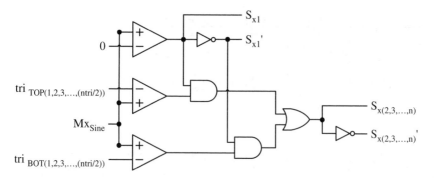

Figure 13.3 Proposed modulation technique for *n*-cell *mL*-RFCI.

The operational principles of the proposed four-cell 7L-RFCI are achieved using two proposed advanced hybrid sinusoidal PWM schemes, named *phase-shifted carrier phase-opposition-disposition PWM* (PSC-POD-PWM) and *phase-shifted carrier phase-disposition PWM* (PSC-PD-PWM), as shown in Fig. 12.4. Generally, for all carrier-based PWM (*mL* − 1), number of triangular carriers (n_{tri}) would be required for any *mL* modulation scheme.

Similarly, for four-cell 7L-RFCI, six n_{tri} are used and divided into two sets equally, where one set level-shifted above zero reference magnitude ($\text{tri}_{TOP(1,2,3)}$) and the other set below the zero reference magnitude ($\text{tri}_{BOT(1,2,3)}$). The top and bottom carriers are 120° phase-shifted evenly, based on equation (12.3).

According to the switching states given in Table 13.1, S_{x1} and S_{x1}' are always low frequency operated switches, with the frequency matching to the reference signal (Mx_{sine}). Thus, the final gating signals for S_{x1} and S_{x1}' are achieved directly by comparing Mx_{sine} with respect to zero magnitude, as shown in Fig. 13.3. For the remaining final gating signals, the six n_{tri} are compared with the Mx_{sine}, respectively, with the aid of simple comparators.

Thereafter, the output comparator signals of the proposed advanced hybrid sinusoidal PWM schemes are further multiplied with S_{x1} and S_{x1}' through the AND logic gates to extract the desired modulated pulses from the respective positive and negative half-periods of Mx_{sine}. Finally, the modulated pulses after the AND gates are then summed together using the OR logic gates to achieve the desired gating signals for S_{x2} to S_{xn}. Hence, the gating signals for the complementary switches (S_x') are obtained simply through the NOT gates. The final gating signals achieved for four-cell 7L-RFCI can be observed from Fig. 13.4 over two fundamental cycles period.

13.4 Design Considerations

13.4.1 Voltage Characteristics Expressions

For the *n*-cell *mL*-RFCI, the voltage across the respective capacitor clamped at p^{th} position ($C_{xp}{}^{th}$) in Fig. 13.1 can be determined based on the following equation (13.2).

$$V_{Cx(p^{th})}^{RFCI} = \frac{p^{th}}{mL - 1} \times V_{dc} \tag{13.2}$$

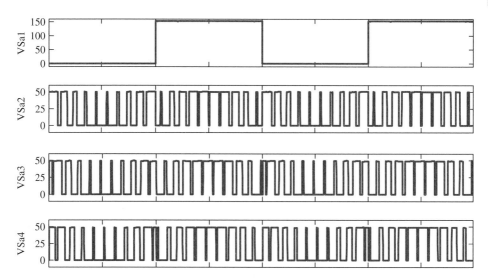

Figure 13.4 Voltage stress across Sa1 (top) to Sa4 (bottom) for four-cell 7L-RFCI based on hybrid PSC-PD-PWM, with Mx = 0.9 and V_{dc} = 300 V.

According to equation (13.2), in the case of four-cell 7L-RFCI, the flying capacitor voltages $V_{Cx}1$, $V_{Cx}2$, and $V_{Cx}3$ are maintained as $V_{dc}/6$, $V_{dc}/3$, and $V_{dc}/2$, respectively, during the steady-state operation. Even though equation (13.2) is expressed similarly for determining the flying capacitor voltages of the classical MFCI, there are two important factors to be taken note of in Table 13.2. It can be clearly observed in the reduced number of flying capacitors and the voltage ratings. The proposed four-cell 7L-RFCI does not only reduced the total number of flying capacitors but also reduced the flying capacitor voltage ratings as compared to the classical four-cell 5L-MFCI.

The voltage stress across the switches of the proposed transformerless RFCI is highly affected by the flying capacitor voltages and is determined using equation (13.3). In the case of the proposed four-cell 7L-RFCI with 300 V_{dc} supplied, the voltage stress across

Table 13.2 Capacitor Voltages for Classical MFCI and Proposed RFCI.

Capacitors Clamped at p^{th} Position	Phase x Capacitor Voltages $V_{Cx,p}{}^{th}$			
	MFCI		RFCI	
	Four-cell (Five-level)	Six-cell (Seven-level)	Three-cell (Five-level)	Four-cell (Seven-level)
1	$V_{dc}/4$	$V_{dc}/6$	$V_{dc}/4$	$V_{dc}/6$
2	$2 V_{dc}/4$	$2 V_{dc}/6$	$2 V_{dc}/4$	$2 V_{dc}/6$
3	$3 V_{dc}/4$	$3 V_{dc}/6$	~	$3 V_{dc}/6$
4	~	$4 V_{dc}/6$	~	~
5	~	$5 V_{dc}/6$	~	~

Here, "x" represents phase a, b and c; and "p^{th}" represents the placing position number of flying capacitors being clamped.

the switches are shown in Fig. 5.6. It is observed that the voltage stress V_{Sx1} is $V_{dc}/2$, while the rest (V_{Sx2}, V_{Sx3}, and V_{Sx4}) are equally distributed and maintained at $V_{dc}/6$. According to equations (13.2) and (13.3), the blocking voltage capability of the two low operational frequency switches S_{x1} and S_{x1}' in Cell 1 must be able to withstand $V_{dc}/2$ for n-cell mL-RFCI at all times. The same voltage stress calculated from equation (13.3) applies to the respective complementary switches.

$$V_{Sx(cell\ n^{th})}^{RFCI}(t) \approx \begin{cases} -\left(\dfrac{(mL-1)-p^{th}}{mL-1}\right) V_{dc} \times S_{x\ (cell\ n^{th})}(t) & \text{if } cell\ n^{th} = 1 \\ -\left(\dfrac{p^{th}-(p^{th}-1)}{mL-1}\right) V_{dc} \times S_{x\ (cell\ n^{th})}(t) & \text{if } cell\ n^{th} \geq 2 \end{cases}$$

$$(13.3)$$

The same output phase voltage (12.10) and output line-to-line voltage expressions (12.11) are applied for RFCI, respectively, based on the corresponding switching states in Table 13.1 and the flying capacitor voltages obtained from equation (13.2). The output line-to-line voltages of four-cell 7L-RFCI under the two proposed hybrid sinusoidal PWM schemes are shown in Fig. 13.5. One can observe that the overall switching frequency is greatly reduced for the proposed four-cell 7L-RFCI employing hybrid PWM schemes when compared to the classical six-cell 7L-MFCI using PSC-PWM, as shown previously in Fig. 12.14.

13.4.2 Design of Flying Capacitors

The critical flying capacitance value for the proposed n-cell mL-RFCI is estimated using equation (13.4):

$$C_{x,p^{th}\langle critical \rangle} \approx \frac{M x_{sine}\ V_{dc}}{2 \times Z_L(j\omega) \times \Delta V_{Cx,p^{th}} \times f_{Os}} \approx \frac{\hat{I}_x \cos(\varphi)}{\Delta V_{Cx,p^{th}} \times f_{Os}} \tag{13.4}$$

where x ϵ phase {a,b,c}

p^{th} = position of the flying capacitors
$M x_{sine}$ = modulation index of phase {a,b,c}
$Z_L(j\omega)$ = complex load impedance
$\Delta V_{Cx,p}^{th}$ = permissible maximum peak-to-peak flying capacitor ripple voltage
f_{Os} = overall switching frequency (Hz)
$\hat{I}_x \cos(\varphi)$ = output peak phase {a,b,c} current with power factor angle

The flying capacitance is inversely proportional to the overall switching frequency ($f_{Os} = n_{tri} \times f_c / 2$), as well as the maximum permissible flying capacitor ripple voltage ($\Delta V_{Cx,p}^{th}$). Due to the asynchronized gating signals, there are periods when the flying capacitors are connected in series with one another. This can be observed from the charging and discharging states of the flying capacitors corresponding to the particular switching state condition, as shown in Table 13.3. This means that the amount of charging and discharging current (the effect of ripple current) for the respective capacitor depends on the series connection of the flying capacitors at that particular switching state condition of the converter.

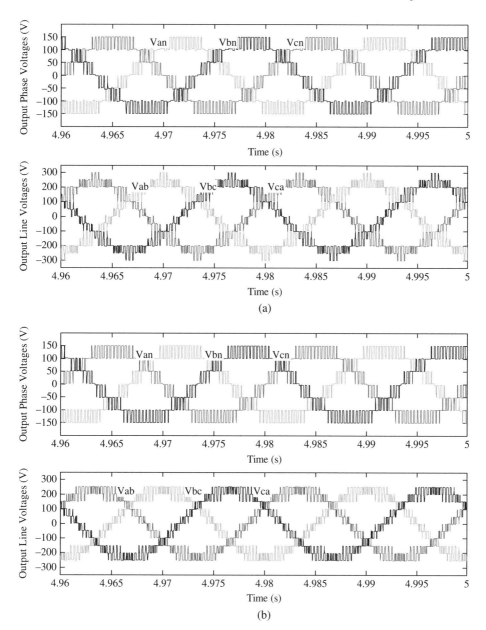

Figure 13.5 Output phase voltages (top) and output line-to-line voltages (bottom) of the proposed four-cell 7L-RFCI using (a) PSC-POD-PWM and (b) PSC-PD-PWM, with Mx = 0.9 and V_{dc} = 300 V.

Thus, both the flying capacitor ripple current and ripple voltage have the same ripple frequency matching to the f_{Os}. It should be noted that the critical flying capacitance value is affected by the desired permissible maximum peak-to-peak flying capacitor ripple voltages based on the f_{Os} operation. Besides that, the critical flying capacitance value is also determined based on the peak magnitude of the output phase current ($\hat{I}_x$), which

Table 13.3 Charging and Discharging States of Flying Capacitors and Corresponding Switching States.

Switching States				Charging/Discharging States			Series Connection of Flying Capacitors
				$I_{Cx}1$	$I_{Cx}2$	$I_{Cx}3$	
0	0	0	0				
0	0	0	1	−			
0	0	1	0	+	−		C_x1, C_x2
0	0	1	1	−			
0	1	0	0	+	−		C_x2, C_x3
0	1	0	1	−	+	−	C_x1, C_x2, C_x3
0	1	1	0	+	−		C_x1, C_x3
0	1	1	1		−		
1	0	0	0			+	
1	0	0	1	−		+	C_x1, C_x3
1	0	1	0	+	−	+	C_x1, C_x2, C_x3
1	0	1	1		−	+	C_x2, C_x3
1	1	0	0		+		
1	1	0	1	−	+		C_x1, C_x2
1	1	1	0	+			
1	1	1	1				

is affected by the power factor angle (φ) due to the output load impedance ($Z_L(j\omega)$). Therefore, it is important to select the flying capacitors appropriately by ensuring that both maximum ripple voltage and peak current are lesser than the rated voltage and current ratings given by the manufacturer datasheet.

13.4.3 Experimental Verification

The studies of both classical six-cell 7L-MFCI using PSC-PWM and the proposed four-cell 7L-RFCI under PSC-PD-PWM are co-simulated between PSIM v9.0.4 Sim-couple and MATLAB R2010b Simulink®. The hardware prototype has been developed to verify the results discussed previously with the aid of dSPACE RT1103 Real-Time Workshop (RTW). All results presented are based on the detailed settings given in Table 13.4.

The experimental results as shown in Figs. 13.6 and 13.7 verified the feasibility of the proposed four-cell 7L-RFCI based on the hybrid PSC-POD-PWM and PSC-PD-PWM, respectively. The output seven-level phase voltage Van (Figs. 13.6(a) and 13.7(a)) and output line-to-line voltage Vab (Figs. 13.6(b) and 13.7(b)) are synthesized correspondingly with their frequency spectrums. The experimental harmonic spectrums of the synthesized output voltages have validated the analytical results shown in Figs. 12.18, 12.19, 12.22 and 12.23. Thus, one can conclude that reduced overall switching frequency by 50% is also achievable for the proposed RFCI with the two hybrid modulation schemes.

Table 13.4 Configuration Settings for Simulation and Hardware Prototype of Four-Cell 7L-RFCI.

System Parameters	Settings
Reference sinusoidal frequency (f_o)	50 Hz
Triangular carrier frequency (f_c)	1 kHz
Modulation amplitude (Ma)	0.9
dc voltage source	300 Vdc
Flying capacitors ($\mathbf{C_{xp}}^{th}$)	2200 µF
Load – resistors / inductors	37.5 Ω / 40 mH
Filter – resistors / capacitors	4 Ω / 25 µF

Sinusoidal output currents are also achieved with low harmonic distortion as shown in Figs. 13.6(c) and 13.7(c) under the hybrid PSC-POD-PWM and PSC-PD-PWM schemes, respectively.

Moreover, the proposed four-cell 7L-RFCI under the hybrid PSC-POD-PWM and PSC-PD-PWM is also evaluated with a low modulation index of 0.3 Ma (under-modulation range) in Figs. 13.8 and 13.9. Despite the under-modulation condition, the flying capacitor voltages in Figs. 13.8(a) and 13.9(a) remain balanced at their desired voltage levels while the output line-to-line stepped voltage waveform is synthesized with lower voltage levels as compared to the result in Figs. 13.6(b) and 13.7(b). In addition to that, the voltage stress across the switches Sa1, Sa2, Sa3, and Sa4 are kept constant due to the balanced flying capacitor voltages as observed in Figs. 13.8(b) and 13.9(b). The results in Figs. 13.8 and 13.9 demonstrate that the inherent natural balancing property is well retained for the proposed RFCI with the aid of RC balance booster filter. Thus, it is proven that the second proposed RFCI approach has better converter stability even under low modulation condition as compared to the first proposed LFCI.

The agreement between the simulation and experimental results have confirmed the practicality of the proposed RFCI and also addressed the issues (a) to (c) faced in the classical MFCI. Therefore, the proposed RFCI synthesizes higher output voltage stepped waveforms with reduced component counts and costs. Besides that, the proposed RFCI has evidently enhanced the performances of the first proposed LFCI and improved the converter reliability as well. Hence, the RFCI is believed to be more attractively applicable for higher medium-voltage drives as compared to the classical MFCI and first proposed LFCI, as well as for applications with space constraint.

13.5 Flying Capacitor Voltage Balancing Control

Since the pole voltage waveforms of the proposed RFCI are synthesized depending on the flying capacitor voltages, it is important to ensure that the voltages across the respective flying capacitors be tracked to their desired reference values based on the given equation (13.2) at all times. In the event of unbalanced flying capacitor voltages, there will be an increase in the voltage stress on the power semiconductors, which will affect the stability of the entire system. Even though this issue can be resolved easily by oversizing the component ratings, this is not a cost-effective solution.

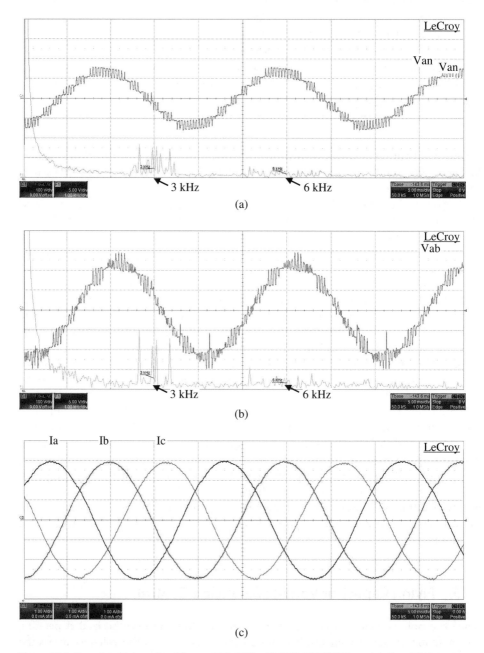

Figure 13.6 Experimental results of four-cell 7L-RFCI with PSC-POD-PWM modulation. (a) Output phase voltages (Van). (b) Output line-to-line voltages (Vab). (c) Output phase currents (Ia, Ib and Ic).

Such undesirable conditions can be prevented with the aid of balance booster filters, owing to the well-retained inherent natural balancing of flying capacitor voltages in the proposed RFCI, as presented previously. However, the converter may take longer to respond to and stabilize any dynamic conditions. Alternatively, a closed-loop flying

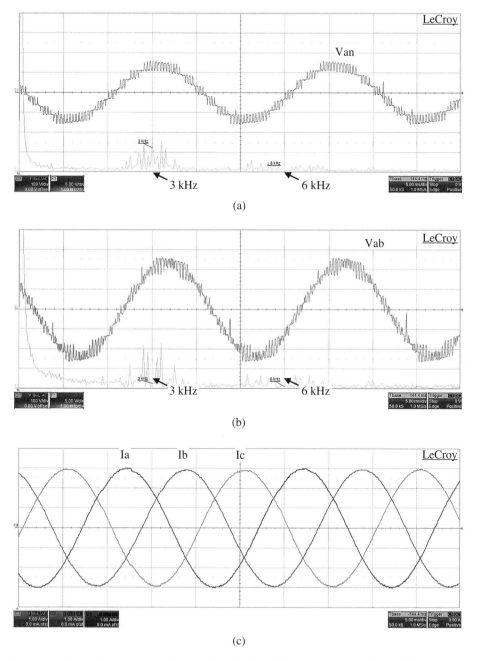

Figure 13.7 Experimental results of four-cell 7L-RFCI with PSC-PD-PWM modulation. (a) Output phase voltages (Van). (b) Output line-to-line voltages (Vab). (c) Output phase currents (Ia, Ib and Ic).

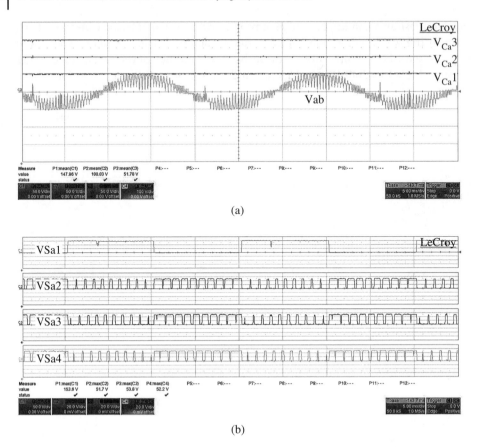

Figure 13.8 Experimental results of four-cell 7L-RFCI under 0.3 Mx$_{sine}$ with PSC-POD-PWM. (a) Output line-to-line voltage (Vab) and flying capacitor voltages ($V_{Ca}1$, $V_{Ca}2$, $V_{Ca}3$). (b) Voltage stress across Sa1 (V_{Sa1}), Sa2 (V_{Sa2}), Sa3 (V_{Sa3}), and Sa4 (V_{Sa4}).

capacitor voltage balancing control is designed for the proposed n-cell mL-RFCI (shown in Fig. 13.10) using the hybrid PSC-PD-PWM scheme.

The voltage across the flying capacitors depends on the current through the respective capacitors as expressed in equation (13.5) and is affected by the switching states of the two adjacent switches, and therefore the flying capacitor voltages can be balanced by varying the duty ratios of the switching pulses.

$$C_{x,p^{th}} \times \frac{dV_{Cx,p^{th}}}{dt} = I_{Cx,p^{th}}(t) = (S_{x(n-p^{th}+1)} - S_{x(n-p^{th})})I_x \tag{13.5}$$

The flying capacitor voltage dynamic in equation (13.6) is expressed in a local average form over a switching period using duty ratio instead of switching states.

$$\frac{\Delta \overline{V}_{Cx,p^{th}}}{\Delta t} = \frac{(\Delta d_{x(n-p^{th}+1)} - \Delta d_{x(n-p^{th})})\overline{I}_x}{C_{x,p^{th}}} \tag{13.6}$$

Thus, the flying capacitor voltages can be regulated by varying the duty ratios of the switching pulses based on the voltage errors between the reference voltages ($\varepsilon_{x(n)}$) in

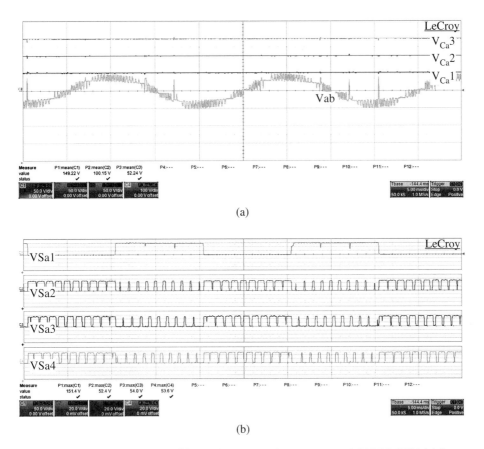

Figure 13.9 Experimental results of four-cell 7L-RFCI under 0.3 Mx$_{sine}$ with PSC-PD-PWM. (a) Output line-to-line voltage (Vab) and flying capacitor voltages (V$_{Ca}$1, V$_{Ca}$2, V$_{Ca}$3). (b) Voltage stress across Sa1 (V$_{Sa1}$), Sa2 (V$_{Sa2}$), Sa3 (V$_{Sa3}$), and Sa4 (V$_{Sa4}$).

Table 13.2 and the measured values. These voltage errors are then minimized using a proportional (P) controller that allows simple implementation as well as easy tuning of the gain to achieve the desired performance.

$$\Delta d_{x(n)} = \text{sign}(I_x)(\varepsilon_{x(n-1)} - \varepsilon_{x(n)})P \tag{13.7}$$

Besides this, the duty cycle ratios are also varied, depending on the directional flow of the phase current, which is defined as 1 and −1 for the respective positive and negative cycles using the sign term. The varied duty ratios will then be added to the reference sinusoidal signal (M$_{Xsine}$) to achieve the modulation sine templates (M$_{Xsine(n)}$), which are required for the proposed hybrid PSC-PD-PWM scheme, as shown in Fig. 12.4(b).

For the case of four-cell 7L-RFCI, a total of six triangular carriers will be divided equally into two sets where one set is level-shifted above zero reference magnitude (tri$_{TOP(1,2,3)}$) and the other set remains in-phase and level-shifted below zero reference magnitude (tri$_{BOT(1,2,3)}$). Both the top and bottom carriers are 120° phase-shifted apart evenly according to equation (12.3).

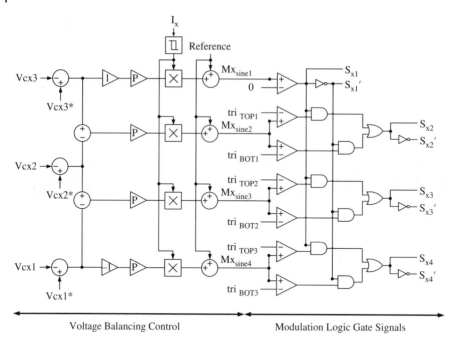

Figure 13.10 Proposed flying capacitor voltage balancing control with modulation logic gate signals.

Based on the switching states given in Table 13.1, S_{x1} and S_{x1}' are low frequency operated switches, with the frequency matching $Mx_{sine(n)}$. Hence, the final gating signals for S_{x1} and S_{x1}' are obtained directly by comparing $Mx_{sine(1)}$ with respect to zero magnitude, as shown in Fig. 13.10. The remaining gating signals are achieved by comparing $Mx_{sine(2,3,4)}$, respectively, with six triangular carriers $tri_{TOP(1,2,3)}$ and $tri_{BOT(1,2,3)}$ using simple comparators. The output comparator signals will be further multiplied with S_{x1} and S_{x1}' through the AND logic gates to extract the desired modulation signals from the respective positive and negative half-cycles of the $Mx_{sine(n)}$. Thereafter, the extracted modulating pulses after the AND gates are summed together using the OR logic gates to achieve the final gating signals for S_{x2} to S_{xn}. Thus, the complementary switches are simply obtained using the NOT gates. The overall voltage balancing control using the hybrid PSC-PD-PWM for the proposed four-cell 7L-RFCI is shown in Fig. 13.10, which can be extended for any *n*-cell *mL*-RFCI configuration as well.

13.6 Experimental Verification

The dynamic performances of the four-cell 7L-RFCI are evaluated with a series of step changes in the modulation index ($Mx_{sine<rectifier>}$) and with the unbalanced load conditions to verify the feasibility of the proposed voltage balancing control. The detailed configuration settings for both simulation and experimental results are presented in Table 13.5.

The proposed 7L-RFCI in Fig. 13.1 is simulated to have the modulation index step changed from 0.9 Mx to 0.3 Mx between t = 2 s to 2.1 s, and 0.9 Mx to 0.6 Mx between

Table 13.5 Configuration Settings for Simulation and Hardware Prototype of Four-Cell 7L-RFCI.

System Parameters	Settings
Reference sinusoidal frequency (f_o)	50 Hz
Triangular carrier frequency (f_c)	1 kHz
Flying capacitors ($C_{xp}{}^{th}$)	2200 μF
dc voltage source	170 Vdc
Load – resistors / inductors	37.5 Ω / 40 mH
Filter – resistors / capacitors	4 Ω / 25 μF

t = 2.3 s to 2.4 s, as shown in Fig. 13.11(a). Meanwhile, an unbalanced load condition occurs during t = 2.15 s to 2.25 s, when the two series-connected resistance-inductors (RLs) of 25 Ω 100 mH and 100 Ω 100 mH are connected, respectively, to Phases B and C. The flying capacitor voltages of 7L-RFCI in Fig. 13.11(b) are observed to be maintained constant at their corresponding levels, even in the events of drastic Mx change and the introduction of unbalanced load conditions.

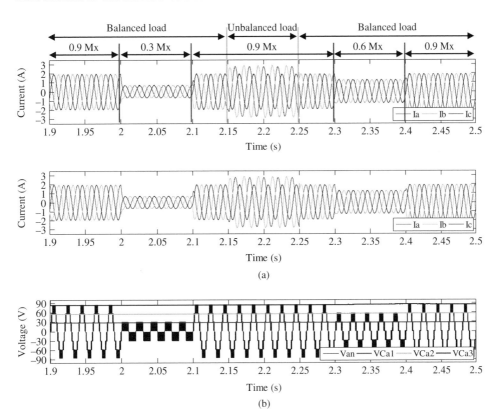

Figure 13.11 Simulated dynamic performance of four-cell 7L-RFCI with Mx$_{sine}$ variation and unbalanced load change. (a) Output currents. (b) Output phase voltage and flying capacitor voltages.

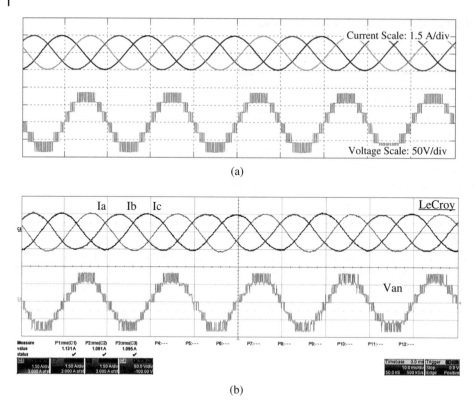

(a)

(b)

Figure 13.12 Results of four-cell 7L-RFCI with 0.9 Mx_{sine} performance based on output currents (Ia, Ib, and Ic) and output phase voltage (Van). (a) Simulation results. (b) Experimental results.

The feasibility of the proposed 7L-RFCI in conjunction with the flying capacitor voltage balancing control is also validated according to the same condition presented in the simulation. The steady-state performances of the inverter operating at 0.9 Mx, 0.6 Mx, and 0.3 Mx are shown in Figs. 13.12–13.17. The seven-level pole voltage (Van) waveform is synthesized with a reduced number of flying capacitors and capacitor voltage ratings, as clearly observed from Fig. 13.8. Besides this, the dynamic performances of

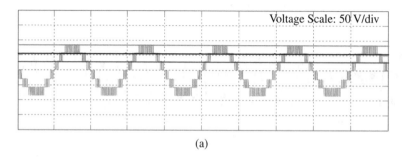

(a)

Figure 13.13 Results of four-cell 7L-RFCI with 0.9 Mx_{sine} performance based on output phase voltage (Van) and flying capacitor voltages ($V_{Ca}1$, $V_{Ca}2$, and $V_{Ca}3$). (a) Simulation results. (b) Experimental results.

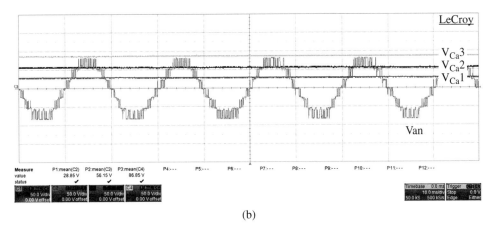

(b)

Figure 13.13 (*Continued*)

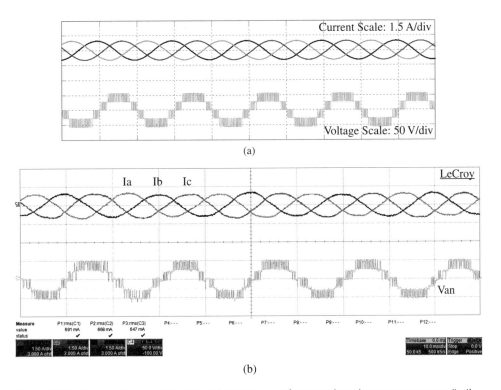

Figure 13.14 Results of four-cell 7L-RFCI with 0.6 Mx$_{sine}$ performance based on output currents (Ia, Ib, and Ic) and output phase voltage (Van). (a) Simulation results. (b) Experimental results.

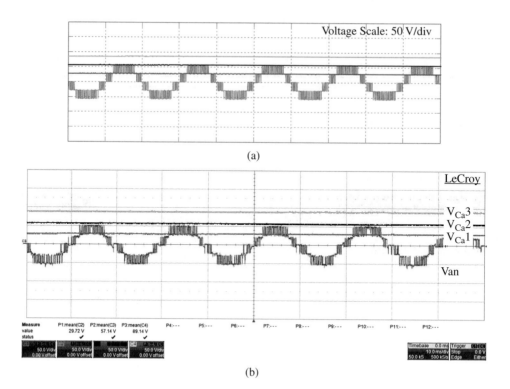

(a)

(b)

Figure 13.15 Results of four-cell 7L-RFCI with 0.6 Mx$_{sine}$ performance based on output phase voltage (Van) and flying capacitor voltages (V$_{Ca}$1, V$_{Ca}$2, and V$_{Ca}$3). (a) Simulation results. (b) Experimental results.

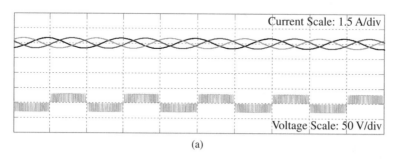

(a)

Figure 13.16 Results of four-cell 7L-RFCI with 0.3 Mx$_{sine}$ performance based on output currents (Ia, Ib, and Ic) and output phase voltage (Van). (a) Simulation results. (b) Experimental results.

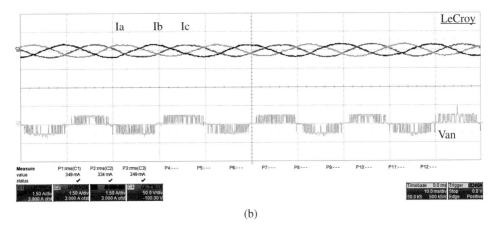

(b)

Figure 13.16 (*Continued*)

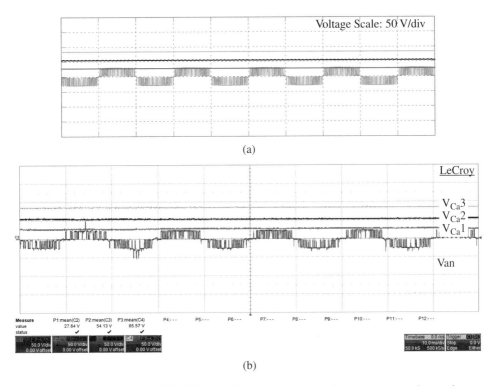

(a)

(b)

Figure 13.17 Results of four-cell 7L-RFCI with 0.3 Mx_{sine} performance based on output phase voltage (Van) and flying capacitor voltages ($V_{Ca}1$, $V_{Ca}2$, and $V_{Ca}3$). (a) Simulation results. (b) Experimental results.

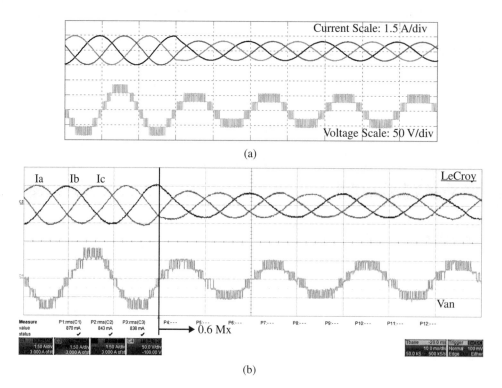

(a)

(b)

Figure 13.18 Results of four-cell 7L-RFCI with modulation index step change from 0.9 to 0.6 Mx$_{sine}$ based on output currents (Ia, Ib, and Ic) and output phase voltage (Van). (a) Simulation results. (b) Experimental results.

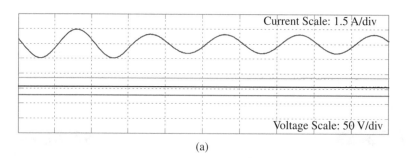

(a)

Figure 13.19 Results of four-cell 7L-RFCI with modulation index step change from 0.9 to 0.6 Mx$_{sine}$ based on output current (Ia) and flying capacitor voltages (V$_{Ca}$1, V$_{Ca}$2, and V$_{Ca}$3). (a) Simulation results. (b) Experimental results.

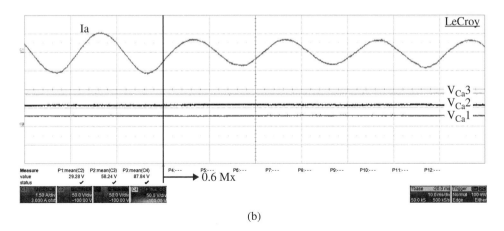

(b)

Figure 13.19 (*Continued*)

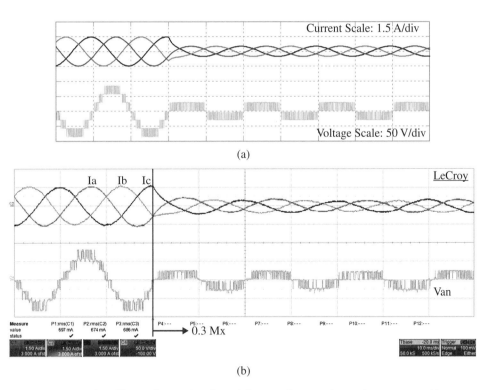

Figure 13.20 Results of four-cell 7L-RFCI with modulation index step change from 0.9 to 0.3 Mx$_{sine}$ based on output currents (Ia, Ib, and Ic) and output phase voltage (Van). (a) Simulation results. (b) Experimental results.

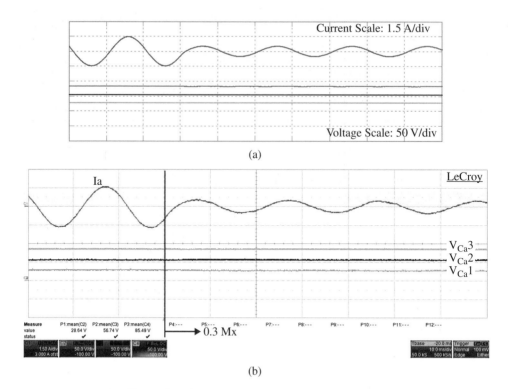

(a)

(b)

Figure 13.21 Results of four-cell 7L-RFCI with modulation index step change from 0.9 to 0.3 Mx_{sine} based on output current (Ia) and flying capacitor voltages ($V_{Ca}1$, $V_{Ca}2$, and $V_{Ca}3$). (a) Simulation results. (b) Experimental results.

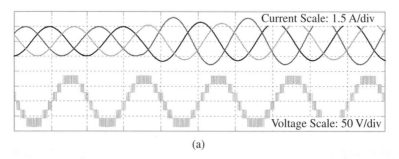

(a)

Figure 13.22 Results of four-cell 7L-RFCI with unbalanced load change based on output currents (Ia, Ib, and Ic) and output phase voltage (Van). (a) Simulation results. (b) Experimental results.

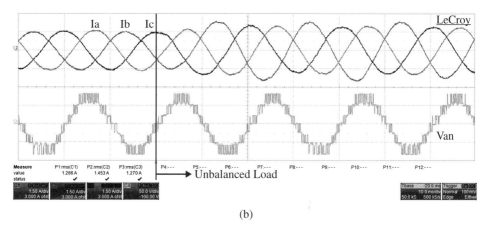

(b)

Figure 13.22 (*Continued*)

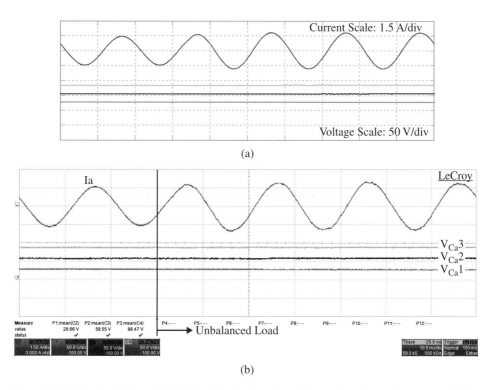

Figure 13.23 Results of four-cell 7L-RFCI with unbalanced load change based on output current (Ia) and flying capacitor voltages ($V_{Ca}1$, $V_{Ca}2$, and $V_{Ca}3$). (a) Simulation results. (b) Experimental results.

the 7L-RFCI are experimented with the sudden step change in the Mx and the unbalanced load conditions, which are presented in Figs. 13.18–13.23. It should be noted that flying capacitor voltages are also experimentally proven to be balanced.

The strong agreements between the simulation results and the experimental results have proven the feasibility of the proposed RFCC with reduced component counts. Good dynamic responses are achieved as well, using closed-loop voltage balancing control.

14

Active Neutral-Point-Clamped Inverter

Ziyou Lim

14.1 Introduction

Multilevel converters have emerged as the most viable solutions and are widely employed in many industries for medium–high-power applications. They are popularly known for their distinguished merits such as having high power-handling capabilities and delivering excellent power quality with the intelligent synthesis of m-level (mL) waveforms.

The commercialized converters available in the markets are the commonly known classical multilevel topologies, which are three-level (3L) neutral-point-clamped inverters (3L-NPC), multilevel flying capacitor inverters (MFCI), and multilevel cascaded H-bridge inverters (MCHB) [1]. Among these, only MCHB is considered for high-voltage drives up to 11 kV, while 3L-NPC and 4L-MFCI are only applicable for the medium-voltage (MV) applications below 6 kV.

In the last decade, an alternative multilevel converter named *five-level active neutral-point-clamped* (5L-ANPC) *inverter* was first proposed by P. Barbosa *et al.* [2]. The mL-ANPC has successfully addressed the drawbacks faced in the aforementioned classical topologies since the converter design is based on the hybrid combination of the NPC and MFCI configurations. Moreover, the costs, size, and weight of the overall mL-ANPC converter are significantly reduced as compared to the classical multilevel topologies [2].

The first 5L-ANPC technology, ACS 2000, has been successfully introduced by ABB Ltd in the industries for voltage drives up to 6.9 kV [3–5]. Over the recent years, several different modulation methods are implemented for the 5L-ANPC such as space vector PWM (SVPWM) [3, 4, 6–9], selective harmonic elimination PWM (SHE-PWM) [5, 10, 11], and carrier-based PWM (CB-PWM) [2, 12–14].

Many previous researches have improved the multilevel converter performances utilizing SV-PWM and SHE-PWM schemes, but CB-PWM is still the one of the most favorable modulation techniques for all multilevel converters in the industries [1]. This may be because of its simplicity and low costs to achieve the desired gating signals. However, CB-PWM is not fully explored and investigated for the ANPC. Currently, level-shifted carrier PWM (LSC-PWM) [2] and phase-shifted carrier PWM (PSC-PWM) [12–14] are the only CB-PWM modulation schemes employed for ANPC.

Advanced Multilevel Converters and Applications in Grid Integration, First Edition.
Edited by Ali I. Maswood and Hossein Dehghani Tafti.
© 2019 John Wiley & Sons Ltd. Published 2019 by John Wiley & Sons Ltd.

Therefore, in this chapter, a new hybrid CB-PWM modulation strategy named *interleaved sawtooth carrier phase-disposition PWM* (ISC-PD-PWM) is proposed for the mL-ANPC. High-resolution waveforms, especially the output line-to-line voltages, are produced employing the proposed ISC-PD-PWM, which makes the ANPC even more attractively suitable for the drive applications. The low-harmonic-profiled ISC-PD-PWM is also proven mathematically against the conventional PSC-PWM used in [12–14].

Nonetheless, it should be noted that the quality of the synthesized output line-to-line voltage waveforms also depends on the operating conditions of the inverter. Since ANPC is configured using both dc-link capacitors and flying capacitors, in the event of experiencing unbalanced capacitor voltages, the output power quality will be greatly affected. Besides that, in practice, ripples are found in the dc-link when the inverters are connected to the non-linear loads or when a front-end rectifier is connected to a weak or unbalanced grid [10, 15]. Thereafter, low-order harmonics are introduced in the output voltages.

Consequently, these mentioned factors deteriorate the output power quality of the ANPC. It is important to address these issues to ensure that good power quality can be delivered at all times. Even though there are various methods reported in the literature, those issues are separately treated using different modulation techniques. Hence, a multiple voltage quantities enhancement (MVQE) control using hybrid ISC-PD-PWM is proposed for mL-ANPC.

Apart from that, the common mode voltage (CMV) reduction technique for the ANPC has not been reported in the literature till date. Since the hybrid configured ANPC is designed for higher MV drive applications, it is also imperative to minimize the common mode voltage, so that premature bearing failure and electromagnetic interference (EMI) can be avoided and reduced. The mentioned CMV issue can be resolved by implementing an extra fourth phase leg in the converter design [16] or by adopting magnetic circuits like common-mode inductors and common-mode transformers [17–19]. As a result, additional cost, losses, and larger installation space are incurred. Alternatively, the CMV can also be reduced or eliminated in an active manner by manipulating the modulation techniques [20–22], but these strategies cannot be directly imposed on the ANPC.

Therefore, a CMV reduction technique is proposed for the mL-ANPC using the manufacturers' most favorable CB-PWM modulation method due to its inherent simplicity and reduced computational efforts when compared to both SHE-PWM and SV-PWM [1]. In this chapter, the proposed CMV reduction technique is applied in conjunction with the alternative new proposed advanced sinusoidal PWM scheme (hybrid ISC-PD-PWM). The performance of the proposed CMV reduction technique using ISC-PD-PWM is also compared with the popularly implemented PSC-PWM schemes discussed in [12–14].

Thus, this chapter focuses on the proposed MVQE control and CMV reduction technique using the advanced hybrid ISC-PD-PWM scheme in Sections 14.4 and 14.5 of this chapter, respectively. The enhanced performances of the ANPC based on the presented proposed principles are verified experimentally with a three-phase seven-level (7L) ANPC hardware prototype.

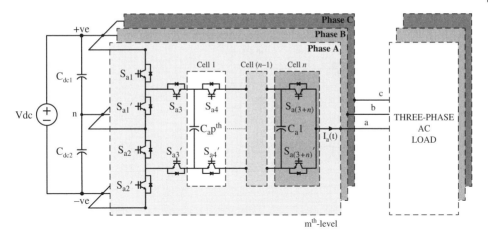

Figure 14.1 Proposed transformerless three-phase n-cell ANPC inverter topology.

14.2 Circuit Configuration

The ANPC topology was fundamentally derived from the three-level (3L) active NPC discussed in [23–25], which is constructed using three pairs of semiconductor switches (S_{x1} to S_{x3}), including their complementary switches (S_{x1}' to S_{x3}'). It was later extended to synthesize higher mL output voltages by clamping a series of modular cells (n_{cells}) onto it. The hybrid configuration of the active NPC and the imbricated cells (Cell 1 to Cell n) results in the design of the mL-ANPC, as shown in Fig. 14.1.

High modularity and flexibility are well retained in the converter, with each modular cell comprising of two switching devices S_x and S_x' along with a capacitor p clamped in between the cells. The number of switches ($i_{switches}$) and flying capacitors ($p_{capacitors}$) can be determined directly from the n_{cells} used. Besides this, the six switches from the basic structure must also be included toward the total switch counts. Because of the hybrid configuration, the mL-ANPC converter does not only address the drawbacks encountered in the classical multilevel topologies but also significantly reduces its cost, size, and weight [2].

14.3 Operating Principles

Having two series-connected modular cells clamped onto the 3L active NPC results in a single-phase seven-level (7L) ANPC inverter, as presented in Fig. 14.2. Each phase leg requires a minimum of 10 $i_{switches}$ (S_{x1} to S_{x5} and S_{x1}' to S_{x5}') and two $p_{capacitors}$ (C_x1 and C_x2), excluding the two dc-link capacitors (C_{dc1} and C_{dc2}), depending on the voltage ratings considered for the converter design.

Hence, a total combination of 16 valid operating modes required for 7L-ANPC is summarized in Table 14.1. It is clearly observed that, among the five pairs of active switches, both switch pairs (S_{x1}–S_{x2}) and (S_{x1}'–S_{x2}') share the same switching sequences

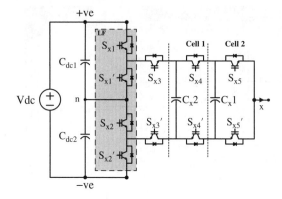

Figure 14.2 Proposed transformerless single-phase two-cell 7L-ANPC.

Table 14.1 Two-Cell 7L-ANPC Voltage Level and Corresponding Switching States.

Switching States	Phase Leg x Voltage
$(S_{x1}, S_{x2}, S_{x3}, S_{x4}, S_{x5})$	V_{xn}
(00000)	$-V_{dc}/2$
(00001) (00010) (00100)	$-V_{dc}/3$
(00011) (00101) (00110)	$-V_{dc}/6$
(00111) (11000)	0
(11001) (11010) (11100)	$+V_{dc}/6$
(11011) (11101) (11110)	$+V_{dc}/3$
(11111)	$+V_{dc}/2$

Note: Here, "x" represents phase a, b and c; and logic "1" and "0" represent the "on" and "off" states, respectively.

at a low frequency (LF) of 50 Hz matching with the modulating frequency (f_o). Meanwhile, the remaining switches S_{x3} to S_{x5} are operating between the modes "000" to "111" corresponding to the carrier frequency (f_c).

All the positive output levels are generated when S_{x1} and S_{x2} conduct, while all negative output levels are achieved when the complementary LF switches S_{x1}' and S_{x2}' conduct. There are two redundant states for the zero output level, and three redundant states for each $\pm V_{dc}/3$ and $\pm V_{dc}/6$ output level. Therefore, the seven-level output stepped voltage waveforms are synthesized accordingly at $\pm V_{dc}/2$, $\pm V_{dc}/3$, $\pm V_{dc}/6$, and zero level with respect to the switching states given in Table 14.1.

Therefore, the flying capacitor voltages V_{Cx1} and V_{Cx2} of 7L-ANPC must be maintained at $V_{dc}/6$ and $V_{dc}/3$, respectively, all the time. The voltage across the corresponding pth capacitor of n-cell mL-ANPC can be determined from Table 14.2 using equation (14.1).

$$V_{Cx(p^{th})}_{ANPC} = \frac{p^{th}}{mL - 1} \times V_{dc} \tag{14.1}$$

Table 14.2 Capacitor Voltages for ANPC.

Capacitors Clamped at p^{th} Position	Phase × Capacitor Voltages ($V_{Cx,p}{}^{th}$)		
	Two-Cell (Seven Levels)	Four-Cell (11 Levels)	Six-Cell (15 Levels)
1	$V_{dc}/6$	$V_{dc}/10$	$V_{dc}/14$
2	$2 V_{dc}/6$	$2 V_{dc}/10$	$2 V_{dc}/14$
3	~	$3 V_{dc}/10$	$3 V_{dc}/14$
4	~	$4 V_{dc}/10$	$4 V_{dc}/14$
5	~	~	$5 V_{dc}/14$
6	~	~	$6 V_{dc}/14$

Note: Here, "x" represents phase a, b and c; and "p^{th}" represents the placing position number of flying capacitor being clamped.

14.4 Design Considerations

A higher number of output voltage stepped levels can be synthesized easily for the modular-structured ANPC by extending the n_{cells}. Hence, the mL-ANPC is expressed in equation (14.2), showing that two additional stepped levels will be produced in the output phase voltage waveforms with every n_{cells} added.

$$(mL)\text{ANPC} = 3 + (2 \times n_{cells}) \tag{14.2}$$

The output phase voltages for mL-ANPC is determined with equation (14.3) based on the corresponding operating modes in Table 14.1, the flying capacitor voltages obtained using equation (14.1), as well as the voltage across both dc-link capacitors C_{dc1} (V_{dc1}) and C_{dc2} (V_{dc2}). According to the voltage divider, V_{dc1} and V_{dc2} should be equivalent to $V_{dc}/2$.

$$
\begin{aligned}
Vxn(t) \approx {}& \frac{V_{dc1}}{2}(t)(S_{x1}(t) + S_{x3}(t)) - \frac{V_{dc2}}{2}(t)(1 - S_{x2}(t) + S_{x3}{}'(t)) \\
& + \sum_{n=1}^{n_{cells}} \frac{V_{Cx((P_{capacitors}+1)-m)}(t)}{2} \left[\begin{pmatrix} S_{x(3+n)}(t) - S_{x(3+n)}{}'(t) \\ -S_{x(3+n-1)}(t) + S_{x(3+n-1)}{}'(t) \end{pmatrix} \right]
\end{aligned}
\tag{14.3}
$$

It is important to determine the voltage stress so that the semiconductor device ratings can be selected appropriately to optimize the converter design. The voltage stress across the switches of the mL-ANPC is heavily dependent on the instantaneous values of dc-link capacitor voltages and flying capacitor voltages, which are clearly expressed in equation (14.4).

$$
V_{Sx(switchi^{th})\atop ANPC}(t) \approx \begin{cases} V_{dc(switchi^{th})}(t) \times (1 - S_{x(switchi^{th})}(t))|_{i<3} \\[2mm] \left[\sum_{q=1}^{2} \left(\frac{p^{th} \times V_{dc}(t)}{mL - 1} - V_{dc(q)}(t) \right) \times ((q-1) - S_{x(q)}(t)) \right] \times (1-S_{x3}(t)) \Big|_{i=3} \\[4mm] \left(\frac{p^{th} - (p^{th} - 1)}{mL-1} \right) V_{dc}(t) \times (1 - S_{x(3+n)}(t)) \Big|_{\substack{i>3 \\ n=i-3}} \end{cases}
\tag{14.4}
$$

Hence, it is also imperative to take note of the peak voltages and inrush currents during the transient period when starting up the converter. Therefore, the pre-charging of these dc-link capacitors and flying capacitors in the ANPC is required to prevent these switching devices from being damaged [26].

Referring to the circuit configuration in Fig. 14.1 and equation (14.4), the blocking voltage capability of the LF switch pairs S_{x1} and S_{x2} for mL-ANPC must be practically able to withstand the instantaneous V_{dc1} and V_{dc2}, respectively. As for the switch pair S_{x3}, which is clamped onto the LF switch pairs S_{x1} and S_{x2}, the voltage stress V_{Sx3} is greatly affected by the voltage across the flying capacitor at p^{th} position $(V_{Cx,p}{}^{th})$ on top of the two dc-link capacitor voltages $(V_{dc1}$ and $V_{dc2})$. Finally, the voltage stress across the remaining switches $(S_{x(3+n)})$ in Cell 1 to Cell n depends solely on the adjacent flying capacitor voltages. The same voltage stress equation (14.4) applies to the respective complementary switches.

14.5 Multiple Voltage Quantities Enhancement Control

The objectives of proposing the MVQE control are to ensure that the (i) dc-link capacitor voltages and (ii) flying capacitor voltages are tracked to their reference values despite any undesirable conditions, as well as for (iii) compensating the dc-link ripples, so that the low-order harmonic components can be eliminated in the output voltages. The delivered output power quality is further enhanced using the proposed hybrid ISC-PD-PWM modulation scheme. The harmonic characteristic of the ISC-PD-PWM is analyzed mathematically as well in the following subsection.

14.5.1 dc-Link Neutral Point Offset Regulator

Despite the proven fact that the topological design of the mL-ANPC (in Fig. 14.1 is favorably preferred in terms of its simplified dc-link configuration as compared to the classical mL-NPC, a potential difference is experienced in the split dc-link voltage across C_{dc1} and C_{dc2}. This is mainly due to the non-identical parameters of the physical capacitors in the dc-link, as well as the presence of non-zero average neutral point current regardless of any modulation scheme used.

Various methods have been reported for neutral point potential balancing using either vector sequence selection based on SVPWM [3, 4, 9] or zero sequence voltage injection technique with CB-PWM [13] for the ANPC. Hence, this section proposes an alternative solution in Fig. 14.3 to balance the neutral point potential by offsetting the dc component in the modulation scheme.

It is clearly understood that the output phase voltage across the respective phase x and the neutral point n will experience an additional dc component $V_o(t)$ when V_{dc1} is not equal to V_{dc2} on top of the high-frequency components generated by the switching devices, as shown in Fig. 14.4. Therefore, the phase voltages of ANPC expressed in equation (14.5) includes the dc component $V_o(t)$.

$$Vx(t) = Vxn(t) + V_o(t) \tag{14.5}$$

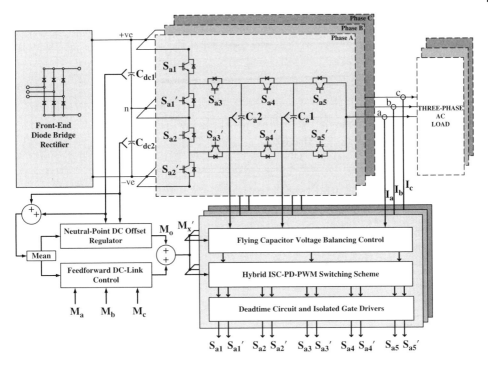

Figure 14.3 Proposed MVQE control scheme for three-phase two-cell 7L-ANPC using hybrid ISC-PD-PWM.

As V_{dc1} and V_{dc2} are not exactly the same due to the tolerance value of the dc-link capacitance, equation (14.3) is expanded into equation (14.6), assuming that all $V_{Cx,p}^{th}$ are ripple free, which can now be equivalently expressed in terms of V_{dc1} and V_{dc2} instead.

$$Vxn(t) \approx \frac{V_{dc1}}{2}(t)(S_{x1}(t) + S_{x3}(t)) - \frac{V_{dc2}}{2}(t)(1 - S_{x2}(t) + S_{x3}'(t))$$

$$+ \sum_{n=1}^{n_{cells}} \left(\begin{array}{c} \dfrac{p^{th} \times V_{dc1}(t)}{mL - 1}[(S_{x(3+n)}(t) - S_{x(3+n-1)}(t))] \\ + \dfrac{p^{th} \times V_{dc2}(t)}{mL - 1}[(-S_{x(3+n)}'(t) + S_{x(3+n-1)}'(t))] \end{array} \right) \tag{14.6}$$

In this case, the switching functions for 7L-ANPC in equation (14.6) are replaced with the modulating functions given in equations (14.7) and (14.8).

$$S_{x1}(t) + \frac{2}{6}S_{x5}(t) + \frac{2}{6}S_{x7}(t) + \frac{2}{6}S_{x9}(t) = Mx(t) + 1 \tag{14.7}$$

$$2 - S_{x1}(t) - \frac{2}{6}S_{x5}(t) - \frac{2}{6}S_{x7}(t) - \frac{2}{6}S_{x9}(t) = 1 - Mx(t) \tag{14.8}$$

Based on equations (14.6)–(14.8), the resulting modulation signal $Mx^{*}(t)$ is derived in equation (14.9), which also equates to equation (14.10). It is noticed that there is the

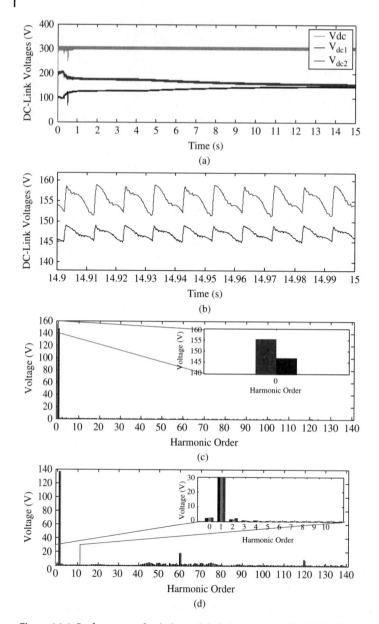

Figure 14.4 Performance of unbalanced dc-link capacitors 7L-ANPC without regulator: (a) dc-link voltages; (b) magnified view of dc-link capacitor voltages; (c) harmonic spectrum of dc-link voltages; and (d) harmonic spectrum of output phase voltages.

presence of an additional dc component Mo(t) in the expressions when both $V_{dc1}(t)$ and $V_{dc2}(t)$ are diverged from one another.

$$\frac{2Vxn(t)}{V_{dc}(t)} = Mx(t) + \frac{V_{dc1}(t) - V_{dc2}(t)}{V_{dc}(t)} \tag{14.9}$$

$$Mx^*(t) = Mx'(t) + Mo(t) \tag{14.10}$$

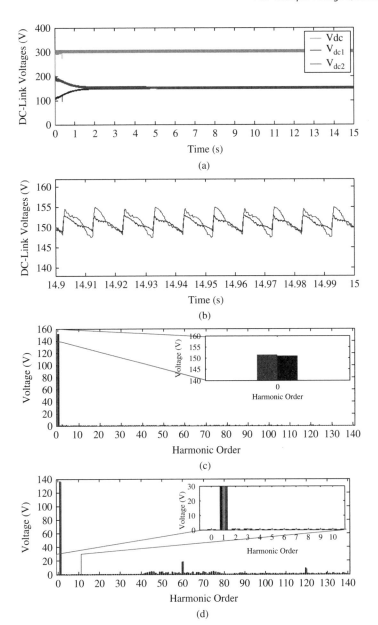

Figure 14.5 Performance of unbalanced dc-link capacitors in 7L-ANPC with dc-link neutral point offset regulator: (a) dc-link voltages; (b) magnified view of dc-link capacitor voltages; (c) harmonic spectrum of dc-link voltages; and (d) harmonic spectrum of output phase voltages.

As a result, the neutral point of the 7L-ANPC is balanced by adding a dc offset to the dc component term from the modulating signal $M_x(t)$. This can be observed in Fig. 14.5.

14.5.2 Feedforward dc-Link Ripple Compensation

Practically, significant ripples are experienced in the dc-link, mainly when the front-end rectifier is fed by a weak or unbalanced ac source, or when the inverter output is connected to nonlinear or unbalanced loads [15]. In such cases, the negative sequence causes the fundamental harmonics to be unbalanced, and low-order harmonics are generated in the output voltages, as observed in Fig. 14.6.

The ripple compensation method for ANPC has been proposed in [10] using SHE-PWM, but it is not applicable for CB-PWM. This chapter proposes a simple dc-link ripple compensation, and yet it is different from [15], which is based on the modified feedforward of modulating functions according to equation (14.11). The normalized dc-link voltage quantity $\widetilde{V}_{dc}$ is achieved with the instantaneous $V_{dc}(t)$ over the average V_{dc} value.

$$Mx'(t) = \frac{Mx(t)}{\widetilde{V}_{dc}} = \frac{Mx(t)}{V_{dc}(t)/V_{dc}} \tag{14.11}$$

With the proposed dc-link ripple compensator, the low-order harmonics found in Fig. 14.6 are being mitigated to high orders. But the amplitude of those mitigated harmonics in Fig. 14.7 is comparatively low with those in Fig. 14.6, which have no significant impact on the output voltage total harmonic distortion (THD). However, a slight improvement is noted in the output line voltage THD using the proposed compensation method due to the elimination of low-order harmonics in Fig. 14.7.

14.5.3 Flying Capacitor Voltage Balancing Control

As the $p_{capacitors}$ increases for higher mL-ANPC, the flying capacitor voltages are likely to be unbalanced, as shown in Fig. 14.8, especially when there is more than one $p_{capacitors}$ in each phase leg. Such consequences lead to an increasingly high voltage stress being imposed on the switching devices, thus greatly affecting the stability of the entire system. Even though the issue can be overcome easily by oversizing the device ratings, it is nevertheless not a cost-effective solution.

Therefore, it is very important to ensure that the flying capacitor voltages of the ANPC are always tracked to their desired reference values based on equation (14.1). Hence, flying capacitor voltage balancing control is required to increase converter reliability. There are some reported control methods – such as modifying the switching pulse width of PSC-PWM through a PI regulator [14], and using redundant switching states based on SHE-PWM [11] or the nearest-level modulator [7, 8]. However, the computational efforts of these are increased, especially for higher mL-ANPC.

Alternatively, a flying capacitor voltage balancing is developed using a simple proportional (P) gain. The current through the respective flying capacitor in equation (14.12) is dependent on the switching condition of the two adjacent switches.

$$C_{x,p^{th}} \times \frac{dV_{Cx,p^{th}}(t)}{dt} = I_{Cx,p^{th}}(t) = (S_{x(3+n-p^{th})}(t) - S_{x(3+n-p^{th}+1)}(t))I_x(t) \tag{14.12}$$

The time domain flying capacitor current in equation (14.12) can be expressed in local average form by replacing those switching state functions with the duty ratio as follows:

$$\frac{\Delta \overline{V}_{Cx,p^{th}}}{\Delta t} = \frac{(\Delta d_{x(3+n-p^{th})} - \Delta d_{x(3+n-p^{th}+1)})\overline{I}_x}{C_{x,p^{th}}} \tag{14.13}$$

$$\Delta d_{x(n)} = sign(I_x)(\varepsilon_{x(n-1)} - \varepsilon_{x(n)})P \tag{14.14}$$

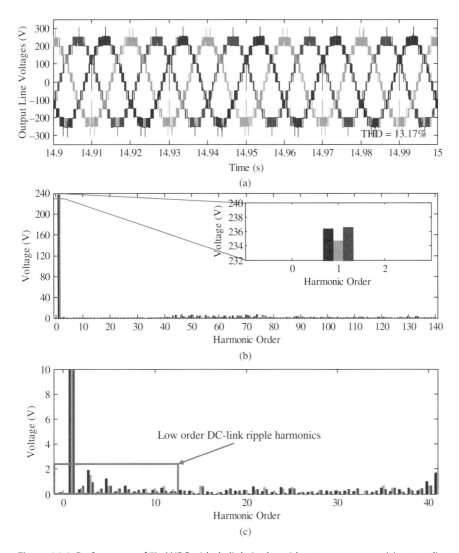

Figure 14.6 Performance of 7L-ANPC with dc-link ripples without compensator: (a) output line-to-line voltages; (b) harmonic spectrum of output line-to-line voltages; and (c) magnified view of harmonic spectrum in (b).

Therefore, the flying capacitor voltages can now be regulated by varying the pulse width of the switching pulses according to the voltage errors ($\varepsilon_{x(n)}$) between the reference capacitor voltage – equation (14.1) – and the instantaneous measured values. These errors are then minimized through the P controller, which can be tuned to obtain the desired performance.

The flying capacitor voltages in Fig. 14.9 are observed to be in well-maintained balance at their corresponding voltage levels with the proposed balancing control. Moreover, the output phase voltage THD of Fig. 14.9 is significantly improved as compared to those in Fig. 14.8 due to the clear synthesis of the seven-level stepped voltage waveforms.

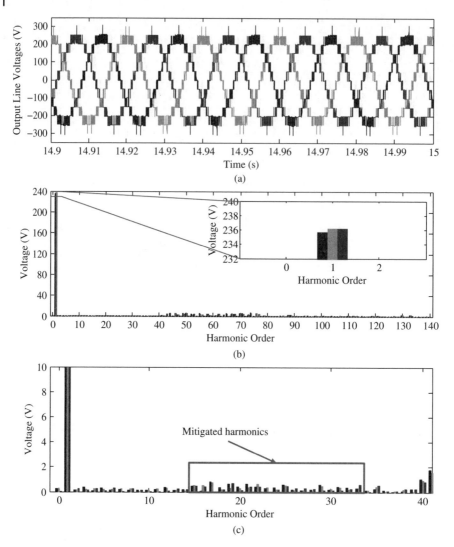

Figure 14.7 Performance of 7L-ANPC with dc-link ripples with compensator: (a) output line-to-line voltages; (b) harmonic spectrum of output line-to-line voltages; and (c) magnified view of harmonic spectrum in (b).

The duty cycle $\Delta d_{x(n)}$ is varied according to the directional flow of the instantaneous output phase current, which is defined as "1" and "−1" for positive and negative cycles, respectively. The modified duty ratio is then added to the resulting $Mx^*(t)$ of equations (14.10) and (14.11) to obtain the finalized modulation signal $Mx_{sine(n)}$ as shown in Fig. 14.10.

14.5.4 Interleaved Sawtooth Carrier Phase-Disposition PWM

The operational principles of the two-cell 7L-ANPC are achieved using a proposed hybrid CB-PWM named the *interleaved sawtooth carrier phase-disposition PWM*

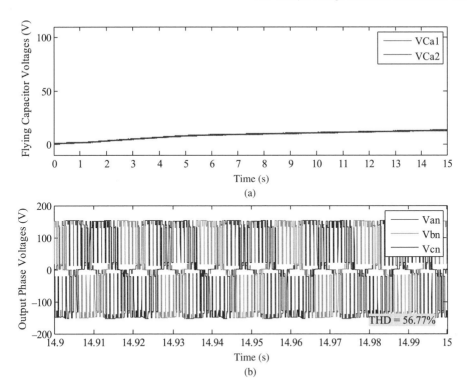

Figure 14.8 Performance of 7L-ANPC flying capacitor voltages without voltage balancing control: (a) phase A flying capacitor voltages; (b) output phase voltages.

(ISC-PD-PWM), shown in Fig. 14.11. Generally, for any carrier-based mL modulation scheme, $(mL - 1)$ number of sawtooth carriers (n_{saw}) would be needed.

In the case of the two-cell 7L-ANPC, six n_{saw} are required, which is then divided equally into two sets. One set is level-shifted above the zero reference magnitude ($saw_{TOP(1,2,3,...,(nsaw/2))}$), and the other set level-shifted below zero reference magnitude ($saw_{BOT(1,2,3,...,(nsaw/2))}$). Both top and bottom sawtooth carriers are 120° phase-shifted apart evenly, based on equation (14.15).

$$\Delta_{phase} = \frac{360°}{(n_{saw}/2)}(\text{deg}) = \frac{2\pi}{(n_{saw}/2)}(\text{rad}) \quad n_{saw} = even > 2 \qquad (14.15)$$

The final gating signals for the respective switching devices are achieved directly by comparing $Mx_{sine(n)}$ with the corresponding top and bottom phase-shifted sawtooth carriers using simple comparators and logic gates, as shown in Fig. 14.10. Under the proposed hybrid modulating technique, the switching states in Table 14.1 are achieved based on the gating signals in Fig. 14.12.

It can be observed that the difference between the proposed hybrid ISC-PD-PWM in this chapter and the commonly employed conventional PSC-PWM approach in [27–29] is the cross-degree of freedom for the three-phase reference wave signals (Mx_{sine}) during the positive and negative cycle period. Thus, the modulating gate signals in this case can be produced without the complicated need to rebuild the Mx_{sine} for the comparative logic stage.

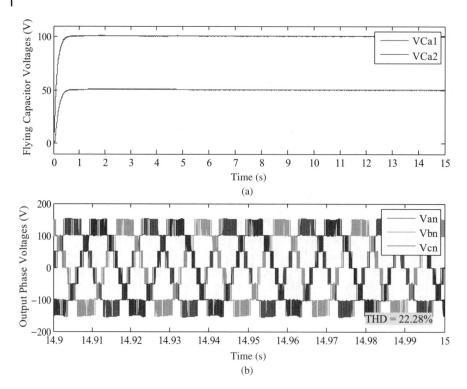

Figure 14.9 Performance of 7L-ANPC flying capacitor voltages with voltage balancing control: (a) phase A flying capacitor voltages; (b) output phase voltages.

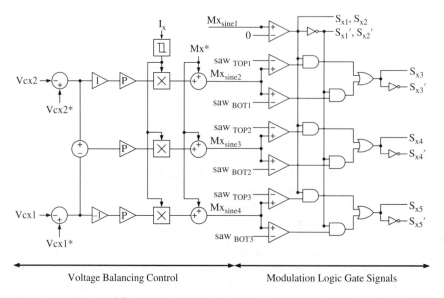

Figure 14.10 Proposed flying capacitor voltage balancing control with modulation logic gate signals for two-cell 7L-ANPC.

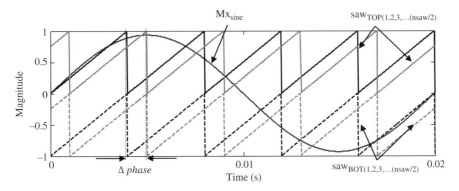

Figure 14.11 Proposed advanced hybrid sinusoidal PWM – ISC-PD-PWM.

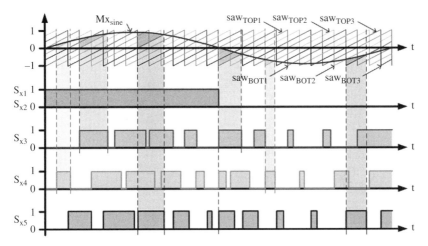

Figure 14.12 Proposed hybrid ISC-PD-PWM modulating waveforms with the corresponding switching gate signals for the two-cell 7L-ANPC.

As a result, the PSC-PWM in [27–29] can also be redefined as a hybrid carrier-based PWM, since another set of phase-shifted triangular carriers can be level-shifted below the zero reference magnitude instead of having the negative cycle of Mx_{sine} shifted up based on the peak magnitude of triangular carrier ($\hat{V}_{tri}$). In fact, the PSC-PWM in [27–29] is equivalent to the presented advanced hybrid PSC-PD-PWM, as analyzed previously in Chapter 5.

Even though similarity is found between the proposed ISC-PD-PWM and the classical PSC-PWM (PSC-PD-PWM) in [27–29], the geometric characteristic of the carriers are totally different. Subsequently, the switching sequences are changed as well. Because of that, the performances of ANPC are greatly affected in terms of the voltage harmonic distortion, which will be discussed in the following subsection.

14.5.5 Harmonic Characteristics of Proposed Hybrid ISC-PD-PWM

A similar approach of analyzing carrier-based PWM schemes using the double Fourier series is applied to the alternatively new advanced hybrid ISC-PD-PWM of Fig. 14.11.

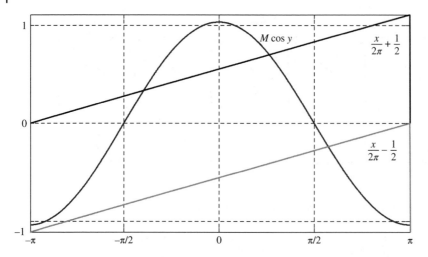

Figure 14.13 Reference sinusoidal signal with two fundamental level-shifted sawtooth carrier ISC-PD-PWM.

Table 14.3 Switching Function of ISC-PD-PWM.

$f(x, y)$	$-\pi < x < \pi$
$+V_{dc}$	$M \cos y > \dfrac{x}{2\pi} + \dfrac{1}{2}$
0	$\dfrac{x}{2\pi} - \dfrac{1}{2} < M \cos y < \dfrac{x}{2\pi} + \dfrac{1}{2}$
$-V_{dc}$	$M \cos y < \dfrac{x}{2\pi} - \dfrac{1}{2}$

The harmonic spectrum analysis for the proposed ISC-PD-PWM is simplified using the single-phase reference sinusoidal signal and the two in-phase level-shifted sawtooth carriers, as shown in Fig. 14.13.

The switching function of ISC-PD-PWM is expressed in Table 14.3, which is derived based on Fig. 14.13. Since two level-shifted sawtooth carriers are used in Fig. 14.13 for analysis, there are only three conditions generated, which are $+V_{dc}$, 0 V, and $-V_{dc}$, following the switching condition expressed in equation (14.1).

In this case, the boundary conditions for the two in-phase level-shifted sawtooth carriers are different from the previous advanced hybrid PSC-POD-PWM and PSC-PD-PWM. Thus, the boundary conditions for the transitions between 0 V to $+V_{dc}$ and $+V_{dc}$ to 0 V are given in equations (14.15) and (14.16), respectively.

$$x_{LOWER} = -\pi \tag{14.16}$$

$$x_{UPPER} = 2\pi \left(M \cos y - \frac{1}{2} \right) \tag{14.17}$$

While achieving the $-V_{dc}$ level, the reference sinusoidal signal must be lesser than both upper and lower sawtooth carriers. Since the carrier used in the ISC-PD-PWM is a ramped (asymmetrical) waveform unlike those symmetric triangular carriers in the other advanced hybrid sinusoidal PWM schemes, the lower and upper boundary conditions for the transitions between 0 V to $-V_{dc}$ and $-V_{dc}$ to 0 V are the same for both the upper carrier and lower carrier, which are given in equations (14.18) and (14.19), respectively.

$$x_{LOWER} = 2\pi \left(M \cos y + \frac{1}{2} \right) \tag{14.18}$$

$$x_{UPPER} = \pi \tag{14.19}$$

Hence, by substituting the boundary condition equations (14.16), (14.17), (14.18), and (14.19) into equation (14.6), the complex form of double Fourier analysis for ISC-PD-PWM modulation can now be expressed in equation (14.20).

$$
C_{mn} = \frac{1}{2\pi^2}
\begin{bmatrix}
\displaystyle\int_{-\frac{\pi}{2}}^{\frac{\pi}{2}} \int_{-\pi}^{2\pi\left(M\cos y - \frac{1}{2}\right)} V_{dc}\, e^{j(mx+ny)}\, dx\, dy \\[2mm]
-\displaystyle\int_{-\pi}^{-\frac{\pi}{2}} \int_{2\pi\left(M\cos y + \frac{1}{2}\right)}^{\pi} V_{dc}\, e^{j(mx+ny)}\, dx\, dy \\[2mm]
-\displaystyle\int_{\frac{\pi}{2}}^{\pi} \int_{2\pi\left(M\cos y + \frac{1}{2}\right)}^{\pi} V_{dc}\, e^{j(mx+ny)}\, dx\, dy
\end{bmatrix}
$$

$$
= \frac{V_{dc}}{2\pi^2} \left[\int_{-\frac{\pi}{2}}^{\frac{\pi}{2}} \int_{-\pi}^{2\pi\left(M\cos y - \frac{1}{2}\right)} e^{j(mx+ny)}\, dx\, dy + \int_{\frac{\pi}{2}}^{\frac{3\pi}{2}} \int_{\pi}^{2\pi\left(M\cos y + \frac{1}{2}\right)} e^{j(mx+ny)}\, dx\, dy \right] \tag{14.20}
$$

In order to determine the dc offset component based on the general spectrum analysis equation (14.5), the m and n in equation (14.20) must be equal to 0 ($m = n = 0$). The dc offset component for the ISC-PD-PWM can be determined as in equation (14.21).

$$
A_{00} + jB_{00} = \frac{V_{dc}}{2\pi^2} \left[\int_{-\frac{\pi}{2}}^{\frac{\pi}{2}} \int_{-\pi}^{2\pi\left(M\cos y - \frac{1}{2}\right)} e^{-j(0+0)}\, dx\, dy + \int_{\frac{\pi}{2}}^{\frac{3\pi}{2}} \int_{\pi}^{2\pi\left(M\cos y + \frac{1}{2}\right)} e^{j(0+0)}\, dx\, dy \right]
$$

$$
= \frac{V_{dc}}{2\pi^2} \left[\int_{-\frac{\pi}{2}}^{\frac{\pi}{2}} [2\pi M \cos y]\, dy + \int_{\frac{\pi}{2}}^{\frac{3\pi}{2}} [2\pi M \cos y]\, dy \right]
$$

$$
= \frac{M V_{dc}}{\pi} \left[\sin\left(\frac{\pi}{2}\right) - \sin\left(-\frac{\pi}{2}\right) \right] + \frac{M V_{dc}}{\pi} \left[\sin\left(\frac{3\pi}{2}\right) - \sin\left(\frac{\pi}{2}\right) \right]
$$

$$
= 0 \tag{14.21}
$$

By letting $m = 0$ and $n > 0$ for equation (14.20), the baseband harmonic of the proposed ISC-PD-PWM scheme is determined in equation (14.22).

$$
A_{0n} + jB_{0n} = \frac{V_{dc}}{2\pi^2} \left[\int_{-\frac{\pi}{2}}^{\frac{\pi}{2}} \int_{-\pi}^{2\pi\left(M\cos y - \frac{1}{2}\right)} e^{j(ny)}\, dx\, dy + \int_{\frac{\pi}{2}}^{\frac{3\pi}{2}} \int_{\pi}^{2\pi\left(M\cos y + \frac{1}{2}\right)} e^{j(ny)}\, dx\, dy \right]
$$

$$
= \frac{M V_{dc}}{\pi} \left[\int_{-\frac{\pi}{2}}^{\frac{\pi}{2}} [\cos y] e^{j(ny)}\, dy + \int_{\frac{\pi}{2}}^{\frac{3\pi}{2}} [\cos y] e^{j(ny)}\, dy \right]
$$

$$
= \frac{M V_{dc}}{2\pi} \left[\int_{-\frac{\pi}{2}}^{\frac{\pi}{2}} (e^{jy(n+1)} + e^{jy(n-1)})\, dy + \int_{\frac{\pi}{2}}^{\frac{3\pi}{2}} (e^{jy(n+1)} + e^{jy(n-1)})\, dy \right] \tag{14.22}
$$

$$A_{01} + jB_{01} = \frac{MV_{dc}}{2\pi} \left[\int_{-\frac{\pi}{2}}^{\frac{\pi}{2}} (e^{jy(0)}) dy + \int_{\frac{\pi}{2}}^{\frac{3\pi}{2}} (e^{jy(0)}) dy \right]$$

$$= MV_{dc} \tag{14.23}$$

The same term $e^{j(ny)}$ is found in the integral equation (14.22), which will be equal to zero for any non-zero condition of n. Thus, n can only be equated to 1, since $A_{0n} + jB_{0n} = 0$ for all conditions when $n > 1$. Let $n = 1$ for equation (14.22), and the baseband harmonic of ISC-PD-PWM can be further expressed as in equation (14.23). The carrier harmonic components of ISC-PD-PWM is also expressed in equation (14.24) by having $m > 0$ and $n = 0$ for equation (14.20).

$$A_{m0} + jB_{m0} = \frac{V_{dc}}{2\pi^2} \left[\int_{-\frac{\pi}{2}}^{\frac{\pi}{2}} \int_{-\pi}^{2\pi\left(M\cos y - \frac{1}{2}\right)} e^{j(mx)} dx \, dy + \int_{\frac{\pi}{2}}^{\frac{3\pi}{2}} \int_{\pi}^{2\pi\left(M\cos y + \frac{1}{2}\right)} e^{j(mx)} dx \, dy \right]$$

$$= \frac{V_{dc}}{jm2\pi^2} \left[e^{jm(-\pi)} \int_{-\frac{\pi}{2}}^{\frac{\pi}{2}} [e^{jm(2\pi M \cos y)} - 1] dy + e^{jm(\pi)} \int_{\frac{\pi}{2}}^{\frac{3\pi}{2}} [e^{jm(2\pi M \cos y)} - 1] dy \right]$$

$$\tag{14.24}$$

Equation (14.24) can be further simplified to equation (14.26) using the Bessel function integral relationships, as expressed in equation (14.25).

$$\int e^{j(\beta)\cos\tau} d\tau = 2\pi J_0(\beta) \tag{14.25}$$

$$A_{m0} + jB_{m0} = \frac{V_{dc}}{jm2\pi^2} [e^{jm(-\pi)}(\pi J_0(m2\pi M) - \pi) + e^{jm(\pi)}(\pi J_0(m2\pi M) - \pi)]$$

$$= \frac{V_{dc}}{jm\pi} [J_0(m2\pi M) - 1](\cos(m\pi)) \tag{14.26}$$

The sideband harmonic components of ISC-PD-PWM are determined by having $m > 0$ and $n \neq 0$ for equation (14.20), which is expressed in equation (14.27).

$$A_{mn} + jB_{mn} = \frac{V_{dc}}{2\pi^2} \left[\int_{-\frac{\pi}{2}}^{\frac{\pi}{2}} \int_{-\pi}^{2\pi\left(M\cos y - \frac{1}{2}\right)} e^{j(mx+ny)} dx \, dy + \int_{\frac{\pi}{2}}^{\frac{3\pi}{2}} \int_{\pi}^{2\pi\left(M\cos y + \frac{1}{2}\right)} e^{j(mx+ny)} dx \, dy \right]$$

$$= \frac{V_{dc}}{jm2\pi^2} \left[e^{jm(-\pi)} \int_{-\frac{\pi}{2}}^{\frac{\pi}{2}} e^{jny} [e^{jm(2\pi M \cos y)} - 1] dy + e^{jm(\pi)} \int_{\frac{\pi}{2}}^{\frac{3\pi}{2}} e^{jny} [e^{jm(2\pi M \cos y)} - 1] dy \right]$$

$$= \frac{V_{dc}}{\pi^2} \left[\begin{array}{l} \frac{1}{m}[\pi J_n(m2\pi M)]\cos(m\pi)\sin\left(n\frac{\pi}{2}\right) \\ -j\frac{1}{m}[\pi J_n(m2\pi M)]\cos(m\pi)\cos\left(n\frac{\pi}{2}\right) \\ +\frac{j}{mn}\left(e^{jm(-\pi)}\sin\left(n\frac{\pi}{2}\right) + e^{jm(\pi)}\sin\left(n\frac{3\pi}{2}\right)\right) \end{array} \right]$$

$$= \frac{V_{dc}}{\pi} \left[\frac{1}{m} J_n(m2\pi M) \right] \cos(m\pi) \left(\sin\left(n\frac{\pi}{2}\right) - j\cos\left(n\frac{\pi}{2}\right) \right) \tag{14.27}$$

It should be noted that the n in $\sin(n\pi/2)$ and $\sin(n3\pi/2)$ must be odd values, while the n in $\cos(n\pi/2)$ must be even values, so that these terms will not be nulled. The final spectrum analysis for the switching function of the ISC-PD-PWM can be achieved by substituting equations (14.21), (14.23), (14.26), and (14.27) back into the fundamental equation (14.5). The same approach of phase-shifting the carrier and sideband harmonics is also applied for the hybrid ISC-PD-PWM using equations (14.21) and (14.22). Hence, the final expression for the output mL phase voltage spectrum analysis is given in equation (14.28).

$$
\begin{aligned}
Vxn_{ISCPDPWM-xL}(t) = {} & \frac{MV_{dc}}{2}\cos(\omega_o t + \theta_0) \\[6pt]
& + \frac{V_{dc}}{2\pi}\left(\sum_{m=1}^{\infty}\frac{1}{(Nm)}[J_0((Nm)2\pi M) - 1](\cos((Nm)\pi)) \atop \times \sin((Nm)[\omega_c t + \theta_c]) \right) \\[6pt]
& + \frac{V_{dc}}{2\pi}\left(\begin{array}{c} \displaystyle\sum_{m=1}^{\infty}\frac{1}{(Nm)}\sum_{\substack{n=-\infty \\ (n\neq 0)}}^{\infty} J_n((Nm)2\pi M)(\cos((Nm)\pi)) \\[6pt] \times\left[\begin{array}{l} \sin\left(n\frac{\pi}{2}\right)\times\cos((Nm)[\omega_c t + \theta_c] + (n)[\omega_o t + \theta_0]) \\[4pt] - \cos\left(n\frac{\pi}{2}\right)\times\sin((Nm)[\omega_c t + \theta_c] + (n)[\omega_o t + \theta_0]) \end{array}\right] \end{array} \right)
\end{aligned}
\tag{14.28}
$$

Here, N = the total number of sawtooth carriers used in ISC-PD-PWM. Similarly, the output line-to-line voltage Vab spectrum analysis for the proposed hybrid ISC-PD-PWM is expressed in equation (14.29).

$$
\begin{aligned}
Vab_{ISCPDPWM-xL}(t) = {} & \frac{\sqrt{3}MV_{dc}}{2}\cos\left(\omega_o t + \frac{\pi}{2}\right) \\[6pt]
& + \frac{V_{dc}}{\pi}\sum_{m=1}^{\infty}\frac{1}{(Nm)}\left(\begin{array}{c} \displaystyle\sum_{\substack{n=-\infty \\ (n\neq 0)}}^{\infty} J_n((Nm)2\pi M)(\cos((Nm)\pi))\left(\sin\left(n\frac{\pi}{3}\right)\right) \\[6pt] \times\left(\begin{array}{l} \sin\left(n\frac{\pi}{2}\right)\times\cos\left((Nm)[\omega_c t] + (n)\left[\omega_o t - \frac{\pi}{3}\right] + \frac{\pi}{2}\right) \\[4pt] - \cos\left(n\frac{\pi}{2}\right)\times\sin\left((Nm)[\omega_c t] + (n)\left[\omega_o t - \frac{\pi}{3}\right] + \frac{\pi}{2}\right) \end{array}\right) \end{array} \right)
\end{aligned}
\tag{14.29}
$$

Equation (14.28) describes that the carrier harmonics are occurred at all multiples of f_c for the ISC-PD-PWM, unlike those in PSC-PD-PWM (in Chapter 5). Moreover, the magnitudes of these ISC-PD-PWM carrier harmonics are also greater, owing to the Bessel function of order zero (J_0) term. It is also worth noting that an infinite number of sideband harmonics occur around the carrier harmonics for ISC-PD-PWM. However, the magnitudes of these sideband harmonics are much lower as compared to those in the hybrid PSC-PD-PWM, owing to the rapid roll-off of the Bessel function (J_n) term as compared to the (J_{2k-1}) term carried in the PSC-PD-PWM.

As for the output line-to-line voltage spectrum analysis in equation (14.29), both the triplen sidebands and carrier harmonics terms in equation (14.28) are eliminated by their nature due to the crossover of 60° between the adjacent phase leg reference sinusoidal signals. Thus, the magnitudes of the remaining sideband harmonics are

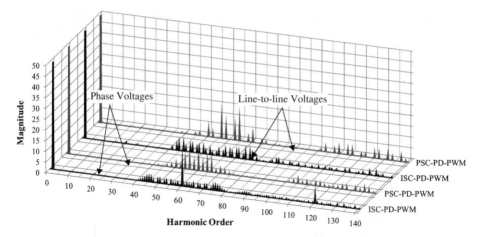

Figure 14.14 Magnified view of the output voltages spectrum synthesized using the PSC-PD-PWM and ISC-PD-PWM schemes.

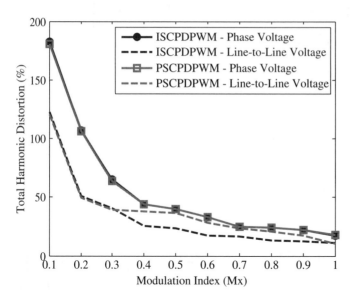

Figure 14.15 Comparison of output voltages THD between the hybrid PSC-PD-PWM and ISC-PD-PWM modulation schemes.

insignificantly smaller. The spectrum analysis of the newly proposed ISC-PD-PWM can be clearly observed in Fig. 14.14 and compared against the classical PSC-PWM (alternatively equivalent to PSC-PD-PWM) used for ANPC.

The THD comparison between both ISC-PD-PWM and PSC-PD-PWM can also be observed in Fig. 14.15. The mathematical analysis of the ISC-PD-PWM harmonic characteristics has evidently proven that the performance of the output line-to-line voltage THD is greatly enhanced for the newly proposed hybrid ISC-PD-PWM as compared to the PSC-PD-PWM, even in the low modulation range. Such attractively low THD performances make ANPC even more beneficially suitable for any drive applications.

The proposed ISC-PD-PWM in this chapter is also applicable to the proposed reduced modular cells converter discussed in Chapter 5.

14.5.6 Experimental Verification

The performances of the two-cell 7L-ANPC based on the proposed MVQE control under hybrid ISC-PD-PWM is investigated using the co-simulation between PSIM v9.0.4 Simcouple and MATLAB R2010b Simulink®. The hardware prototype is developed to verify the presented studies with the aid of the dSPACE RT1103 Real-Time Workshop (RTW).

In order to verify the feasibility of the proposed MVQE in conjunction with the hybrid ISC-PD-PWM, an uncontrolled three-phase diode rectifier is connected at the front-end of the two-cell 7L-ANPC (Fig. 14.3), so that ripples can be generated in the dc-link. Besides that, the two dc-link capacitance values are different as well to create the unbalanced neutral point effects. All the results are presented based on the detailed settings given in Table 14.4.

It is clearly observed from the experimental results that the average dc-link capacitor voltages V_{dc1} and V_{dc2} in Fig. 14.16(b) are closely tracked using the proposed regulator in the MVQE control, as compared to those without control in Fig. 14.16(a). Ripples are found in the dc-link because of the front-end connected rectifier, which has caused the low-order harmonics in Fig. 14.17(a). Thus, the low-order harmonics in Fig. 14.17(a) are being mitigated away in Fig. 14.17(b) using the dc-link ripple compensator in the MVQE control scheme.

The steady-state conditions as well as the dynamic performances are evaluated in Fig. 14.18. Despite the ripples experienced in the dc-link in Fig. 14.18(a) and the sudden load changes in Fig. 14.18(b), the flying capacitor voltages remained balanced at their desired voltage levels. As a result, both output line-to-line voltage and output phase voltage waveforms are clearly synthesized. Hence, better stability is achieved for the ANPC from the evidenced results in Fig. 14.18, with the flying capacitor voltages tracked to the desired values at all times using the proposed voltage balancing method.

Both experimental harmonic spectrums of output stepped phase voltage and line-to-line voltage in Figs. 14.19(a) and 14.19(b), respectively, have validated the analytical results in Fig. 14.14. One can conclude that high-resolution waveforms,

Table 14.4 Configuration Settings for Simulation and Hardware Prototype of the Two-Cell 7L-ANPC.

System Parameters	Simulation	Experiment
Reference sinusoidal frequency (fo)	50 Hz	50 Hz
Triangular carrier frequency (fc)	1 kHz	1 kHz
AC voltage source (Vrms)	170, 185, 155	100, 115, 85
Output dc-link voltage reference (Vdc)	300	250
Flying capacitors (Cx,pth)	2200 μF	2200 μF
dc-link capacitors (Cdc1, Cdc2)	3000 μF, 1000 μF	2400 μF, 1000 μF
Load – resistors, inductors	25 Ω, 122 mH	37.5 Ω, 40 mH

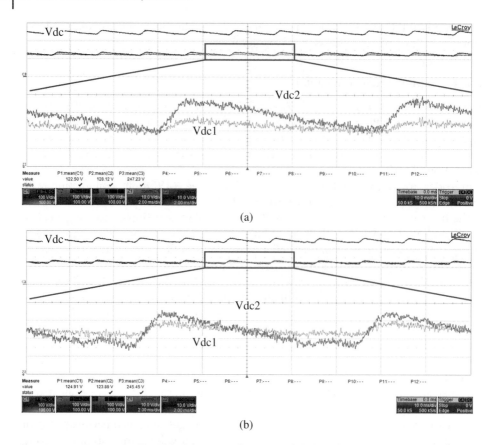

Figure 14.16 Experimental results of the two-cell 7L-ANPC dc-link capacitor voltages: (a) without dc-link neutral-point offset regulator; (b) with regulator.

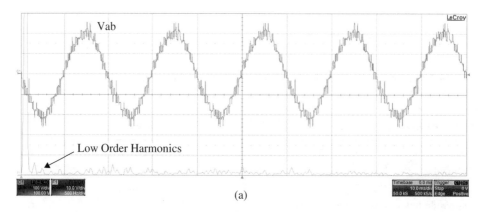

Figure 14.17 Experimental results of the two-cell 7L-ANPC output line-to-line voltage and the FFT spectrum: (a) without dc-link compensator; (b) with compensator.

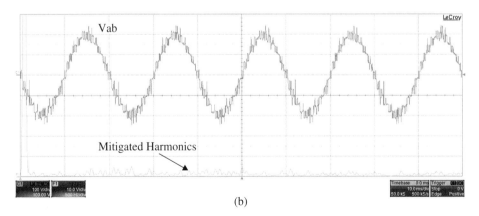

(b)

Figure 14.17 (*Continued*)

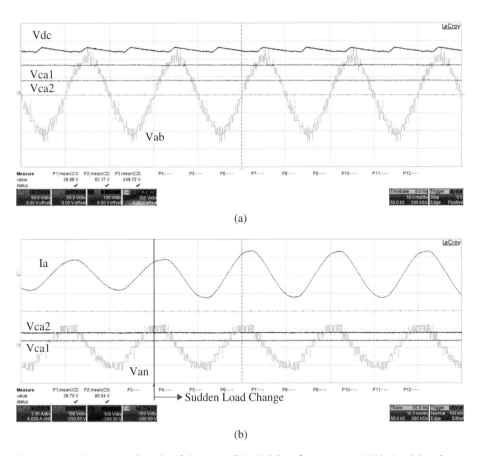

Figure 14.18 Experimental results of the two-cell 7L-ANPC performances at 0.9 Mxsine: (a) under steady-state condition (output line-to-line voltage and flying capacitor voltages); (b) dynamic condition (output phase current, phase voltage, and flying capacitor voltages).

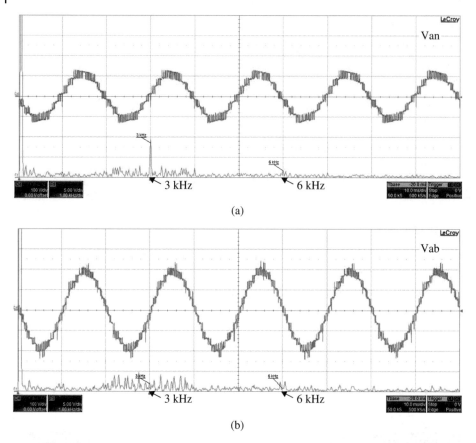

Figure 14.19 Experimental results of the harmonic spectrum under the proposed ISC-PD-PWM: (a) output phase voltage; (b) output line-to-line voltage.

especially the output line-to-line voltages, are synthesized using the proposed hybrid ISC-PD-PWM scheme for the ANPC in comparison with the PSC-PD-PWM (the typically known PSC-PWM implemented for the ANPC [12–14].

The agreement between the simulation and experimental results has confirmed the feasibility of the proposed MVQE control, which has successfully addressed the voltage quality issues (i) to (iii). Therefore, the overall reliability of the converter can be increased, and better power quality can be achieved with the lower harmonic distorted voltage waveforms (especially the output line-to-line voltages) using the proposed ISC-PD-PWM. As such, the perceived merits of the ANPC under the proposed MVQE control with ISC-PD-PWM make it attractively applicable for its well-known medium-voltage drive systems.

14.6 Common Mode Reduction

Common mode modulation techniques are proposed for the classical multilevel topologies as reported in [20–22], but these strategies cannot be applied directly for ANPC, since a hybrid synthesis method is required. Therefore, the second objective of this chapter is to offer an alternative technique to reduce the CMV using hybrid carrier-based PWMs (conventional PSC-PWM and proposed ISC-PD-PWM) for mL-ANPC.

The CMV is typically the potential difference between the neutral point of the three-phase load (o) and the dc-link neutral point (n), as presented in equation (14.30), according to the output phase voltages of the inverter.

$$\text{CMV(Von)} = \frac{\text{Van} + \text{Vbn} + \text{Vcn}}{3} \tag{14.30}$$

As observed from equation (14.30), the CMV is usually created due to the differential phase voltage levels that are synthesized by the respective phase leg under normal modulation technique. Moreover, only the odd mL output phase voltage that contains the states of zero voltage level can be controlled to achieve complete CMV cancellation, unlike those even mL inverters. Since zero output voltage level is produced in the 7L-ANPC (see Table 14.1, CMV elimination can only be guaranteed if the three-phase voltages in equation (14.30) are summed to zero at all times.

In order to produce zero CMV for the seven-level inverter, $7^3 = 343$ possible combinations are now restricted to the 37 switching combinations in Fig. 14.20 without any redundancy. For example, the state indexes of zero CMV $\{S_A, S_B, S_C\} = \{5, 1, 3\}$ is referred to the respective phase voltage levels $\{\text{Van} = +V_{dc}/3, \text{Vbn} = -V_{dc}/3, \text{Vcn} = 0\}$ at that particular switching instant. To nullify the summation of the phase voltages in equation (14.30), the zero CMV vectors for any mL inverter must satisfy the condition in equation (14.31) through the combinational value of three-phase switching states.

$$\sum \{S_A, S_B, S_C\} = \frac{3(mL - 1)}{2} \tag{14.31}$$

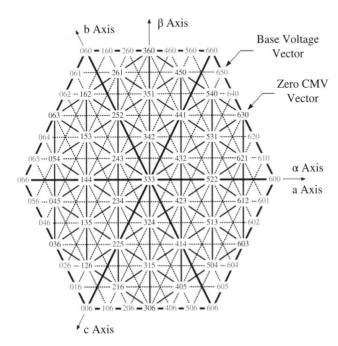

Figure 14.20 Vector diagram of a seven-level inverter with zero CMV states.

It is worth noting in Fig. 14.20 that the entire zero CMV vector diagram (blue lines) has been 30° phase displaced from the outer voltage hexagon lattice of a seven-level inverter (black lines). According to [21], the zero CMV vector diagram of the mL inverter (e.g. 7L-ANPC) is, in fact, equivalent to the simplified vectorial representation of a conventional $((mL+1)/2)$-level (i.e. four-level) inverter comprising only the exclusive switching states.

By associating the normal carrier-based modulation strategy with the simplified vector diagram as described to achieve the common mode reduction, the zero CMV vectors are determined basically from the differences between the adjacent phase leg switching indexes $\{S_A, S_B, S_C\} = \{S_P - S_Q, S_Q - S_R, S_R - S_P\}$, where $\{S_P, S_Q, S_R\}$ are referred to as the *equivalent switching indexes* of $\{S_A, S_B, S_C\}$. This essential approach results in a 30° phase displacement between the two vector diagrams.

For instance, for the 7L-ANPC inverter in this chapter, three four-level modulators are required principally without any exceptional techniques. This means that only half of the $(mL - 1)$ carrier bands (i.e. three reduced carrier bands C_1, C_2, and C_3) are always used. The carrier bands can be sawtooth waves for the proposed ISC-PD-PWM, or triangular waves for the classical PSC-PWM (PSC-PD-PWM). On account of the hybrid carrier-based PWM schemes as described in Section 14.5.4, it should be noted that the magnitude of these reduced carriers is now normalized between -1 and $+1$. Nevertheless, the reduced carriers should remain evenly interleaved among them.

Hence, normal comparing logical rules remain the same for the four-level modulators. The equivalent counterpart switching indexes $\{S_P, S_Q, S_R\}$ are generated independently by comparing the respective three-phase command references (V_P, V_Q, and V_R), as given in equation (14.32), with the three phase-shifted carrier bands C_1, C_2, and C_3.

$$V_P = Ma_{sine} \cos(\omega t)$$
$$V_Q = Mb_{sine} \cos(\omega t - 2\pi/3)$$
$$V_R = Mc_{sine} \cos(\omega t + 2\pi/3) \tag{14.32}$$

Subsequently, the subtracted $\{S_{PQc}, S_{QRc}, S_{RPc}\}$ are obtained simply by the abstracting the differences between the counterpart state indexes of $\{S_P, S_Q, S_R\}$ and its adjacent phase, with respect to c as referred to the same carrier band. However, the output zero CMV vectors of mL-ANPC cannot be explicitly achieved as desired according to the above-mentioned description $\{S_A, S_B, S_C\} \neq \{S_{PQc}, S_{QRc}, S_{RPc}\}$.

Additional modulation techniques are required to modify $\{S_{PQc}, S_{QRc}, S_{RPc}\}$ in order to achieve the zero CMV vectors. This modification is necessary due to the working principles of mL-ANPC, which involves a low-frequency operation across the switches S_{x1}, S_{x2}, and their complementary ones. Second, the subtracted $\{S_{PQc}, S_{QRc}, S_{RPc}\}$ creates three-level PWM pulses, as observed in Fig. 14.21, which cannot be served as the final gating signals.

The three-level PWM pulses (e.g. S_{PQ1} in Fig. 14.21 must be split into two sets of train pulses, where one set is the positive switching pulses (PS_{PQ1}) between 0 and 1, and the other set is the negative switching pulses (NS_{PQ1}) between -1 and 0. The extraction method is done by comparing S_{PQ1} with respect to a zero reference. Intrinsically, PS_{PQ1} is obtained when S_{PQ1} is greater than zero. However, the switching pulses (NS_{PQ1}), which are being abstracted when S_{PQ1} is less than zero, must be inverted again in order to attain the desired NS_{PQ1}' that is similar to those switching transitions between -1 and 0.

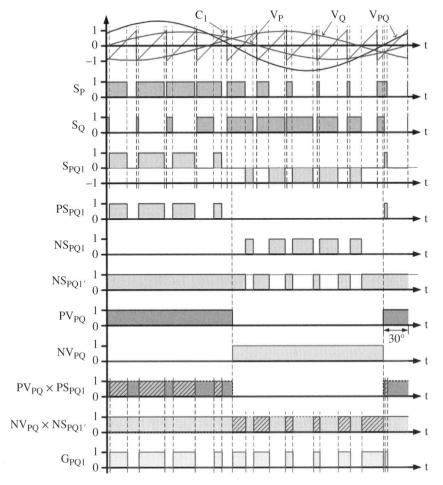

Figure 14.21 Switching pulses of equivalent counterparts (SP and SQ), as well as the derivation of the gating signal (GPQ1), with respect to the command references VP and VQ.

It can be clearly observed from the S_{PQ1} in Fig. 14.21 that both PS_{PQ1} and NS_{PQ1} are now 30° phase-shifted with reference to V_p due to the adjacent phase subtraction V_{PQ} = ($V_P - V_Q$). Before combining PS_{PQ1} and NS_{PQ1} to achieve the final gating signal G_{PQ1} for the switch S_{a3} in the 7L-ANPC, the 30° phase-shifted switching transitions of PS_{PQ1} and NS_{PQ1} must be extracted out from the respective positive and negative half-periods of V_{PQ}. This can be done simply by multiplying the resulting logics (PV_{PQ} and NV_{PQ}) of comparing V_{PQ} with the zero reference. The same approach is applied for the remaining two interleaved carrier bands C_2 and C_3 to determine the final gating signals G_{PQ2} and G_{PQ3}, respectively, for the switches S_{a4} and S_{a5}.

The output logic of V_{PQ} and zero reference comparisons is indeed a 30° phase-shifted 50 Hz squared pulse. These low-frequency squared pulses shall be the final gating signal G_A for S_{a1} and S_{a2} to satisfy the switching conditions described in Table 14.1. Meanwhile, the gating signals for the complementary switches (S_x') can be obtained through the inverter "NOT" gates.

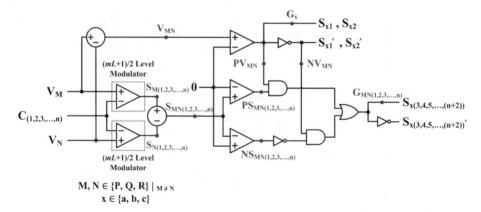

Figure 14.22 Proposed CMV reduction technique using hybrid carrier-based PWM modulation schemes for mL-ANPC.

The described CMV reduction technique is hence achieved using the logic gates as shown in Fig. 14.22, which is applicable for any of the two mentioned hybrid carrier-based modulation strategies (classical PSC-PWM and proposed ISC-PD-PWM). The switching signal waveforms obtained after each stage of the proposed CMV reduction technique in Fig. 14.22 can be better visualized from those switching pulses in Fig. 14.21. Ultimately, the zero CMV vectors for 7L-ANPC are now achieved: $\{S_A, S_B, S_C\} = \{(G_A + G_{PQ1} + G_{PQ2} + G_{PQ3}), (G_B + G_{QR1} + G_{QR2} + G_{QR3}), (G_C + G_{RP1} + G_{RP2} + G_{RP3})\}$.

Even though the CMV magnitudes developed under the proposed hybrid ISC-PD-PWM in Fig. 14.23(b) is slightly higher than when using the conventional PSC-PWM scheme in Fig. 14.23(a), they are greatly reduced, as observed in Fig. 14.24, using the proposed CMV reduction technique. Therefore, this approach can be extended for any mL-ANPC despite utilizing any hybrid carrier-based modulation schemes to reduce the CMV.

14.6.1 Vector Equivalence Mapping

In the three-phase 7L-ANPC, the modulating reference signals V_X (X = P, Q, and R), according to the working principle of the proposed CMV reduction strategy, can be divided into the following four different parts: part I: $0.33 < V_X < Mx_{sine}$; part II: $0 < V_X < 0.33$; part III: $-0.33 < V_X < 0$; and part IV: $-Mx_{sine} < V_X < -0.33$. With regard to that, the synthesis of the output line-to-line voltage can be analyzed using the arbitrary two-phase command references. Hence, the combinations of any two adjacent phases are given as follows: (II, IV), (I, IV), (I, III), (I, II), and (III, IV).

For simplification of the analysis using the vector representation, the combination (I, IV) is used as an example here for both the conventional PSC-PWM in [12–14] and the proposed hybrid ISC-PD-PWM, as shown in Figs. 14.25 and 14.27, respectively. Under the condition of (I, IV), V_P (red line) is greater than the three interleaved carriers C_1, C_2, and C_3, while V_Q (green line) is lesser than C_1, C_2, and C_3.

By observing the synthesized three-phase voltage PWM pulses (Van, Vbn, and Vcn) in both Figs. 14.25 and 14.27, multiple steps (the overlapping of layers over a carrier period T_C) are produced in the phase voltage waveforms of Fig. 14.25 by the symmetrical

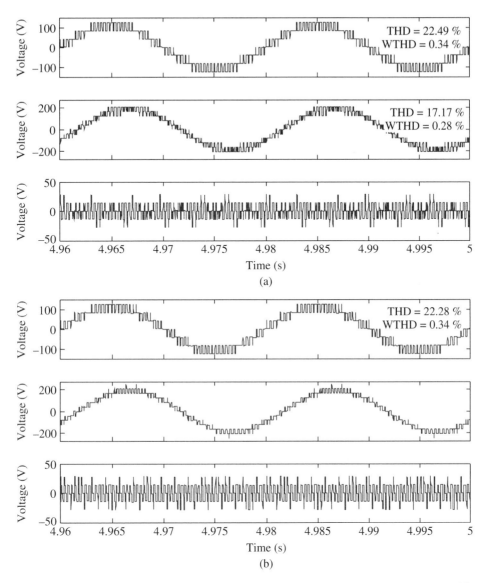

Figure 14.23 Performance of 7L-ANPC without CMV reduction – (top) output phase voltage; (middle) output line-to-line voltage; (bottom) common mode voltage: (a) conventional PSC-PWM (PSC-PD-PWM); (b) proposed hybrid ISC-PD-PWM.

geometric characteristics of the triangular carriers used in the conventional PSC-PWM, as compared to those in Fig. 14.27 using the proposed hybrid ISC-PD-PWM. When multiple stepped levels are generated in the waveforms over a carrier period, undesirable higher-order harmonics distortion is also introduced.

On top of that, the switching states of the conventional PSC-PWM in Fig. 14.25 employing the proposed CMV reduction technique are as follows: 513 – 522 – 612 – 513 – 612 – 522 – 531 – 522 – 612 – 513 – 612 – 531 – 621 – 612 – 513 – 612 – 621 – 531. Therefore, a total of five vectors – 513, 522, 612, 531, and 621 – are created within the

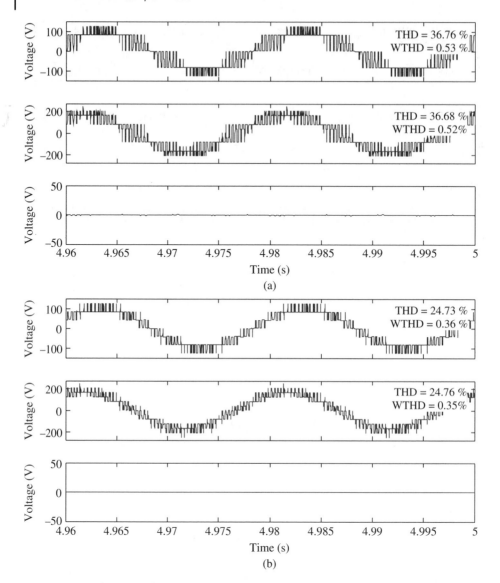

Figure 14.24 Performance of 7L-ANPC using the proposed CMV reduction – (top) output phase voltage; (middle) output line-to-line voltage; (bottom) common mode voltage: (a) conventional PSC-PWM (PSC-PD-PWM); (b) proposed hybrid ISC-PD-PWM.

carrier period, as shown in Fig. 14.26, using the vector representation. Even though zero CMV vectors are achieved, the sequences of these states under the classical PSC-PWM violate the nearest three vectors (NTV) technique.

However, it is clearly noted in Fig. 14.28 that the number of vectors utilized within a carrier period using the proposed ISC-PD-PWM scheme is greatly reduced from five to three (513, 522, and 612). Because of the asymmetrical geometric characteristics of the sawtooth carriers, the PWM pulses in Fig. 14.27 are produced according to the four switching points of the carrier period (0, $T_C/3$,

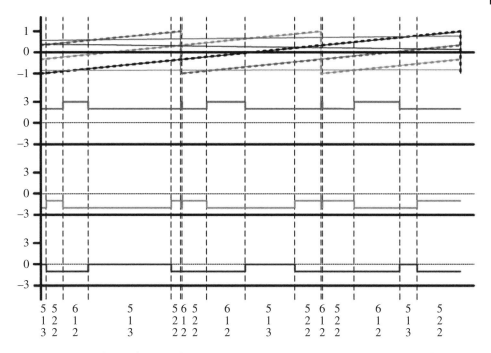

Figure 14.25 Synthesis of output phase voltages using CMV reduction under the proposed hybrid ISC-PD-PWM scheme.

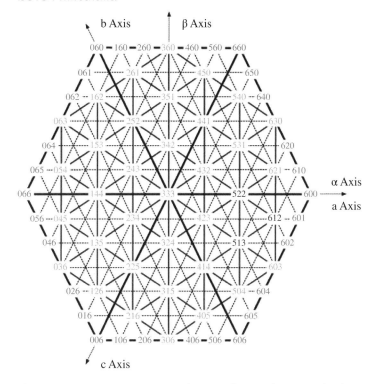

Figure 14.26 Vector representation diagram of CMV reduction under the proposed hybrid ISC-PD-PWM scheme.

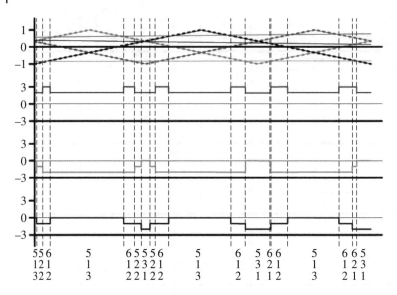

Figure 14.27 Synthesis of output phase voltages using CMV reduction under conventional PSC-PWM (PSC-PD-PWM) scheme [12–14].

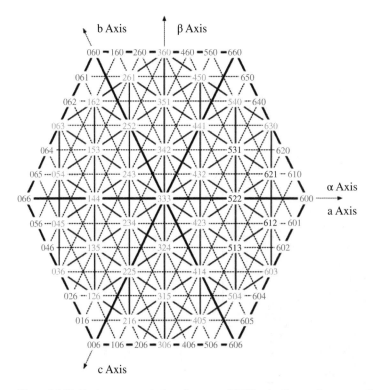

Figure 14.28 Vector representation diagram of CMV reduction under conventional PSC-PWM (PSC-PD-PWM) scheme [12–14].

$2T_C/3$, and T_C). As such, the switching states are in the following sequences: $513 - 522 - 612 - 513 - 522 - 612 - 522 - 612 - 513 - 522 - 612 - 522 - 612 - 513 - 522$. These sequences have conformed the NTV approach with respect to the three zero CMV vectors 513, 522, and 612, as shown in the optimized vector diagram of Fig. 14.28.

14.6.2 Spectral Harmonic Characteristics

The classical PSC-PWM scheme is popularly used in many multilevel converters, especially those designs that involve flying capacitors. Not surprisingly, this scheme is also utilized for mL-ANPC, as reported in [12–14]. However, the typically employed conventional PSC-PWM for mL-ANPC is an alternative hybrid carrier-based PWM scheme, as described previously, which is equivalent to the hybrid PSC-PD-PWM scheme. In the event of normal modulation, only half of the ($mL - 1$) carriers are required and evenly interleaved apart, which is much alike the proposed hybrid ISC-PD-PWM scheme.

Because of the phase-shifted characteristics [30, 31] in both classical PSC-PWM and the proposed ISC-PD-PWM, natural balancing for the flying capacitors can be achieved easily by connecting a balance booster filter to the load side without the need of any control scheme. Nevertheless, the phase-shifting features will cause the first group of sideband harmonics to be produced further away from the fundamental component. Basically, it is affected by the number of carriers being phase-shifted and the carrier frequency. This theoretical concept applies not only to normal modulation, but also to the proposed CMV reduction technique.

It is proven with the spectrum analysis of the output phase voltages and line-to-line voltages, as expressed in equations (14.33) and (14.34) for the classical PSC-PWM (alternatively PSC-PD-PWM), and equations (14.35) and (14.36) for the proposed ISC-PD-PWM. Due to the N term in equations (14.33)–(14.36), it can be observed in Fig. 14.29 that the first group sideband harmonics of both hybrid carrier-based PWM schemes occur around the 60th harmonic order (3 kHz), since three phase-shifted carriers of 1 kHz are used for the four-level modulators in the CMV reduction technique.

$$Vxn_{CMVR-PSCPDPWM-mL}(t) = \left(\frac{\sqrt{3}MV_{dc}}{4} \right) \cos\left(\omega_o t + \theta_0 + \frac{\pi}{6} \right)$$

$$+ \frac{2V_{dc}}{\pi} \sum_{m=1}^{\infty} \frac{1}{(Nm)} \sum_{n=-\infty}^{\infty} \begin{bmatrix} J_n\left((Nm)\frac{\pi}{2}M \right) \sin\left(\frac{n\pi}{3} \right) \sin\left(\frac{\pi}{2}[(Nm) + n] \right) \\ \times \cos\left((Nm)[\omega_c t + \theta_c] + (n)\left[\omega_o t + \theta_0 - \frac{\pi}{3} \right] + \frac{\pi}{2} \right) \end{bmatrix}$$

$$(14.33)$$

$$Vab_{CMVR-PSCPDPWM-mL}(t) = \left(\frac{3MV_{dc}}{4} \right) \cos\left(\omega_o t + \frac{\pi}{3} \right)$$

$$+ \frac{4V_{dc}}{\pi} \sum_{m=1}^{\infty} \frac{1}{(Nm)} \sum_{n=-\infty}^{\infty} \begin{bmatrix} J_n\left((Nm)\frac{\pi}{2}M \right) \sin^2\left(\frac{n\pi}{3} \right) \sin\left(\frac{\pi}{2}[(Nm) + n] \right) \\ \times \cos\left((Nm)[\omega_c t] + (n)\left[\omega_o t - \frac{\pi}{3} \right] + \pi \right) \end{bmatrix}$$

$$(14.34)$$

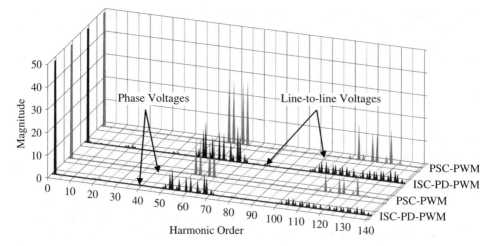

Figure 14.29 Magnified view of output voltages harmonic spectrum synthesized using classical PSC-PWM [12–14] and the proposed hybrid ISC-PD-PWM.

$$Vxn_{CMVR-ISCPDPWM-mL}(t) = \left(\frac{\sqrt{3}MV_{dc}}{4}\right)\cos\left(\omega_o t + \theta_0 + \frac{\pi}{6}\right)$$

$$+\frac{V_{dc}}{\pi}\sum_{m=1}^{\infty}\frac{1}{(Nm)}\sum_{n=-\infty}^{\infty}\begin{bmatrix} J_n((Nm)\pi M)\sin\left(\frac{n\pi}{3}\right) \\ \times\begin{pmatrix} \sin\left(n\frac{\pi}{2}\right)\times\cos\left((Nm)[\omega_c t + \theta_c] + (n)\left[\omega_o t + \theta_0 - \frac{\pi}{3}\right] + \frac{\pi}{2}\right) \\ +\cos\left(n\frac{\pi}{2}\right)\times\sin\left((Nm)[\omega_c t + \theta_c] + (n)\left[\omega_o t + \theta_0 - \frac{\pi}{3}\right] - \frac{\pi}{2}\right) \end{pmatrix} \end{bmatrix}$$

(14.35)

$$Vab_{CMVR-ISCPDPWM-mL}(t) = \left(\frac{3MV_{dc}}{4}\right)\cos\left(\omega_o t + \frac{\pi}{3}\right)$$

$$+\frac{2V_{dc}}{\pi}\sum_{m=1}^{\infty}\frac{1}{(Nm)}\sum_{n=-\infty}^{\infty}\begin{bmatrix} J_n((Nm)\pi M)\sin^2\left(\frac{n\pi}{3}\right) \\ \times\begin{pmatrix} \sin\left(n\frac{\pi}{2}\right)\times\cos\left((Nm)[\omega_c t] + (n)\left[\omega_o t - \frac{\pi}{3}\right] + \pi\right) \\ -\cos\left(n\frac{\pi}{2}\right)\times\sin\left((Nm)[\omega_c t] + (n)\left[\omega_o t - \frac{2\pi}{3}\right] - \pi\right) \end{pmatrix} \end{bmatrix}$$ (14.36)

According to the double Fourier series expressed in equations (14.33)–(14.36), all the carrier harmonics terms are being removed when obtaining the (S_{PQc}, S_{QRc}, S_{RPc}). Besides, the crossover intersection of these adjacent phase reference signals will also cancel away the triple sideband harmonics naturally, which explains the sin ($n\pi/3$) terms carried in these expressions.

However, the alterations in the geometry of the carrier will generate the sideband harmonics differently as well. For the classical PSC-PWM (alternatively equivalent PSC-PD-PWM), both equations (14.33) and (14.34) describe that odd sideband harmonics occur around even N multiples of f_c; and, vice versa, even sideband harmonics occur around odd N multiples of f_c. On the contrary, an infinite number of sideband

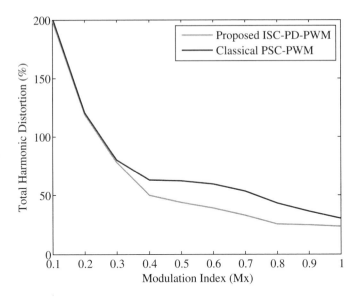

Figure 14.30 Comparison of output voltage THD performance under CMV reduction technique.

harmonics are found at all N multiples of f_c for the proposed ISC-PD-PWM in equations (14.35) and (14.36).

Despite of such differences, it is necessarily to highlight that the magnitudes of the sideband harmonics in the PSC-PWM are twice greater than those in the proposed hybrid ISC-PD-PWM scheme. Together with the cancellation of the triplen sideband harmonics, the harmonic profile is proven to be greatly enhanced for the proposed ISC-PD-PWM modulation as compared to the classical PSC-PWM (PSC-PD-PWM) (see Fig. 14.30).

Based on the proposed CMV reduction approach, the THD of the respective output phase voltages and line-to-line voltages are identical, since both voltages share the same spectral harmonic characteristics. Given the proven facts that higher-resolution voltage waveforms can be produced using ISC-PD-PWM, the input filter size can be further reduced. The THD is popularly used as a figure of merit for filter designing.

In the case of motor drive applications, it is necessary to analyze the current harmonic performances based on the two hybrid modulation schemes used for the ANPC. Hence, the current THD can be predicted using the load-independent metric, which is the weighted total harmonic distortion (WTHD). The comparative result in Fig. 14.31 shows that the CMV reduction technique under the proposed ISC-PD-PWM achieves much lower output current THD except for the case of under-modulation conditions ($\text{Mx}_{\text{sine}} < 0.33$).

14.6.3 Switching Frequency Reduction

When implementing the symmetrical geometry of the triangular carriers in both the normal modulation technique and the CMV reduction strategy, it should be noted that the next switching state will remain the same as the previous conditions whenever a

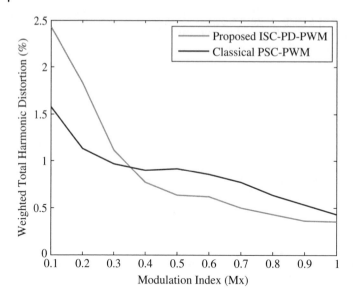

Figure 14.31 Comparison of output voltage weighted THD performance under CMV reduction technique.

carrier period is over. The continuous switching transitions can be clearly observed in Fig. 14.32 for the triangular PWM modulation.

However, these continuous switching transitions can be eliminated by changing the carrier shape to asymmetrical (sawtooth). Hence, the switching transitions for the sawtooth PWM modulation are now between consecutive carrier periods. Therefore, the number of switching transitions is significantly reduced using the sawtooth carrier. The reduction of switching transitions would be more obvious every time the reference signal Mx_{sine} crosses over the zero reference magnitude (see Fig. 14.32).

As such, the average switching frequency of the converter is determined by taking the mean of the total switching frequency across the switches S_{x1}, S_{x2}, S_{x3}, S_{x4}, and S_{x5} in the case of 7L-ANPC. The comparative result between the classical PSC-PWM and the proposed ISC-PD-PWM schemes employed in the CMV reduction technique is shown in Fig. 14.33. It is evidently proven that the proposed hybrid ISC-PD-PWM had remarkably reduced the average switching frequency, especially for the low modulation indices.

14.6.4 Power Device Losses

The total distributed losses (total switching loss and total conduction loss) of the discrete components are shown in Fig. 14.34 based on the low rated power conditions given in Table 14.5. The loss comparison is evaluated between the two modulation schemes (classical PSC-PWM and the proposed hybrid ISC-PD-PWM) under the CMV reduction method based on different modulation indices. The results clearly show that both conduction loss and switching loss are notably decreased for the ANPC employing the proposed ISC-PD-PWM modulation scheme.

It is worth noting that the differences in the total distributed losses between ISC-PD-PWM and PSC-PWM are relatively smaller, especially during the

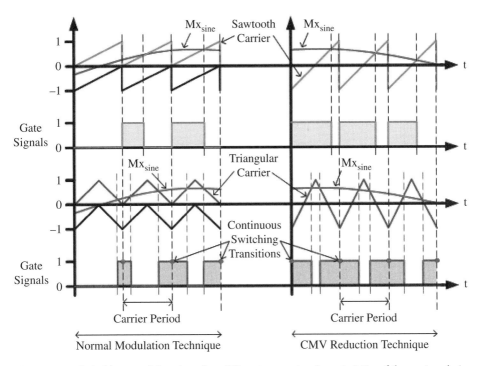

Figure 14.32 Switching conditions based on different geometry characteristics of the carriers during modulation.

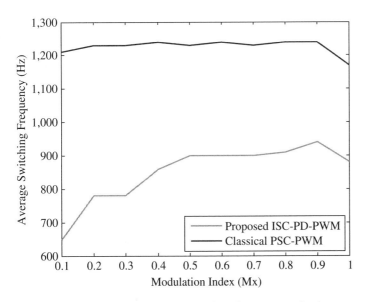

Figure 14.33 Comparison of average switching frequency under the proposed CMV reduction technique.

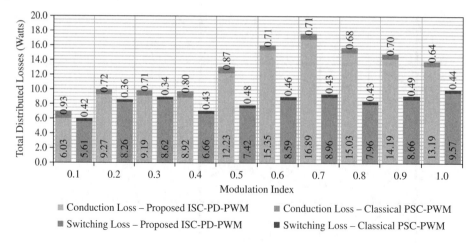

Figure 14.34 Total distributed losses comparison of a 7L-ANPC phase leg under the proposed CMV reduction technique.

Table 14.5 Configuration Settings for Simulation and Hardware Prototype of the Two-Cell 7L-ANPC.

System Parameters	Settings
Reference sinusoidal frequency (f_o)	50 Hz
Triangular carrier frequency (f_c)	1 kHz
Modulation amplitude (Mx_{sine})	0.9
dc voltage source	250 Vdc
dc-link and flying capacitors ($C_{x,p}{}^{\text{th}}$)	2200 μF
Load – resistors / inductors	37.5 Ω / 40 mH
Filter – resistors / capacitors	4 Ω / 25 μF

under-modulation condition ($\text{Mx}_{\text{sine}} < 0.33$). This is due to the high conduction loss incurred in this case, since it is greatly affected by the expectedly poor current harmonic performance based on the voltage WTHD (see Fig. 14.31).

Nevertheless, the total distributed losses for ISC-PD-PWM are still considerably lower because of the greatly reduced switching loss. It is clearly owing to its asymmetrical wave that the continuous switching transitions are eliminated. Hence, the ANPC converter employing the proposed hybrid ISC-PD-PWM features higher efficiency than the classical PSC-PWM modulation scheme (alternatively equivalent PSC-PD-PWM).

14.6.5 Experimental Verification

The proposed common mode reduction technique using both classical PSC-PWM and the proposed ISC-PD-PWM is applied to the three-phase 7L-ANPC. The performance is co-simulated between PSIM v9.0.4 Simcouple and MATLAB R2010b Simulink®, as

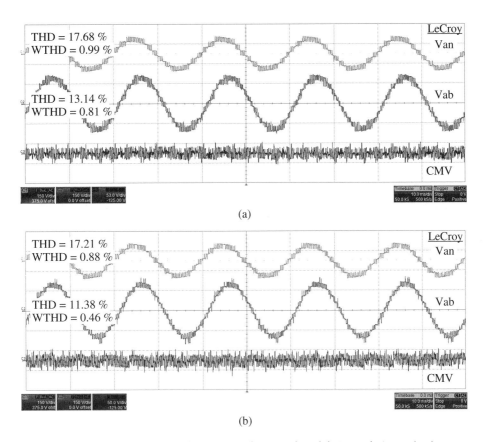

Figure 14.35 Experimental results of 7L-ANPC under normal modulation technique – (top) output phase voltage; (middle) output line-to-line voltage; (bottom) common mode voltage: (a) conventional PSC-PWM; (b) proposed hybrid ISC-PD-PWM.

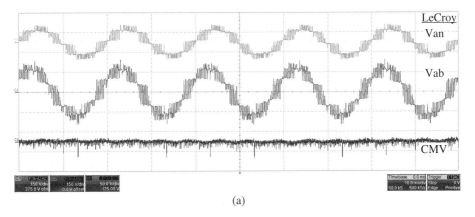

Figure 14.36 Experimental results of 7L-ANPC using the proposed CMV reduction technique – (top) output phase voltage; (middle) output line-to-line voltage; (bottom) common mode voltage: (a) conventional PSC-PWM; (b) proposed hybrid ISC-PD-PWM.

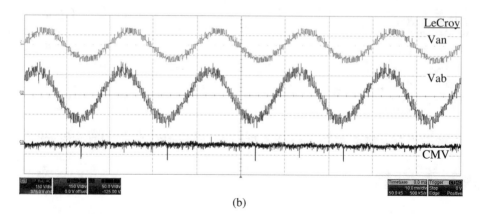

(b)

Figure 14.36 (*Continued*)

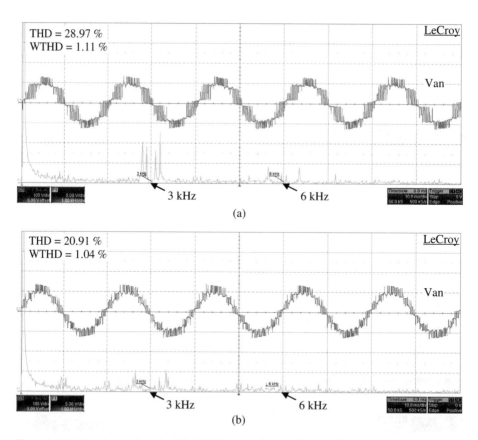

Figure 14.37 Experimental result of 7L-ANPC output phase voltage and its FFT spectrum: (a) conventional PSC-PWM; (b) proposed hybrid ISC-PD-PWM.

well as validated experimentally with the aid of dSPACE RT1103 Real-Time Workshop (RTW). All the results are presented based on the detailed settings given in Table 14.5.

The output voltages of 7L-ANPC in Fig. 14.35 are clearly synthesized using the normal modulation technique based on both discussed hybrid PWM schemes. The CMV under the proposed ISC-PD-PWM in Fig. 14.35(b) is slightly distorted as compared to the CMV under the classical PSC-PWM in Fig. 14.35(a). However, the common mode voltages are significantly reduced in Fig. 14.36 by applying the proposed common mode reduction technique to the ANPC. It is also noted that the CMV in Fig. 14.36(b) is now much reduced as compared to the one in Fig. 14.36(a). Even though the CMV is supposed to be totally eliminated as simulated in Fig. 14.24, those spikes as seen in Fig. 14.36 are caused by the dead-time interval effects.

The harmonic spectrum of the output phase voltages and line-to-line voltages for the 7L-ANPC employing the CMV reduction approach is shown in Figs. 14.36 and 14.37, respectively, under different PWM schemes. The magnitudes of the sideband harmonics in the classical PSC-PWM, as shown in Figs. 14.36(a) and 14.37(a), are approximately twice greater than those in the proposed ISC-PD-PWM, shown in Figs. 14.36(b) and 14.37(b).

Moreover, the experimental harmonic spectrums are exactly matched with the analytical results in Fig. 14.38. Thus, one can conclude that high-resolution waveforms are

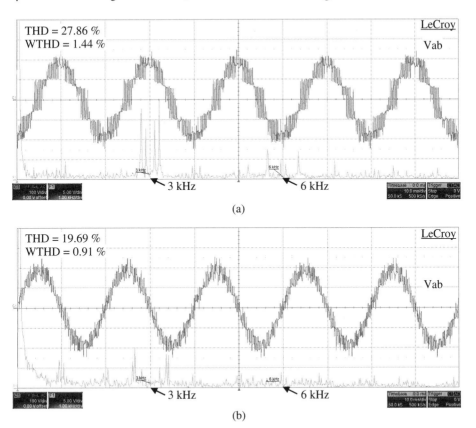

(a)

(b)

Figure 14.38 Experimental result of 7L-ANPC output line-to-line voltage and its FFT spectrum: (a) conventional PSC-PWM; (b) proposed hybrid ISC-PD-PWM.

synthesized for the ANPC when employing the proposed ISC-PD-PWM scheme in conjunction with the CMV reduction technique. The agreement between the simulation and the experimental results has also confirmed the feasibility of the proposed CMV reduction technique, as well as that of the enhanced THD and WTHD performances achieved using the proposed ISC-PD-PWM.

In addition to that, the asymmetrical geometry of the sawtooth carriers eliminates the continuous switching transitions, which results a great reduction in the switching frequency of the power devices. Because of these features, the efficiency of ANPC is further improved as well, with the total device power losses significantly decreased, especially under the low rated power condition. As such, the perceived merits of employing the proposed technique with hybrid ISC-PD-PWM modulation makes ANPC more attractively fitted for its well-known MV drive systems.

References

1 S. Kouro, M. Malinowski, K. Gopakumar, J. Pou, L. G. Franquelo, W. Bin *et al.*, "Recent advances and industrial applications of multilevel converters," *Industrial Electronics, IEEE Transactions on*, vol. 57, pp. 2553–2580, 2010.

2 P. Barbosa, P. Steimer, J. Steinke, M. Winkelnkemper, and N. Celanovic, "Active-neutral-point-clamped (ANPC) multilevel converter technology," in *Power Electronics and Applications, 2005 European Conference on*, pp. 10, 2005.

3 F. Kieferndorf, M. Basler, L. A. Serpa, J. H. Fabian, A. Coccia, and G. A. Scheuer, "ANPC-5L technology applied to medium voltage variable speed drives applications," in *Power Electronics Electrical Drives Automation and Motion (SPEEDAM), 2010 International Symposium on*, pp. 1718–1725, 2010.

4 F. Kieferndorf, M. Basler, L. A. Serpa, J. H. Fabian, A. Coccia, and G. A. Scheuer, "A new medium voltage drive system based on ANPC-5L technology," in *Industrial Technology (ICIT), 2010 IEEE International Conference on*, pp. 643–649, 2010.

5 J. Meili, S. Ponnaluri, L. Serpa, P. K. Steimer, and J. W. Kolar, "Optimized pulse patterns for the 5-level ANPC converter for high speed high power applications," in *IEEE Industrial Electronics, IECON 2006 – 32nd Annual Conference on*, pp. 2587–2592, 2006.

6 E. Burguete, J. Lopez, and M. Zabaleta, "New five-level active neutral-point-clamped converter," *Industry Applications, IEEE Transactions on*, vol. 51, pp. 440–447, 2015.

7 J. I. Leon, L. G. Franquelo, S. Kouro, W. Bin, and S. Vazquez, "Novel modulator for the hybrid two-cell flying-capacitor based ANPC converter," in *Power Engineering, Energy and Electrical Drives (POWERENG), 2011 International Conference on*, pp. 1–6, 2011.

8 J. I. Leon, L. G. Franquelo, S. Kouro, W. Bin, and S. Vazquez, "Simple modulator with voltage balancing control for the hybrid five-level flying-capacitor based ANPC converter," in *Industrial Electronics (ISIE), 2011 IEEE International Symposium on*, pp. 1887–1892, 2011.

9 T. Guojun, D. Qingwei, and L. Zhan, "An optimized SVPWM strategy for five-level active NPC (5L-ANPC) converter," *Power Electronics, IEEE Transactions on*, vol. 29, pp. 386–395, 2014.

10 S. R. Pulikanti, G. Konstantinou, and V. G. Agelidis, "DC-link voltage ripple compensation for multilevel active-neutral-point-clamped converters operated with SHE-PWM," *Power Delivery, IEEE Transactions on,* vol. 27, pp. 2176–2184, 2012.

11 S. R. Pulikanti and V. G. Agelidis, "Control of neutral point and flying capacitor voltages in five-level SHE-PWM controlled ANPC converter," in *Industrial Electronics and Applications, 2009. ICIEA 2009. 4th IEEE Conference on,* pp. 172–177, 2009.

12 S. R. Pulikanti, G. S. Konstantinou, and V. G. Agelidis, "Generalisation of flying capacitor-based active-neutral-point-clamped multilevel converter using voltage-level modulation," *Power Electronics, IET,* vol. 5, pp. 456–466, 2012.

13 W. Kui, Z. Zedong, L. Yongdong, L. Kean, and S. Jing, "Neutral-point potential balancing of a five-level active neutral-point-clamped inverter," *Industrial Electronics, IEEE Transactions on,* vol. 60, pp. 1907–1918, 2013.

14 W. Kui, X. Lie, Z. Zedong, and L. Yongdong, "Capacitor voltage balancing of a five-level ANPC converter using phase-shifted PWM," *Power Electronics, IEEE Transactions on,* vol. 30, pp. 1147–1156, 2015.

15 S. Kouro, P. Lezana, M. Angulo, and J. Rodriguez, "Multicarrier PWM with DC-link ripple feedforward compensation for multilevel inverters," *Power Electronics, IEEE Transactions on,* vol. 23, pp. 52–59, 2008.

16 C. Seung-Jun, K. Sanggi, K. Hyeon-Sik, and S. Seung-Ki, "Common-mode voltage reduction of three-level four-leg PWM converter," *Industry Applications, IEEE Transactions on,* vol. 51, pp. 4006–4016, 2015.

17 M. M. Swamy, K. Yamada, and T. Kume, "Common mode current attenuation techniques for use with PWM drives," *Power Electronics, IEEE Transactions on,* vol. 16, pp. 248–255, 2001.

18 C. Xiyou, X. Dianguo, L. Fengchun, and Z. JianQiu, "A novel inverter-output passive filter for reducing both differential- and common-mode dv/dt at the motor terminals in PWM drive systems," *Industrial Electronics, IEEE Transactions on,* vol. 54, pp. 419–426, 2007.

19 C. Wenjie, Y. Xu, X. Jing, and F. Wang, "A novel filter topology with active motor CM impedance regulator in PWM ASD system," *Industrial Electronics, IEEE Transactions on,* vol. 61, pp. 6938–6946, 2014.

20 A. Videt, P. Le Moigne, N. Idir, P. Baudesson, and X. Cimetiere, "A new carrier-based PWM providing common-mode-current reduction and DC-bus balancing for three-level inverters," *Industrial Electronics, IEEE Transactions on,* vol. 54, pp. 3001–3011, 2007.

21 L. Poh Chiang, D. G. Holmes, Y. Fukuta, and T. A. Lipo, "Reduced common-mode modulation strategies for cascaded multilevel inverters," *Industry Applications, IEEE Transactions on,* vol. 39, pp. 1386–1395, 2003.

22 Z. Haoran, A. von Jouanne, D. Shaoan, A. K. Wallace, and W. Fei, "Multilevel inverter modulation schemes to eliminate common-mode voltages," *Industry Applications, IEEE Transactions on,* vol. 36, pp. 1645–1653, 2000.

23 D. Floricau, E. Floricau, and M. Dumitrescu, "Natural doubling of the apparent switching frequency using three-level ANPC converter," in *Nonsinusoidal Currents and Compensation, 2008. ISNCC 2008. International School on,* pp. 1–6, 2008.

24 D. Floricau, E. Floricau, L. Parvulescu, and G. Gateau, "Loss balancing for Active-NPC and Active-Stacked-NPC multilevel converters," in *Optimization of Electrical and Electronic Equipment (OPTIM), 2010 12th International Conference on*, pp. 625–630, 2010.

25 L. Jin, L. Jinjun, and D. Boroyevich, "A simplified three phase three-level zero-current-transition active neutral-point-clamped converter with three auxiliary switches," in *Applied Power Electronics Conference and Exposition (APEC), 2010 Twenty-Fifth Annual IEEE*, pp. 1521–1526, 2010.

26 W. Kui, L. Yongdong, Z. Zedong, X. Lie, and M. Hongwei, "Self-precharge of floating capacitors in a five-level ANPC inverter," in *Power Electronics and Motion Control Conference (IPEMC), 2012 7th International*, pp. 1776–1780, 2012.

27 F. Forest, T. A. Meynard, S. Faucher, F. Richardeau, J. J. Huselstein, and C. Joubert, "Using the multilevel imbricated cells topologies in the design of low-power power-factor-corrector converters," *Industrial Electronics, IEEE Transactions on*, vol. 52, pp. 151–161, 2005.

28 M. L. Heldwein, S. A. Mussa, and I. Barbi, "Three-phase multilevel PWM rectifiers based on conventional bidirectional converters," *Power Electronics, IEEE Transactions on*, vol. 25, pp. 545–549, 2010.

29 K. A. Corzine and J. R. Baker, "Reduced-parts-count multilevel rectifiers," *Industrial Electronics, IEEE Transactions on*, vol. 49, pp. 766–774, 2002.

30 R. H. Wilkinson, T. A. Meynard, and H. Du Toit Mouton, "Natural balance of multicell converters: The general case," *Power Electronics, IEEE Transactions on*, vol. 21, pp. 1658–1666, 2006.

31 B. P. McGrath and D. G. Holmes, "Natural capacitor voltage balancing for a flying capacitor converter induction motor drive," *Power Electronics, IEEE Transactions on*, vol. 24, pp. 1554–1561, 2009.

15

Multilevel Z-Source Inverters

Muhammad M. Roomi

15.1 Introduction

Power electronics technology is utilized to regulate the flow of electric energy, by supplying voltages and currents in a form that is optimally suited for user loads [1]. The energy regulation system consists of a power controller and a power converter interfaced by sensors and semiconductor driving circuits. The power controller performs operations including maximum power point tracking, grid synchronization, and fast dynamics for the system. The power converter then receives the signal and performs the conversion of magnitude, frequency, and phase of the electrical quantities before the smooth grid connection is affected. Inverters represent the converter operation, where the average power flow is from the dc to the ac side. High-performance voltage and current-fed inverters are usually used for dc-ac conversion.

The two traditional inverters, i.e. voltage source inverters (VSIs) and current source inverters (CSIs), are classified on the basis of the directions of power flow. In a typical VSI, the dc voltage source is supported by a large capacitor. This inverter comprises six switches, and each switch has an anti-parallel diode to provide current flow in both directions and unidirectional voltage blocking capacity. Although these inverters are frequently employed, there are some limitations, e.g. the dc input voltage needs to be greater than the ac output voltage. Therefore, this inverter operates as either buck or boost depending on the type of conversion required. In some applications where the dc input voltage is limited, an additional boost converter needs to be incorporated, which increases cost and lowers efficiency. In addition, in this traditional VSI, the switches in the upper and lower phase of the same leg preclude "on" status simultaneously. This results in a shoot-through (ST), which would eventually destroy the devices. In addition, an LC filter should be used to achieve sinusoidal voltage output. In a CSI, the dc current source is a relatively large inductor, which is fed by the dc voltage sources. This inverter consists of six switches with a diode connected in series, permitting current flow in one direction while blocking voltage in both directions. The CSI requires additional converters for boosting or bucking operation that may increase cost and lower efficiency, which is similar to the VSI. In addition, the ac voltage has to be greater than that of the dc input voltage for the CSI. To achieve this operation, the CSI acts as a boost or a buck converter depending on the application. Another major drawback of this inverter

Advanced Multilevel Converters and Applications in Grid Integration, First Edition.
Edited by Ali I. Maswood and Hossein Dehghani Tafti.
© 2019 John Wiley & Sons Ltd. Published 2019 by John Wiley & Sons Ltd.

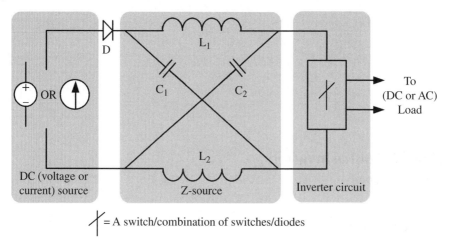

= A switch/combination of switches/diodes

Figure 15.1 Typical connection of ZSI to the load.

is that any one of the switches in the upper and lower legs has to be maintained in the on position continuously. When this condition is not maintained, an open circuit of the inductor causes destruction of the device. These traditional inverters operate usually as either a boost or a buck converter. In addition, they are sensitive to electromagnetic interference (EMI) noises. The foregoing problems are overcome by using a Z-source inverter (ZSI), proposed by F.Z. Peng in [2]. Fig. 15.1 shows the typical connection of a ZSI to the load.

Traditional VSIs and CSIs use a capacitor and an inductor, respectively. In contrast, the ZSI consists of a unique impedance network that connects the source and the load or another converter. The ZSI consists of two inductors, L_1 and L_2, and two capacitors, C_1 and C_2. The inductors and capacitors are connected in X-shape, thereby forming an impedance network that connects the source with the load or another converter. The source used for this converter may be a dc voltage source or a dc current source. The source can be a fuel cell, an inductor, a capacitor, a battery, a rectifier, or a combination of these sources. The switches used in the ZSI can be either a combination of semiconductor devices and an anti-parallel diode similar to the VSI or a series combination of switching devices and the diode as in the CSI. Thus, the drawbacks of the VSI and CSI main circuits become alike and can be overcome by an impedance inverter. The inductors used to form the impedance network can be either a split inductor or two different inductors. The added advantage of this inverter is that it can be used for all ac-to-dc, dc-to-ac, ac-to-ac, and dc-to-dc conversions.

The ZSI is an attractive option for the power conversion given that the theoretical output voltage can be any value regardless of the input. This feature ensures that the ZSI outperforms the traditional VSI and CSI. In addition, traditional converters have only eight switching states (six active states when the dc voltage source is connected to the load and two zero states when the load is shorted by either the upper or lower switches in the three legs). The ZSI provides an additional state when the load is shorted, enabling this inverter to possess nine switching states. The additional state, which is the ST state, can be obtained in three ways:

- Shorting the switches in any one-phase leg
- Shorting the switches in any two-phase leg
- Shorting all the devices in the three-phase leg

This additional state provides the ZSI with its unique buck-boost characteristic.

15.2 Two-Level ZSI

The ZSI can be configured into a VSI or CSI depending on the input supply. The network structure is identical for both the configurations, with the impedance network connected between the source and the traditional VSI bridge. This impedance network will provide the required voltage boost capability, while at the same time performing the voltage buck operation depending on the application. Even though both the voltage- and current-type converters are introduced simultaneously, voltage-type ZSIs are increasingly preferred to current-type ZSIs. Therefore, the modulation techniques, dynamics, and sizing of voltage-type inverters have been thoroughly investigated recently. This inverter is commonly utilized in applications including motor drives, solar techniques, fuel cells, and electric vehicles. The operation of the inverter is unchanged irrespective of the application. The operation involves the charging of two inductors, which results in voltage boosting. The advantage of the Z-source over traditional two-stage inverter is attributable to its ability to perform the boosting operation without an additional switch. The function of the additional switch is compensated by an extra state in the ZSI, at which both the switches in a single leg are gated on. The topology of the voltage-type ZSI is illustrated in Fig. 15.2.

In traditional two-stage inverters, the shorting of the two switches in the same leg damages the circuit. However, in ZSI, this damage is prevented due to the restricting action by the inductors. In the normal and traditional converters, dead-time protection is needed to make sure the active switches are all gated "off" before any other switch is gated on. This happens for a few microseconds. During this period, a fly-back or flywheel diode is connected anti-parallel to every switch and provides a path for the current. In the ZSI, gating switches on or off in the same leg does not affect the circuit. Therefore, inverters may be operated without the encumbrances of dead-time protection. Subsequently, inverters are generally less sensitive to electromagnetic noises. The ZSI operates in two modes: ST state and non-shoot-through (N-ST) state. The operations of voltage-type ZSIs in these two states are explained below.

The analysis of the ZSI is based on the assumption that the inductors, L_1 and L_2, and the capacitors, C_1 and C_2, are identical. If both the inductance and capacitance values are identical, the impedance network attains symmetry. Therefore, with a symmetrical structure (illustrated in Fig. 15.2), the voltages across the inductors and capacitors are:

$$v_{L1} = v_{L2} = v_L \tag{15.1}$$

$$V_{C1} = V_{C2} = V_C \tag{15.2}$$

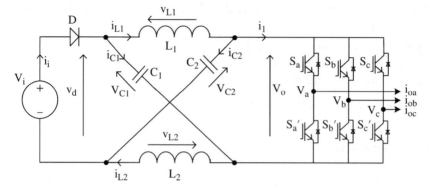

Figure 15.2 Voltage-type ZSI topology.

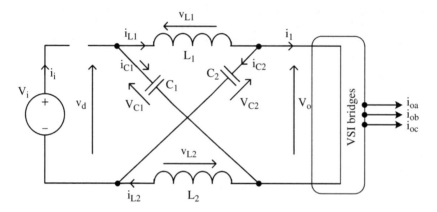

Figure 15.3 Equivalent circuit of voltage-type ZSI during ST state.

a) *ST State*

During the ST state, both of the switches in any one of the three-phase legs are gated on. The equivalent circuit of the inverter in this state is shown in Fig. 15.3. During this state, the diode is reverse biased (open circuit), and a short circuit represents the inverter bridge.

For a switching cycle T, the inverter operates in ST mode for a period T_0. From the equivalent circuit shown in Fig. 15.3, it is given that:

$$v_L = V_C \tag{15.3}$$

$$v_d = 2V_C \tag{15.4}$$

$$v_0 = 0 \tag{15.5}$$

b) *Non-ST State*

This operation of the ZSI during this state is similar to the operation of the traditional inverters. This state represents any one of the six active states or the two null states of the inverter bridge. Fig. 15.4 depicts the equivalent circuit of this state. During this state, the diode is forward biased and conducting, and the inverter bridge operates in any one of the traditional eight states.

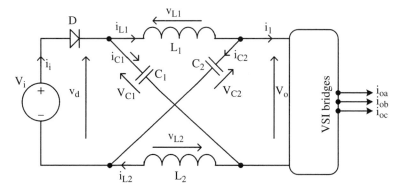

Figure 15.4 Equivalent circuit of voltage-type ZSI during N-ST state.

The inverter operates in N-ST mode for a period T_1. Thus, from the equivalent circuit of this mode as shown in Fig. 15.4:

$$v_L = V_i - V_C \tag{15.6}$$

$$v_d = V_i \tag{15.7}$$

$$v_0 = V_C - v_L = 2V_C - V_i \tag{15.8}$$

Here, v_L is the voltage across the inductor, V_C is the voltage across the capacitor, V_i is the voltage source, and $T = T_1 + T_0$.

For a period T, the average value of the inductor is given by:

$$V_L = \overline{v_L} = \frac{T_0 \cdot V_C + T_1(V_i - V_C)}{T} \tag{15.9}$$

In the steady state, the average value of the inductor is zero. Therefore:

$$V_L = \frac{T_0 \cdot V_C + T_1(V_i - V_C)}{T} = 0 \tag{15.10}$$

Hence:

$$\frac{V_C}{V_i} = \frac{T_1}{T_1 - T_0} \tag{15.11}$$

Similarly, the average value of the dc-link voltage can be given by:

$$V_0 = \frac{T_1}{T_1 - T_0}V_i = V_C \tag{15.12}$$

From equation (15.8), the peak value of the dc-link voltage can be modified as:

$$\widehat{v_0} = 2V_C - V_i = \frac{T}{T_1 - T_0}V_i = B \cdot V_i \tag{15.13}$$

Here, $B = \frac{T}{T_1 - T_0} \geq 1$ is the boost factor obtained from the ST state, and $\widehat{v_0}$ is the peak value of the dc-link voltage.

Multiplying the modulation index by half the peak value of the dc-link voltage generates the peak value of the ac output voltage. The peak value is given by equation (15.14):

$$\widehat{v_{ac}} = M \cdot \frac{\widehat{v_0}}{2} \tag{15.14}$$

Here, M is the modulation index. By substituting the value of $\widehat{v_0}$, equation (15.14) becomes:

$$\widehat{v_{ac}} = M \cdot \frac{B \cdot V_i}{2} \tag{15.15}$$

From equation (15.15), it is determined that the ac output voltage can be stepped up by adjusting the boost factor, which is absent in the traditional VSI. The product of the modulation index and the boost factor determines the buck-boost factor.

$$B_B = M \cdot B \tag{15.16}$$

The boost factor B is determined by the modulation index M and can be controlled by the ST duty cycle of the zero state over the N-ST duty cycle of the active states of the pulse width modulation (PWM) inverters. As the load terminal of the inverter is subjected to the same zero voltage during the ST state, this ST state does not disrupt the switching control of the inverter.

15.3 Three-Level ZSI

Three-level neutral-point-clamped (NPC) inverters possess notable advantages like higher output waveform quality, lower voltage stress, lower common mode voltage, and lower semiconductor losses. In addition, the power semiconductor devices need to block half of the dc-link voltage, when compared to a two-level inverter. Similar advantages apply to the Z-source NPC inverter. The principle of boosting the voltage in the three-level Z-source NPC inverter is achieved by utilizing the ST states, in which the upper two switches and lower two switches are gated on. However, the main challenge is to connect the neutral point of the inverter. In order to achieve that, various topologies have been proposed. Some of these topologies are the single Z-source network NPC (SZSN NPC) inverter and the dual Z-source network NPC (DZSN NPC) inverter. Figs. 15.5 and 15.6 depict the different topologies of single Z-source network NPC inverter. Fig. 15.5 shows the traditional way of creating the neutral point using a split capacitor bank [3]; whereas Fig. 15.6 indicates that the neutral point is created between the split input dc sources [4, 5]. The operation of the circuits at active and zero states, and the corresponding voltage representation are provided below.

15.3.1 Three-Level Single Z-Source Network with Neutral Point Connected to Split Capacitor Bank Inverter

The equivalent circuit of the SZSN NPC inverter using split capacitor banks is illustrated in Fig. 15.7, where the current source represents the inverter circuit and the load. In the two-level inverter, there is only one ST state. However, for three-level inverters, there are three ST states: (1) full shoot-through (FST), where all the switches (e.g. S_{a1}, S_{a2}, $S_{a1'}$, $S_{a2'}$) in a single leg are gated on, (2) upper shoot-through (UST) state, where the upper three switches are gated on, and (3) lower shoot-through (LST) state, where the lower three switches are gated on. The analysis of the ZSI is based on the assumption that the inductors, L_1 and L_2, and the capacitors, C_1 and C_2 possess the same values, represented as L and C, respectively. This assumption ensures the symmetry of the impedance network, which is explained by equations (15.1) and (15.2). The dc-link voltage levels are

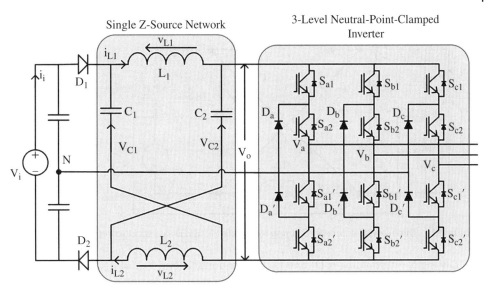

Figure 15.5 Topology of single Z-source network with neutral-point connected to split capacitor bank.

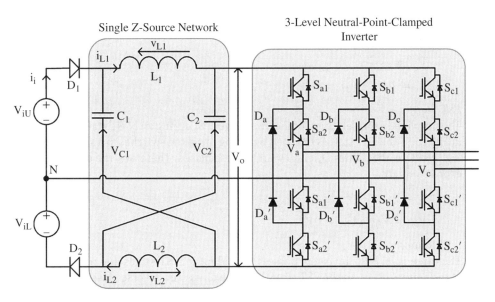

Figure 15.6 Topology of single Z-source network with neutral-point connected to split input dc source.

represented by $V_{(+)}$, V_N, and $V_{(-)}$. The equivalent circuits and the mode of operation for different states are explained below.

a) Non-ST State

The dc-link capacitors are represented as V_{d1} and V_{d2}, respectively. Fig. 15.7 shows the simplified representation of the three-level ZSI during the N-ST state.

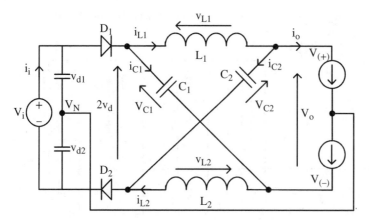

Figure 15.7 Three-level Z-source NPC during N-ST state (D_1 and D_2 are conducting).

During N-ST, the diodes D_1 and D_2 are conducting, and $V_{d1} = V_{d2} = v_d$. Therefore, the voltage expressions for this state are given as:

$$v_L = 2v_d - V_C \tag{15.17}$$

$$V_{(+)} = +\frac{V_0}{2} \tag{15.18}$$

$$V_N = 0V \tag{15.19}$$

$$V_{(-)} = -\frac{V_0}{2} \tag{15.20}$$

$$V_0 = 2(V_C - v_d) \tag{15.21}$$

b) *FST State*

The simplified representation of the three-level ZSI during an N-ST state is illustrated in Fig. 15.8. The FST state is initiated by gating on all the switches in a single leg. During this state, the diodes D_1 and D_2 are reverse biased. Therefore, the corresponding expressions are:

$$v_L = V_C \tag{15.22}$$

$$V_{(+)} = V_N = V_{(-)} = 0V \tag{15.23}$$

c) *UST State*

The three-level NPC inverters possess two additional ST states due to the presence of the neutral point. In the FST state, both the diodes are reverse biased. However, in the additional states, one of the diodes is blocking, while the other is conducting. One of the two additional states is the UST state, which is achieved by shorting the upper three switches of the inverter. In this state, diode D_2 is reverse biased, and diode D_1 is conducting. The equivalent circuit of the state is depicted in Fig. 15.9.

During this state, the voltage mathematical expressions are given by:

$$v_{L1} = V_{d1} \tag{15.24}$$

$$V_{(+)} = 0, V_N = 0 \tag{15.25}$$

$$V_{(-)} = V_{d1} - V_{C1} \tag{15.26}$$

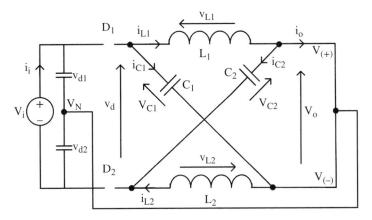

Figure 15.8 Three-level Z-source NPC during FST state (D₁ and D₂ are blocking).

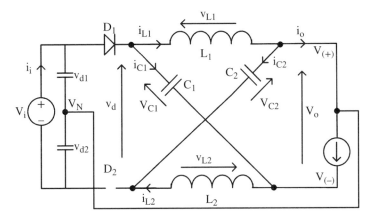

Figure 15.9 Three-level Z-source NPC during UST state (D₁ conducting and D₂ blocking).

d) LST State

Another additional state is the LST state, where the lower three switches are shorted to provide a voltage boost. In this state, the diode D1 is reverse biased while diode D2 is conducting, which can be deduced from the equivalent circuit depicted in Fig. 15.10.

During this state, the voltage expressions are:

$$v_{L2} = V_{d2} \tag{15.27}$$

$$V_{(+)} = -V_{d2} + V_{C2} \tag{15.28}$$

$$V_{(-)} = 0, V_N = 0 \tag{15.29}$$

If the UST, LST and N-ST are chosen for the inverter operation, the averaged inductor voltage for a period of T is given by:

$$V_C = \frac{2v_d^*(1 - T_0/2T)}{1 - T_0/T} \tag{15.30}$$

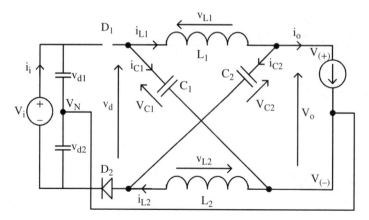

Figure 15.10 Three-level Z-source NPC during LST (D$_2$ conducting and D$_1$ blocking).

Here, T_0 is the common ST period for both UST and LST states. Using equation (15.30), the peak dc-link voltage is given by:

$$\hat{v}_0 = V_C - v_L = \frac{2v_d}{1 - T_0/T} \tag{15.31}$$

Therefore, the peak ac voltage is derived as:

$$\widehat{v_{ac}} = M \cdot \frac{\hat{v}_0}{2} \tag{15.32}$$

15.3.2 Three-Level Single Z-Source Network with Neutral Point Connected to Split Input dc Source Inverter

The split dc source SZSN NPC inverter is similar to the split capacitor bank except for the neutral-point formation. In this topology, the neutral point is formed between the two input dc sources (V_{iU} and V_{iL}) as depicted in Fig. 15.6. The mode of operation includes N-ST, FST, UST, and LST states. To achieve voltage boosting, the inverter can be operated in two combinations: (1) N-ST state with FST state and (2) N-ST state with UST and LST states.

a) Non-ST State

The simplified representation of the three-level ZSI during the N-ST state is illustrated in Fig. 15.11.

Considering $V_{iU} = V_{iL} = V_i$ and during N-ST, diodes D$_1$ and D$_2$ are conducting. Therefore, the voltage expressions for this state are given as:

$$v_L = 2V_i - V_C \tag{15.33}$$

$$V_{(+)} = +\frac{V_0}{2} \tag{15.34}$$

$$V_N = 0V \tag{15.35}$$

$$V_{(-)} = -\frac{V_0}{2} \tag{15.36}$$

$$V_0 = 2(V_C - V_i) \tag{15.37}$$

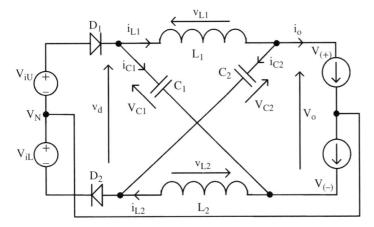

Figure 15.11 Three-level Z-source NPC during N-ST state (D_1 and D_2 are conducting).

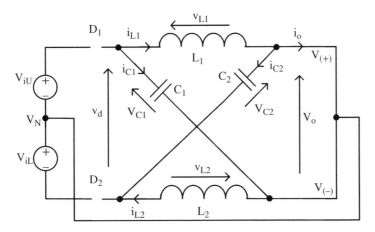

Figure 15.12 Three-level Z-source NPC during FST state (D_1 and D_2 are reverse biased).

b) FST State

The simplified representation of the three-level ZSI during an N-ST state is shown in Fig. 15.12. The FST state is initiated by gating on all the switches in a single leg. During this state, the diodes D_1 and D_2 are reverse biased. Therefore, the corresponding expressions can be given by:

$$v_L = V_C \tag{15.38}$$

$$V_{(+)} = V_N = V_{(-)} = 0V \tag{15.39}$$

c) UST State

The three-level NPC inverters possess an additional two ST states due to the presence of a neutral point. In the FST state, both the diodes are reverse biased. However, in the additional states, one diode is blocking while the other is conducting. One of the two additional states is the UST state, which is achieved by shorting the upper three

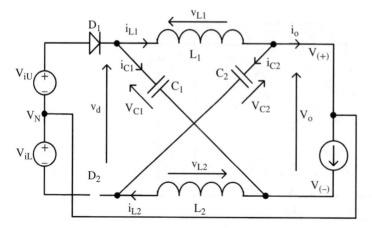

Figure 15.13 Three-level Z-source NPC during UST state (D_1 forward and D_2 reverse biased).

switches of the inverter. In this state, the diode D_2 is reverse biased, and the diode D_1 is conducting. The equivalent circuit of this state is shown in Fig. 15.13.

During this state, the voltage mathematical expressions are given by:

$$v_{L1} = V_{iU} \tag{15.40}$$

$$V_{(+)} = 0, V_N = 0 \tag{15.41}$$

$$V_{(-)} = V_{iU} - V_{C1} \tag{15.42}$$

d) LST State

The other additional ST state is the LST state, in which the lower three switches are shorted to provide the voltage boost. In this state, diode D_1 is reverse biased and diode D_2 is conducting, which can be determined from the equivalent circuit represented by Fig. 15.14.

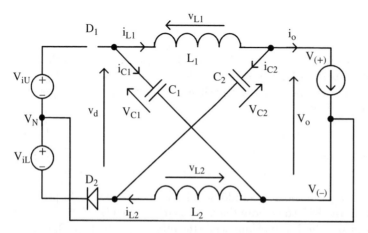

Figure 15.14 Three-level Z-source NPC during LST (D_2 forward and D_1 reverse biased).

In this state, the voltage expressions are:

$$v_{L2} = V_{iL} \tag{15.43}$$

$$V_{(+)} = -V_{iL} + V_{C2} \tag{15.44}$$

$$V_{(-)} = 0, V_N = 0 \tag{15.45}$$

If the FST and N-ST are chosen for the inverter operation, then averaging the inductor voltage for a period of T yields:

$$V_C = \frac{2V_i \cdot (1 - T_0/T)}{1 - 2T_0/T} \tag{15.46}$$

Here, T_0 is the ST period for the FST state. From equation (15.46), the peak dc-link voltage can be calculated as:

$$\widehat{v_0} = 2V_C - 2V_i = \frac{2V_i}{1 - 2T_0/T} \tag{15.47}$$

Therefore, the peak phase voltage is derived as:

$$\widehat{v_{ac}} = M \cdot \frac{\widehat{v_0}}{2} = M \cdot \frac{V_i}{1 - 2T_0/T} = M \cdot B \cdot V_i \tag{15.48}$$

Here, M is the modulation index and $B = 1/(1 - 2T_0/T)$ is the boost factor. B should be greater than 1 for boost operation.

15.3.3 Three-Level Dual Z-Source NPC Inverter

The most common topology for designing a Z-source NPC is by connecting two Z-source networks instead of the capacitor banks in traditional NPC inverters. Fig. 15.15 shows a dual Z-source network NPC (DZSN NPC) inverter with two Z-source networks (upper and lower) connected to the dc-link. The two Z-source networks are supplied by two isolated dc sources. The negative dc terminal of the upper Z-source network is connected to the positive dc terminal of the lower Z-source network. The inverter dc-link is attached to the middle point of the upper and lower networks, represented by point N in Fig. 15.15, thus forming a neutral point with a potential of 0 V. The employed impedance network results in voltage buck-boost and exhibits greater reliability. This is attributed to the insertion of ST state, and it would not cause any semiconductor to fail. In addition, as the ST allows the inverter to operate without any dead-time protection, the quality of the output is better. The only limitation of the DZSN NPC inverter is its cost, as it requires two passive Z-source networks.

With the neutral point forming between the upper and lower Z-source networks, the voltage boosting of both the networks should be simultaneous. For the upper Z-source network, the dc-link voltage boosting can be achieved by gating on the upper three switches of any phase leg (e.g. S_{a1}, S_{a2}, and $S_{a1'}$) with the diodes D_1 and $D_{a'}$ being forward biased and D_2 reverse biased. Similarly, the voltage boosting for the lower Z-source network is stimulated by gating on the lower three switches of any phase leg (e.g. S_{a2}, $S_{a1'}$, and $S_{a2'}$) with diodes D_2 and D_a in conducting mode and D_1 in blocking mode. While performing dc-link voltage boosting, an important factor that needs to be considered is

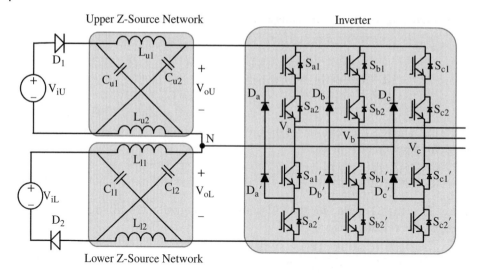

Figure 15.15 Topology of three-level dual Z-source NPC inverter.

that both the Z-source networks should not be boosted simultaneously. Simultaneous boosting short-circuits the full dc-link, which degrades the performance of the harmonics. Therefore, boosting must be performed at equal time intervals to ensure that an unbalanced dc-link voltage is avoided. Therefore, while implementing the modulation method for dual Z-source network, FST state (achieved by gating on all switches in a single-phase leg) should be avoided. The following section deals with the different modulation methods for Z-source NPC inverters.

15.4 Modulation Methods for Three-Level Z-Source NPC Inverter

Among the three topologies, the single Z-source network with a split dc source NPC inverter and a dual Z-source NPC inverter are preferred. Voltage boosting in the Z-source NPC inverter is achieved by utilizing the zero states in the traditional inverter as ST states. These ST states should be carefully inserted using the appropriate modulation method. In traditional two-level inverters, the full dc-link is shorted to provide voltage boosting. As discussed in the previous section, for three-level NPC inverters, voltage boosting is achieved by changing the zero states into the FST state (all switches are gated on), UST state (top three switches in any leg are gated on), and LST states (bottom three switches in any leg are gated on). The different operating states of the inverter, and the corresponding switches and diodes that are gated on during these operating states for a single-phase (phase "a") are listed in Table 15.1.

For traditional NPC inverters, the two most common carrier-based modulation methods are phase opposition disposition level-shifted PWM (POD LSPWM) and In-phase disposition level-shifted PWM (I-PD LSPWM). These carrier comparison methods are subsequently used for the three-level ZSI with the ST state inserted by performing an additional comparison. Nevertheless, sufficient care should be taken to ensure that the

Table 15.1 Switching States of a Three-level ZSI.

Operating States	Gated on switches	Gated on diodes	Output voltage
Non-shoot-through (N-ST)	S_{a1}, S_{a2}	D_1, D_2	$+v_i/2$
Non-shoot-through (N-ST)	$S_{a2}, S_{a1'}$	D_1, D_2 (D_a or $D_{a'}$)	0
Non-shoot-through (N-ST)	$S_{a1'}, S_{a2'}$	D_1, D_2	$-v_i/2$
Full shoot-through (FST)	$S_{a1}, S_{a2}, S_{a1'}, S_{a2'}$	—	0
Full shoot-through (FST)	$S_{a1}, S_{a2}, S_{a1'}, S_{c2}, S_{c1'}, S_{c2'}$	$D_{a'}, D_c$	0
Upper shoot-through (UST)	$S_{a1}, S_{a2}, S_{a1'}$	$D_{a'}, D_1$	—
Lower shoot-through (LST)	$S_{a2}, S_{a1'}, S_{a2'}$	D_a, D_2	—

normalized volt-second average is maintained at all times, whether the FST state or a combination of UST and LST states is used for dc-link voltage boosting.

15.4.1 Phase Opposition Disposition Modulation Method

In this method, the three-phase sinusoidal reference waves ($V_a{}^*$, $V_b{}^*$, and $V_c{}^*$) are compared with two triangular carrier waves ($V_{carr}+$ and $V_{carr}-$) that are phase shifted by 180° as depicted in Fig. 15.16. This comparison produces two individual gate signals for the upper two switches in each phase, and their corresponding NOT signals are fed

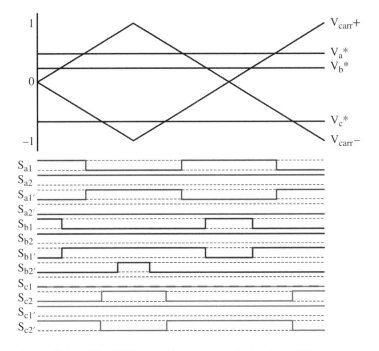

Figure 15.16 POD LSPWM switching sequence for traditional NPC.

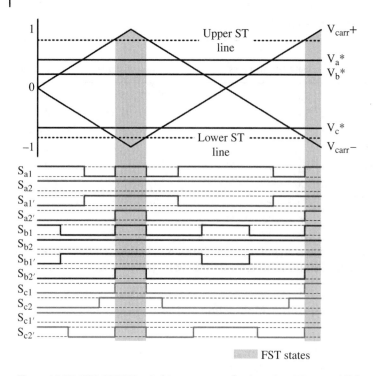

Figure 15.17 POD LSPWM switching sequence for three-level Z-source NPC.

to the lower two switches. This modulation method used for the traditional three-level inverters can be extended to the three-level Z-source because the rear end of both the configurations is an NPC circuit. However, if no modification is made to the traditional switching sequence, this modulation method leads to voltage buck operation when used with ZSI.

Therefore, in order to boost the voltage using the ZSI, insertion of ST states has to be carried out without modification of the normalized volt-second average appearing across the load. Consequently, two additional signals, UST line and LST line as represented in Fig. 15.17, are utilized. The UST line is compared to one of the triangular carriers ($V_{carr}+$), and the LST line is compared to the other triangular carrier ($V_{carr}-$) to determine the insertion of the ST states. From Fig. 15.17, it can be deduced that all switches are gated on during the ST state, leading to an FST state in place of a combination of UST and LST states. This is attributable to the fact that the ST due to the two ST lines occurs at the same period. From Table 15.1, it can be noted that the FST states are inserted either by gating on all the switches in a single-phase leg, or by gating on a combination of switches from the two-phase legs.

15.4.2 In-Phase Disposition (I-PD) Modulation Method

The following method was developed on the basis of carrier comparison, and it is similar to the POD modulation method. This I-PD modulation method is known to produce less harmonic content. In this method, two triangular waves that are in phase with each other are compared with three reference sinusoidal waves to produce the switching sequence

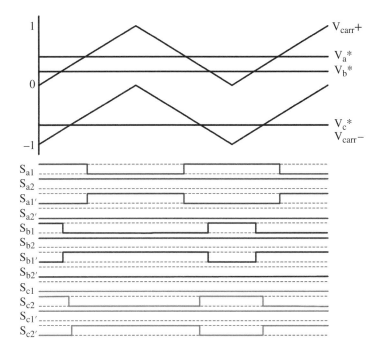

Figure 15.18 I-PD LSPWM switching sequence for three-level traditional NPC.

to the inverter. From Fig. 15.18, it can be inferred that at no point of time are all the switches in a single-phase leg gated on. Therefore, the FST state is not available for maintaining the normalized volt-second average across an externally connected load. Therefore, in order to apply the I-PD modulation method to the ZSI, the only alternative is to insert the ST state by a combination of the UST and LST states as shown in Table 15.1.

A modified switching sequence is developed for inserting ST states to the three-level Z-source NPC. The modified method consists of two additional ST lines that are similar to the ST lines in the phase opposition disposition method. These ST lines are compared with the carrier waves to produce the ST state. As shown in Fig. 15.19, the UST line is compared to the carrier signal ($V_{carr}+$), and subsequently the inverter enters the UST state. The LST line is compared to the carrier signal ($V_{carr}-$), and the inverter operates in the LST state. During these switching states, the control signal for a single-phase leg is "1110" for the UST state and "0111" for the LST state. Undistorted voltage boosting requires balanced insertion, which needs to be performed with high precision.

15.5 Modulation Method for Three-Level Dual Z-Source NPC Inverter

Subsequent to the discussion on the topologies and modulation methods for the three-level Z-source NPC inverter in Sections 15.3 and 15.4, this section presents a modulation method that has been recently proposed [6] for DZSN NPC. The highly

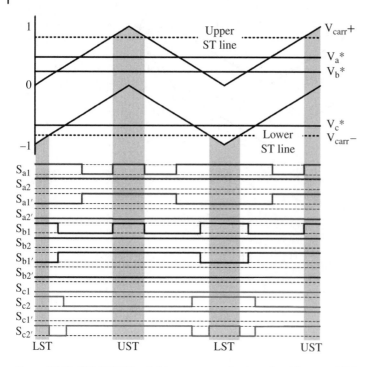

Figure 15.19 I-PD LSPWM switching sequence for three-level Z-source NPC.

preferred DZSN NPC inverter uses two isolated dc sources integrated with two different Z-source networks to power the inverter. The topology is shown in Fig. 15.15. The newly proposed modulation method uses the traditional carrier comparison technique. However, a single carrier is used instead of the conventional two carriers. An important factor when designing any modulation method for DZSN NPC is to avoid the FST state caused by shorting all switches in a single leg. As mentioned earlier, this FST state should be avoided to eliminate the volt-second error. Therefore, the proposed method uses the UST and LST states to provide dc-link voltage boosting. The operation of this method and its validation are discussed in the following subsections.

15.5.1 Reference Disposition Level-Shifted PWM Method

The new modulation method uses two reference signals and one carrier signal, which contrasts with the conventional carrier-based techniques. The two reference signals are level-shifted with respect to each other, which subsequently provides the switching sequence to the inverter when compared with the single carrier signal. Fig. 15.20 illustrates the reference disposition level-shifted PWM (RD LSPWM) method to produce PWM switching without ST. The signals to the switches are provided by comparing the two reference waves with the triangular carrier wave and are shown in Fig. 15.20.

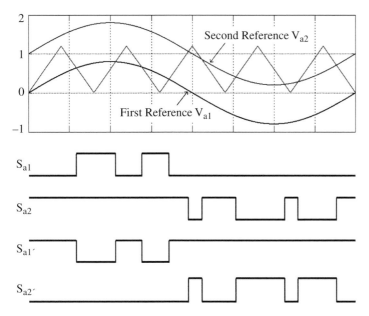

Figure 15.20 New RD LSPWM method without ST signals.

Equations (15.49) and (15.50) express the reference waves V_{a1} and V_{a2}, respectively, where $\theta = \omega t$ and $M-$ modulation ratio:

$$\left. \begin{aligned} V_{a1} &= \sin(\theta) \cdot M \\ V_{b1} &= \sin\left(\theta - \frac{2}{3}\pi\right) \cdot M \\ V_{c1} &= \sin\left(\theta - \frac{4}{3}\pi\right) \cdot M \end{aligned} \right\} \tag{15.49}$$

$$\left. \begin{aligned} V_{a2} &= 1 + [\sin(\theta) \cdot M] \\ V_{b2} &= 1 + \left[\sin\left(\theta - \frac{2}{3}\pi\right) \cdot M\right] \\ V_{c2} &= 1 + \left[\sin\left(\theta - \frac{4}{3}\pi\right) \cdot M\right] \end{aligned} \right\} \tag{15.50}$$

To achieve insertion of ST for the ZSI, traditional PWM strategies require modifications. Several carrier-based modified PWM methods for ZSI based on traditional control methods were proposed. A simple boost control (SBC) method [2] utilizes two lines to control the ST states. In maximum boost control (MBC) method [7], the ST state occurs when the carrier is greater than the reference waves. However, drawbacks like low-frequency ripples, voltage stress on switches, and the operating range of these techniques resulted in an improved maximum constant boost control (MCBC) method [8]. In MCBC, two modified curves greater than or equal to the amplitude of the reference waves are compared with the triangular wave to control the ST. In this method, by main-

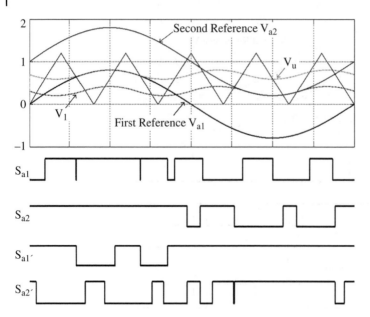

Figure 15.21 New RD LSPWM method with ST signals.

taining the ST duty ratio constant, the low-frequency current ripple can be eliminated, thereby reducing the volume and cost of the network. Furthermore, to reduce stress across the semiconductor switches, a greater voltage boost is required for any modulation index. In addition to a constant duty ratio and voltage stress, neutral-point voltage balancing needs to be considered. Considering these requirements, reference disposition level-shifted PWM (RD LSPWM) is implemented to achieve the maximum voltage gain using the MCBC method as illustrated in Fig. 15.21.

The sketch map of RD LSPWM depicted in Fig. 15.21 comprises eight modulation signals: two sets of three reference signals V_{a1}, V_{b1}, V_{c1} and V_{a2}, V_{b2}, V_{c2}, and two ST signals V_u and V_l. The first set of reference signals (V_{a1}, V_{b1}, V_{c1}) corresponds to the switches S_{x1} and $S_{x1'}$ where x = a, b, c, and the second set of reference signals (V_{a2}, V_{b2}, V_{c2}) corresponds to the switches S_{x2} and $S_{x2'}$ where x = a, b, c. The circuit enters the UST state when the upper modified signal V_u (which is equal to or greater than the peak value of the first set of reference signals) is lower than the carrier signal. Similarly, the circuit enters the LST state when the lower modified signal V_l (which is equal to the lower peak value of the second set of reference signals) is greater than the carrier signal. The switching during the N-ST states for the three-level ZSI follows a pattern identical to that of the traditional carrier-based PWM, with the exception that the two reference signals and the carrier signal are compared.

The UST and LST signals are periodical, and there are two half periods $[0, \pi/3]$ and $[\pi/3, 2\pi/3]$ in a single switching cycle. During the first half period of the switching cycle, the modified curves are given by:

UST curve:

$$V_u = M \left(\sqrt{3} + \sin\left(\theta - \frac{2}{3}\pi\right) \right) \text{ where } \theta \in [0, \pi/3] \tag{15.51}$$

LST curve:

$$V_l = M \left((1/M) + \sin \left(\theta - \frac{2}{3}\pi \right) \right) \text{ where } \theta \in \left[0, \pi/3 \right] \tag{15.52}$$

During the second half period of the switching cycle, the modified signals are as follows:

UST curve:

$$V_u = M \left(\sin(\theta) \right), \text{ where } \theta \in \left[\pi/3, 2\pi/3 \right] \tag{15.53}$$

LST curve:

$$V_l = M \left((1/M) + \sin(\theta) - \sqrt{3} \right), \text{ where } \theta \in \left[\pi/3, 2\pi/3 \right] \tag{15.54}$$

The ST duty ratio for any modulation index is the distance between the upper and lower envelope curves. Equations (15.51)–(15.54) indicate that this distance remains constant. This implies that the ST duty ratio (D_o) is constant. For the single ZSI, the boost factor can be determined from D_o and is given by [5]:

$$B = \frac{1}{1 - 2D_o} \text{ where } D_o = \frac{T_0}{T} \tag{15.55}$$

Now, the DZSN NPC, which is supplied through two individual dc sources, includes two Z-source networks between the source and the NPC inverter circuit. Therefore, two boost factors for the DZSN NPC are defined by the following expressions:

$$B_{up} = \frac{1}{1 - \dfrac{2(T_{oU})}{T}} \tag{15.56}$$

$$B_{low} = \frac{1}{1 - \dfrac{2(T_{oL})}{T}} \tag{15.57}$$

Here, B_{up} and B_{low} are the boost factors of the upper and lower Z-source networks, respectively; T_{oU}/T, T_{oL}/T define the ratio of the ST durations for the upper network and the lower network, respectively. From equations (15.56) and (15.57), the upper and lower dc-link voltages are given by equations (15.58) and (15.59), respectively.

$$V_{oU} = V_{iU} \cdot B_{up} = \frac{V_{iU}}{1 - \dfrac{2(T_{oU})}{T}} \tag{15.58}$$

$$V_{oL} = V_{iL} \cdot B_{low} = \frac{V_{iL}}{1 - \dfrac{2(T_{oL})}{T}} \tag{15.59}$$

At certain instances when the voltage levels are different, the neutral-point balancing will be affected. Hence, to achieve balancing, some modifications to the ST duty duration for both the upper and lower Z-source networks are required, and are shown in equations (15.60) and (15.61):

$$V_{oU} = V_{iU} \cdot B_{up} = \frac{V_{iU}}{1 - \dfrac{2(T_{oU} + T_T)}{T}} \tag{15.60}$$

$$V_{oL} = V_{iL} \cdot B_{low} = \frac{V_{iL}}{1 - \dfrac{2(T_{oL} + T_T)}{T}} \tag{15.61}$$

Here, V_{iU} and V_{iL} are the individual dc sources connected to the upper and lower Z-source networks. V_{oU} and V_{oL} are the corresponding output dc-link voltages, and T_T is the common ST duration for both the upper and lower Z-source networks. Ideally, both the dc-link sources are maintained at the same voltage levels. The boost factors determine the output voltage of the dc-links, which should be of identical magnitude to achieve low output total harmonic distortion (THD). Therefore, the peak ac voltage for the dual network can be expressed by equation (15.62), in which V_i represents the value of the dc sources:

$$\widehat{v_{ac}} = M \cdot B \cdot V_i \tag{15.62}$$

15.5.2 Switching States

From the topology depicted in Fig. 15.15, the two Z-source networks draw power from individual dc power sources. The negative dc line of the upper Z-source network is connected to the positive dc line of the lower Z-source network. The midpoint of these lines is connected to the dc-link of the inverter, thus forming a neutral point with a potential of 0 V. Furthermore, the ZSI has several unique ST states in addition to the active and zero states of the conventional inverters. As the Z-source NPC inverter boosts the dc-link voltage during these ST states, the ST interval needs to be generated carefully in order to equally distribute the boosted inductive energy into the Z-source network. Table 15.2 summarizes the switching states of the inverter during the two operating modes: ST and N-ST states.

According to Table 15.2, it can be surmised that the Z-source NPC inverter possesses three ST states, namely, FST, UST, and LST. FST occurs when all the switches of the single leg, or any two legs, or all three legs are shorted. UST represents the condition when the upper three switches (S_{x1}, S_{x2}, $S_{x1'}$ where x = a, b, c) in a leg are gated on with $D_{a'}$ being forward biased, and similarly LST implies that the bottom three switches (S_{x2}, $S_{x1'}$, $S_{x2'}$ where x = a, b, c) in a leg are gated on with D_a being forward biased.

During these ST states, the output voltage of the inverter is boosted. However, care should be taken that the boosting period of the upper and lower networks should be identical. This ensures that shorting of the full dc-link will be avoided. This also ensures balanced voltage boosting. As discussed earlier, by gating on all switches in three legs, the circuit enters FST mode, which also aids in voltage boosting. However, gating on all inverter switches leads to a "volt-second error" [5] in the inverter output. Therefore, the new method RD LSPWM uses UST and LST states to achieve voltage boosting. The diodes (D_1 and D_2) in Fig. 15.15, which are connected to the Z-source networks, ensure unidirectional operation. If the diodes were replaced with switches, then the appropriate switches must be gated off to disconnect the source during voltage boosting. In the chosen experimental setup, the diodes will reverse bias depending on the shorted Z-source network.

During the ST states, the inverter terminals are shorted, changing the output voltage of the inverter to 0 V. The inductive energy stored during ST is delivered to the ac load

Table 15.2 Switching States of the DZSN NPC Inverter, Illustrated in Fig. 15.15 (ST ≡ Shoot-Through).

Phase	Switching States	Switches (ON)	Diodes (ON)	Output
a	Non-ST	S_{a1}, S_{a2}	D_1, D_2	$+V_i$
a	Non-ST	$S_{a2}, S_{a1'}$	D_1, D_2, D_a (or $D_{a'}$)	0
a	Non-ST	$S_{a1'}, S_{a2'}$	D_1, D_2	$-V_i$
a	Upper-ST	$S_{a1}, S_{a2}, S_{a1'}$	-	0
a	Lower-ST	$S_{a2}, S_{a1'}, S_{a2'}$	-	0
a	Full-ST	$S_{a1}, S_{a2}, S_{a1'}, S_{a2'}$	-	0
b	Non-ST	S_{b1}, S_{b2}	D_1, D_2	$+V_i$
b	Non-ST	$S_{b2}, S_{b1'}$	D_1, D_2, D_b (or $D_{b'}$)	0
b	Non-ST	$S_{b1'}, S_{b2'}$	D_1, D_2	$-V_i$
b	Upper-ST	$S_{b1}, S_{b2}, S_{b1'}$	-	0
b	Lower-ST	$S_{b2}, S_{b1'}, S_{b2'}$	-	0
b	Full-ST	$S_{b1}, S_{b2}, S_{b1'}, S_{b2'}$	-	0
c	Non-ST	S_{c1}, S_{c2}	D_1, D_2	$+V_i$
c	Non-ST	$S_{c2}, S_{c1'}$	D_1, D_2, D_c (or $D_{c'}$)	0
c	Non-ST	$S_{c1'}, S_{c2'}$	D_1, D_2	$-V_i$
c	Upper-ST	$S_{c1}, S_{c2}, S_{c1'}$	-	0
c	Lower-ST	$S_{c2}, S_{c1'}, S_{c2'}$	-	0
c	Full-ST	$S_{c1}, S_{c2}, S_{c1'}, S_{c2'}$	-	0
x, y = combination of any two legs	Full-ST	$S_{x1}, S_{x2}, S_{x1'}$, $S_{y2}, S_{y1'}, S_{y2'}$	D_{x2}, D_{y1}	0
A, B, C	Full-ST	$S_{a1}, S_{a2}, S_{a1'}, S_{a2'}, S_{b1}, S_{b2}$, $S_{b1'}, S_{b2'}, S_{c1}, S_{c2}, S_{c1'}, S_{c2'}$	-	-

during any of the N-ST states mentioned in Table 15.2, where the switches are assembled in complementary pairs for each phase. From the table, the output voltages of the inverter during the N-ST states are determined to be $+V_i$, 0, $-V_i$.

15.5.3 Validation

The performance of the RD LSPWM method for the dual Z-source network is validated through simulations that were performed using PSIM with MATLAB/Simulink coupler. *The parameter values are chosen on the basis of similar researches conducted earlier.* The individual dc voltage sources V_{iU} and V_{iL} supplying the upper and lower Z-source networks are maintained at 100 V. Similarly, the Z-source network parameters for both networks are L = 2 mH and C = 1000 μF with a switching frequency of 10 kHz. A three-phase RL circuit serves as the load of the ZSI.

Identical dc Sources
In this section, the theoretical calculations of the dual Z-source network implementing the reference disposition method are compared with the simulation outputs, and

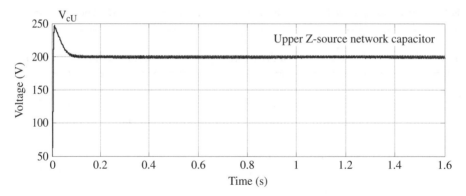

Figure 15.22 Capacitor voltage of the upper Z-source network with M = 0.85 and B = 2.1 (refer to Fig. 15.15).

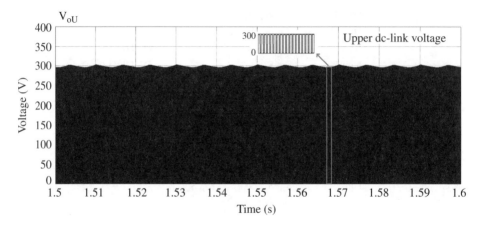

Figure 15.23 Voltage across the upper dc-link with M = 0.85 and B = 2.1 (refer to Fig. 15.15).

the results are discussed. For the evaluation, the modulation index was set to M = 0.85, with which the boost factor of the Z-Source in the MCBC method is calculated using the equation $B = 1/(\sqrt{3}M - 1)$. The value of boost factor was determined as 2.1. Thus, the ST duty cycle is set to $T_0/T = 0.26$. The output of the upper dc voltage is induced by the Z-source capacitors in the upper network. In regard to the circuit configuration given in Fig. 15.15, key waveforms for the configuration are presented in Figs. 15.22–15.30. Figs. 15.22 and 15.23 depict the voltage across the upper capacitor and upper dc-link, respectively. According to Fig. 15.23, the simulated average value of the upper dc-link voltage is computed to be 200 V. The computed value is similar to the theoretically calculated value of the upper dc-link, and is shown in equation (15.63):

$$V_{oU} = V_{iU} \cdot B_{up} = 100*2.1 = 210 \, V \tag{15.63}$$

Similarly, for the lower Z-source network, which is fed by a 100 V dc source, the capacitor voltage and the dc-link voltage are illustrated in Figs. 15.24 and 15.25. The average value of the lower dc-link voltage through simulation in Fig. 15.25 is computed

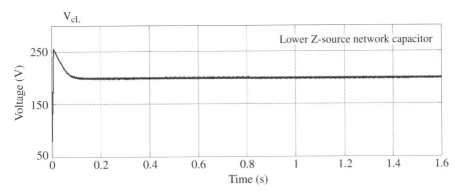

Figure 15.24 Capacitor voltage of the lower Z-source network with M = 0.85 and B = 2.1 (refer to Fig. 15.15).

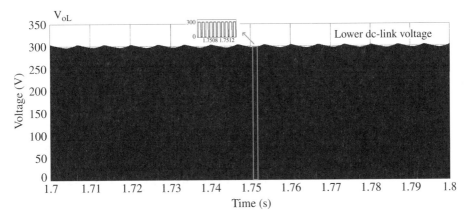

Figure 15.25 Voltage across the lower dc-link with M = 0.85 and B = 2.1 (refer to Fig. 15.15).

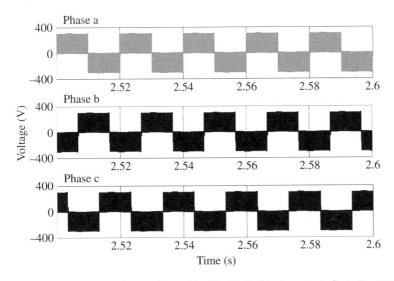

Figure 15.26 Output phase voltages (V_{aN}, V_{bN}, V_{cN}) of the inverter (refer to Fig. 15.15).

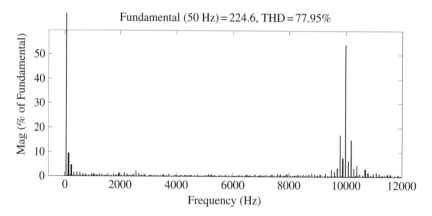

Figure 15.27 THD of output phase voltage of the inverter (refer to Fig. 15.15).

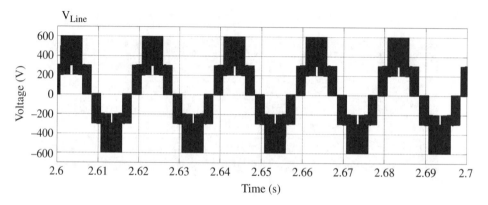

Figure 15.28 Output line voltage (V_{ab}) of the inverter (refer to Fig. 15.15).

as 202 V, which is similar to the expected theoretical value of 210 V calculated using equation (15.59):

$$V_{oL} = V_{iL} \cdot B_{low} = 100^*2.1 = 210 \text{ V} \tag{15.64}$$

The waveforms shown in Figs. 15.23 and 15.25 confirm that the upper and lower dc-link voltages are almost identical. Furthermore, the capacitor voltages are balanced by the proper insertion of the ST states using this method, which balances the neutral-point potential. Fig. 15.26 illustrates the output phase voltage of the inverter. The THD of the output phase voltage is depicted in Fig. 15.27.

Figs. 15.28 and 15.29 provide the output line voltage and its harmonic distortion waveforms, respectively. The THD of the output line voltage incorporating reference disposition is 47.69%. Fig. 15.30 illustrates the output line currents of the inverter, whose ripple can be eliminated using a proper filter, for the reference disposition method.

Non-identical dc Sources

Generally, the upper and lower Z-source networks are supplied by identical dc sources. However, when applied to a distributed energy resources (DER) system, the dc sources

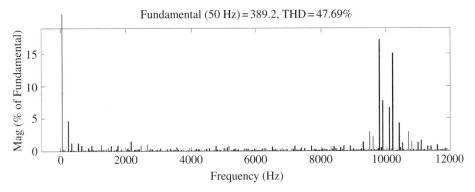

Figure 15.29 Total harmonic distortion (THD) of output line-to-line voltage of the inverter (refer to Fig. 15.15).

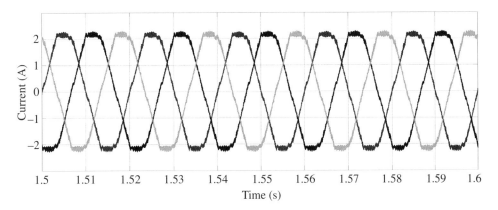

Figure 15.30 Output line currents of the inverter (refer to Fig. 15.15).

would be identical. This section analyzes the performance of the proposed method when the dc sources are not identical. Neutral-point balancing is a major problem that occurs when the sources are unbalanced. To achieve balancing during the non-identical condition, the ST duration of both the upper and lower Z-source networks should be adjusted. The new ST durations of both networks are explained by equations (15.60) and (15.61), respectively. The boost factor changes correspondingly as it is a function of the ST period, and is calculated from equation (15.55).

The key results for the system under non-identical dc sources are presented in this section. The upper (V_{cU}) and lower (V_{cL}) capacitor voltages when supplied by non-identical dc voltage sources are shown in Fig. 15.31. From this figure, it is inferred that the upper capacitor voltage is significantly greater than the lower capacitor voltage. For this output, no modification in the ST duration is introduced. After the modification in the ST, the subsequent output of the capacitor voltages is shown in Fig. 15.32. From the figure, it can be verified that both capacitors are balanced, thereby providing neutral-point balancing.

The capacitor voltages of the upper and lower Z-source network are balanced, and the upper and lower dc-link voltages are shown in Figs. 15.33 and 15.34, respectively. It is

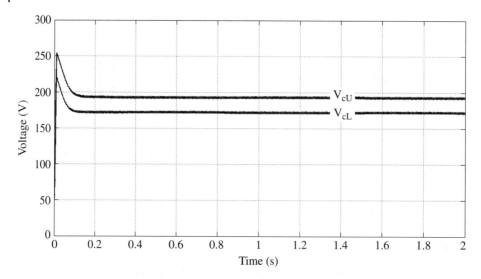

Figure 15.31 Upper and lower capacitor output voltages supplied by non-identical dc sources (without modification in ST duration) (refer to Fig. 15.15).

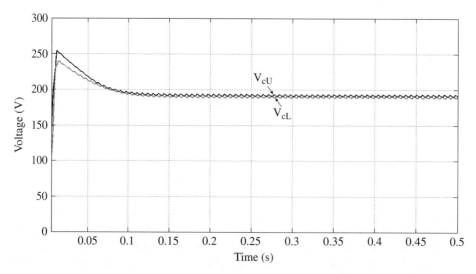

Figure 15.32 Upper and lower capacitor output voltages supplied by non-identical dc sources (with modification in ST duration) (refer to Fig. 15.15).

observed that the magnitudes of the upper and the lower dc-link voltages are identical. From the simulation results, the average values of the upper and lower dc-link voltages are computed as 191 V and 189 V, respectively. Therefore, the neutral-point potential balance is achieved by this proposed modulation method.

In this method, the two individual dc sources are non-identical. However, from Fig. 15.32, it can be determined that the capacitors in both networks are balanced. Therefore, the output phase voltage of the inverter is non-distorted as observed from Fig. 15.35.

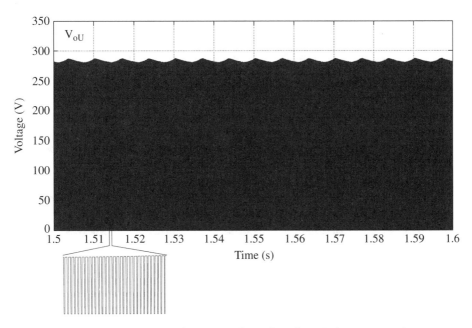

Figure 15.33 Upper dc-link voltage during non-identical condition (refer to Fig. 15.15).

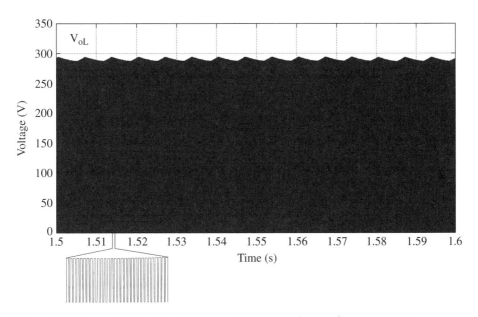

Figure 15.34 Lower dc-link voltage during non-identical condition (refer to Fig. 15.15).

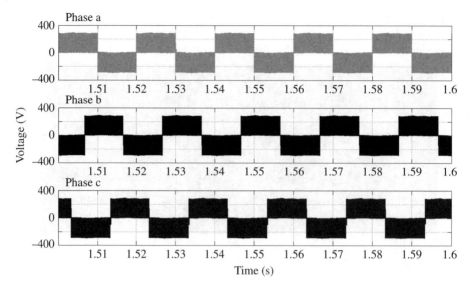

Figure 15.35 Output phase voltages (V_{aN}, V_{bN}, V_{cN}) of the inverter during non-identical condition (refer to Fig. 15.15).

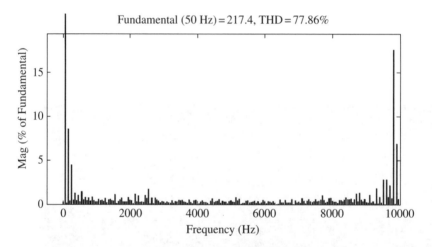

Figure 15.36 Fast Fourier transform (FFT) of the output phase voltage during non-identical condition (refer to Fig. 15.15).

The output voltages of the non-identical dc sources are obtained and plotted in Fig. 15.35. The peak value of the output voltage for the non-identical dc sources was determined as slightly lower than the case when identical dc sources were considered. The adjustment in the ST duration influences the boost factor. This is the primary reason for the variation of magnitudes between the identical and non-identical conditions. The THD of the output phase voltage is illustrated in Fig. 15.36.

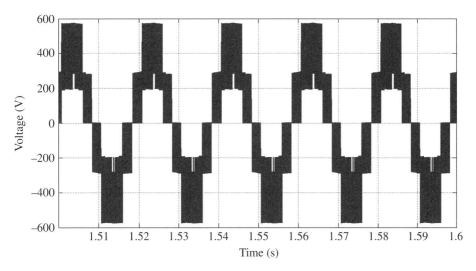

Figure 15.37 Output line voltage (V_{ab}) of the inverter during non-identical condition (refer to Fig. 15.15).

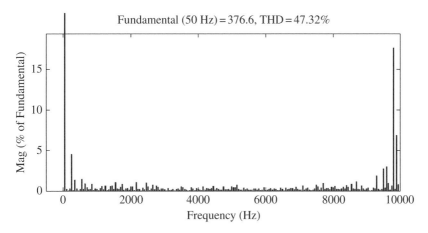

Figure 15.38 Fast Fourier transform (FFT) of output line voltage during non-identical condition (refer to Fig. 15.15).

Figs. 15.37 and 15.38 provide the output line voltage and its harmonic distortion waveforms, respectively. The THD of the output line voltage incorporating reference disposition is 47.32%. The harmonic analysis for both the cases is very similar. Fig. 15.39 depicts the three-phase output currents of the ZSI. There are small spikes that are visible, but these are absent when the dc sources are identical. These can be eliminated by either modifying the modulation index or by using a proper filter.

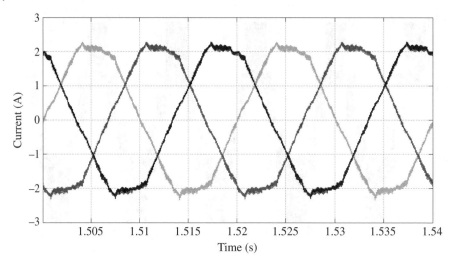

Figure 15.39 Output line currents during non-identical condition (refer to Fig. 15.15).

15.6   Reference Disposition Level-Shifted PWM for Non-ideal Dual Z-Source Network NPC Inverter

Several stages of dc voltages may be used to deliver the desired output voltage for multi-level inverters. Implementing multiple dc voltages helps in achieving a near-sinusoidal output voltage with the multilevel inverter, while adopting fundamental frequency switching. One of the most common multilevel topologies that is widely used is the diode-clamped inverter, commonly known as the NPC inverter. In Section 15.5, the NPC inverter was used along with a ZSI. To form the neutral point for the inverter, two different Z-source networks were implemented in the circuit, thereby forming the DZSN NPC. A new carrier-based modulation method that has been recently proposed in [6] was utilized to incorporate the ST states for the three-level Z-source NPC inverter. The RD LSPWM method is used for the analysis of the non-ideal DZSN NPC.

15.6.1   Non-ideal Circuit Topology

In Section 15.5, the RD LSPWM method was validated for identical and non-identical dc source conditions. The inverter was operated under ideal condition in the DZSN NPC; i.e. no losses in the circuit were considered. In practice, a voltage drop across the components is inevitable. Due to this phenomenon, this section assesses and discusses the performance of the aforementioned new modulation method for the DZSN NPC in real scenarios. Therefore, this section deals with the study of the non-ideal DZSN, where the circuit resistance and their respective voltage drop are considered. The real-time voltage drops are introduced in the simulation study by connecting a resistance to the existing topology (shown in Fig. 15.15). Fig. 15.40 illustrates the circuit that incorporates these modifications. A detailed circuit analysis is presented in the following subsection.

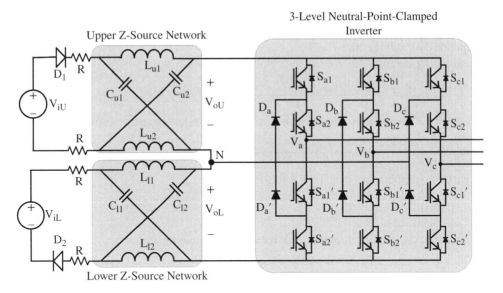

Figure 15.40 Non-ideal circuit of DZSN NPC with two individual dc sources.

The ST to the inverter is inserted through the new modulation method (RD LSPWM) described in Section 15.5.1.

15.6.2 Circuit Analysis

There are two symmetrical single Z-source networks in the DZSN. Since the networks are similar, the circuit analysis is performed for one of the networks. From Fig. 15.40, it can be observed that the upper Z-source network possesses two capacitors and inductors. By assuming that the inductors and the capacitors have the same value of inductance and capacitance, the network attains symmetry. In this case, the voltages across both the capacitors and inductors are represented as V_C and v_L, respectively; and the voltage drop across the resistance is denoted by V_R. Therefore, the operation of the inverter for a switching period of T (ST state for a period T_0 and N-ST states for a period T_1) is expressed by the equations (15.65)–(15.68).

A symmetrical network is assumed, where the inductors L_1 and L_2, and the capacitors C_1 and C_2 possess similar inductance and capacitance values. Therefore, the expressions for inductance and capacitance values are given by the equations (15.1) and (15.2), respectively.

During the ST zero state for a period T_0, the voltages are:

$$v_L = V_C - V_R \tag{15.65}$$

$$V_{oU} = 0 \tag{15.66}$$

During one of the eight N-ST states for a period T_1, the voltages are:

$$v_L = V_{iU} - V_C - V_R \tag{15.67}$$

$$V_{oU} = V_C - v_L = 2V_C - V_{iU} + V_R \tag{15.68}$$

Here, v_L is the voltage across the inductor, V_C is the voltage across the capacitor, V_R is the voltage drop across the resistance, V_{iU} is the upper dc voltage source, V_{oU} is the upper dc-link voltage, and $T = T_1 + T_0$.

Averaging the inductor voltage over a period T gives:

$$V_L = \overline{v_L} = \frac{T_0(V_C - V_R) + T_1(V_{iU} - V_C - V_R)}{T} \tag{15.69}$$

Under the steady-state condition, equation (15.69) equals to zero. Thus:

$$V_L = \frac{T_0(V_C - V_R) + T_1(V_{iU} - V_C - V_R)}{T} = 0 \tag{15.70}$$

Further simplification yields the following:

$$V_C = \frac{TV_R - T_1 V_{iU}}{T_0 - T_1} \tag{15.71}$$

Similarly, the average upper dc-link voltage of the inverter is:

$$V_{oU} = \overline{v_{oU}} = \frac{T_0 \cdot 0 + T_1(2V_C - V_{iU} + V_R)}{T} \tag{15.72}$$

Hence:

$$\overline{v_{oU}} = \frac{T_1}{T}(2V_C - V_{iU} + V_R) \tag{15.73}$$

The peak dc-link voltage is determined as:

$$V_{oU(peak)} = \frac{1}{1 - \frac{2(T_o)}{T}} (V_{iU} - 2V_R) \tag{15.74}$$

Here, T_0 is the ST duration. Equations (15.73) and (15.74) provide the mathematical expressions for the average and peak values of the dc-link voltage, which are used to evaluate the performance of the newly proposed reference disposition PWM for DZSN.

15.6.3 Switching States

As observed from Fig. 15.40, each of the two Z-source networks has its individual dc power supply. One dc output terminal (negative) of the upper Z-source network is connected to one dc output terminal (positive) of the lower Z-source network. The dc-link of the inverter is connected to the midpoint of the upper and the lower networks, represented by the point N in Fig. 15.40. N is the neutral point with a potential of 0 V. In addition, ZSI has several unique ST states, in addition to the active and zero states of the conventional inverters. As the Z-source NPC inverter utilizes these ST states to boost the dc bus voltage, the ST interval has to be generated with precision. This ensures that the boosted inductive energy is equally distributed in the Z-source network. The modulation scheme is explained in Section 15.5.1. Fig. 15.41 shows the control logic of a single-phase for the DZSN NPC inverter. One of the regular three-phase reference signals (V_{a1}) and the level-shifted reference signal (V_{a2}) are compared with the carrier signals, rendering the inverter operation similar to the carrier comparison PWM implemented in the VSI. In addition, V_u and V_l determine the UST and LST duty cycles. Moreover, the comparisons shown in Fig. 15.41 generate the switching signals for the inverter.

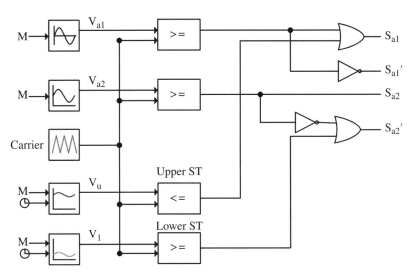

Figure 15.41 ST control logic for phase "a" using RD LSPWM.

Table 15.3 Simplified Switching Representation for a Non-ideal Three-level ZSI ($z = a$, b, orc).

Switching states	Switches (ON)	Diodes (ON)	Output
Non-ST	S_{z1}, S_{z2}	D_1, D_2	$+V_i$
Non-ST	$S_{z2}, S_{z1'}$	$D_1, D_2, D_z (\text{or } D_{z'})$	0
Non-ST	$S_{z1'}, S_{z2'}$	D_1, D_2	$-V_i$
Upper-ST	$S_{z1}, S_{z2}, S_{z1'}$	$D_1, D_{z'}$	0
Lower-ST	$S_{z2}, S_{z1'}, S_{z2'}$	D_2, D_z	0
Full-ST (preferred but not included)	$S_{z1}, S_{z2}, S_{z1'}, S_{z2'}$	—	0
Full-ST (not preferred)	$S_{a1}, S_{a2}, S_{a1'}, S_{a2'}, S_{b1}, S_{b2},$ $S_{b1'}, S_{b2'}, S_{c1}, S_{c2}, S_{c1'}, S_{c2'}$	—	—

Table 15.3 summarizes the switching states of the inverter during the ST and N-ST states. From the table, it can be determined that the Z-source NPC inverter possesses three ST states, namely, FST, UST, and LST. Even though the three ST states are feasible, it is evident that the voltage boosting of the dual Z-source network is achieved by using the UST and LST states from Fig. 15.41.

15.6.4 Validation

The RD LSPWM method is used for providing the switching signals to the inverter. The performance evaluation of non-ideal dual Z-source network incorporating the proposed method is performed using PSIM with a MATLAB/Simulink coupler. The simulation results are discussed in this section, followed by the experimental verification of the setup in Section 15.6.5.

For evaluating the performance of the RD LSPWM modulation method for the non-ideal dual Z-source network, each of the two individual dc voltage sources V_{iU} and V_{iL} supplying the upper and lower Z-source networks are maintained at 40 V. The simulation parameters are chosen on the basis of the experimental parameters. The parameters for both the upper and lower Z-source networks are L = 5 mH and C = 1000 μF with a switching frequency of 1 kHz. Due to limitations in the laboratory equipment, the experiments were performed at a low switching frequency. A three-phase RL circuit serves as the load of the ZSI.

For the designed impedance values, the modulation index is set to M = 0.85 in order to achieve maximum voltage boosting. Having set the M value, the ST duty cycle is calculated as $T_0/T = 0.26$. The simulated results are discussed subsequently. The output of the upper dc-link voltage is induced by the Z-source capacitors in the upper network. Figs. 15.42 and 15.43 depict the voltage across the upper capacitor and the upper dc-link voltage, respectively.

From Fig. 15.43, the simulated average value of the upper dc-link voltage is determined to be 53.1 V, and is similar to the theoretical value, which is 54. The theoretical value is calculated using the mathematical expression given in equation (15.73).

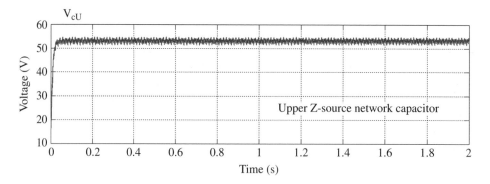

Figure 15.42 Capacitor voltage of the upper Z-source network with M = 0.85 (refer to Fig. 15.40).

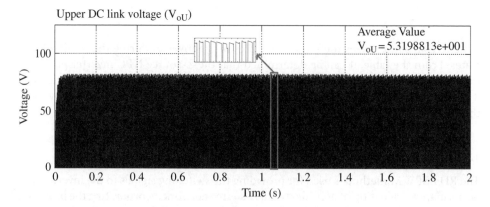

Figure 15.43 Voltage across the upper dc-link with M = 0.85 (refer to Fig. 15.40).

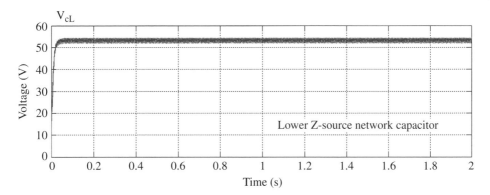

Figure 15.44 Capacitor voltage of the lower Z-source network with M = 0.85 (refer to Fig. 15.40).

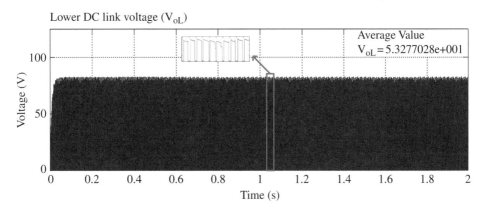

Figure 15.45 Voltage across the lower dc-link with M = 0.85 (refer to Fig. 15.40).

Similarly, for the lower Z-source network, which is fed by another 40 V dc source, the capacitor voltage and the dc-link voltage are shown in Figs. 15.44 and 15.45, respectively. The average value of the lower dc-link voltage through the simulation is observed to be 53.2 V as shown in Fig. 15.45. This is identical to the expected theoretical value of 54 V calculated using equation (15.73).

The waveforms illustrated in Figs. 15.43 and 15.45 show that the upper and lower dc-link voltages are similar, balancing the neutral-point potential. Under ideal conditions, the circuit resistance is not considered. Next, the boost factor of the Z-source in the MCBC method is calculated using the equation $B = 1/(\sqrt{3}M - 1)$ and is computed as 2.1. From the simulation results depicted in Figs. 15.43 and 15.45, the peak value of the dc-link voltage is 80 V, which matches equation (15.74). The peak value obtained by incorporating the MCBC method in the proposed modulation method proves that the boosting of the dc-link voltage matches the "boost factor" value (equation (15.74)) claimed by Shen et al. [8]. Fig. 15.46 illustrates the output phase voltage (V_{aN}) of the inverter. According to the depicted waveform, it is established that the peak value of the voltage is 85 V, and agrees with the theoretical calculation of 88 V, calculated using $\widehat{v_{ac}} = M \cdot B \cdot V_i$. The THD of the output phase voltage is depicted in Fig. 15.47.

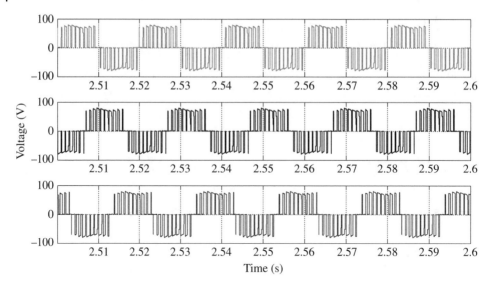

Figure 15.46 Output phase voltage (V_{aN}, V_{bN}, V_{cN}) of the inverter with M = 0.85 (refer to Fig. 15.40).

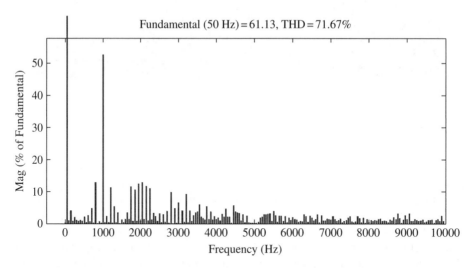

Figure 15.47 Total harmonic distortion (THD) of output phase voltage with M = 0.85 (refer to Fig. 15.40).

Fig. 15.48 provides the output voltage measured between phase "a" and the load neutral. Figs. 15.49 and 15.50 show the output line voltage and its harmonic distortion waveforms, respectively. The THD of the output line voltage incorporating the new reference disposition modulation technique is 41.09%. The output line currents of the inverter for the RD LSPWM, whose ripple can be eliminated using a proper filter, are illustrated in Fig. 15.51.

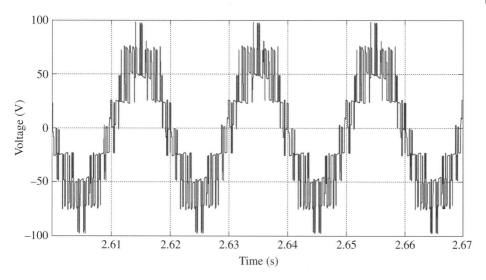

Figure 15.48 Output voltage (line to load neutral) of the inverter with M = 0.85 (refer to Fig. 15.40).

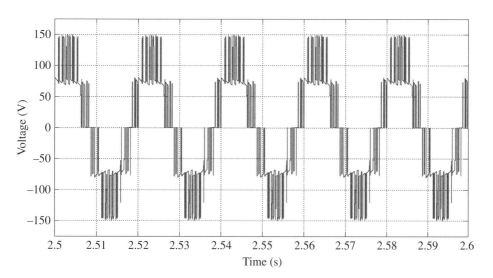

Figure 15.49 Output line voltage (V_{ab}) of the inverter with M = 0.85 (refer to Fig. 15.40).

15.6.5 Experimental Verification

A hardware prototype of the dual Z-source network implementing the proposed reference disposition modulation with the help of dSpace RT1103 is built for this research. The hardware prototype is designed and implemented to verify the simulation results discussed in Section 15.6.4, and it is shown in Fig. 15.52. All the results presented in Section 15.6 are implemented using the parameters tabulated in Table 15.4. The

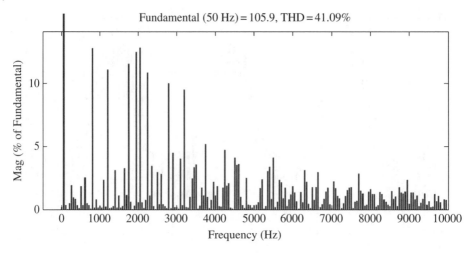

Figure 15.50 Total harmonic distortion (THD) of output line voltage with M = 0.85 (refer to Fig. 15.40).

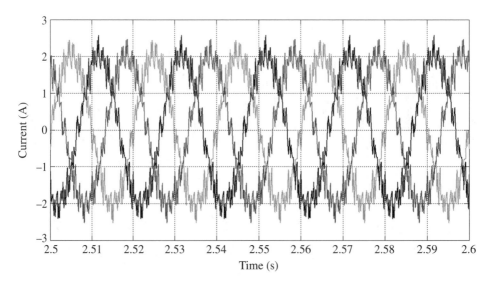

Figure 15.51 Output line currents of the inverter with M = 0.85 (refer to Fig. 15.40).

obtained experimental results match the theoretical calculations and simulation results, confirming the applicability of the proposed modulation control even under a low switching frequency of 1 kHz.

Figs. 15.53 and 15.54 illustrate the experimental values of the upper link capacitor and dc-link voltages, respectively. From Fig. 15.54, the average value of the dc-link is determined as 54.6 V, which is similar to the theoretical calculation of 54 V. The upper link capacitors and the dc-link voltage are similar to the values obtained in the simulation results.

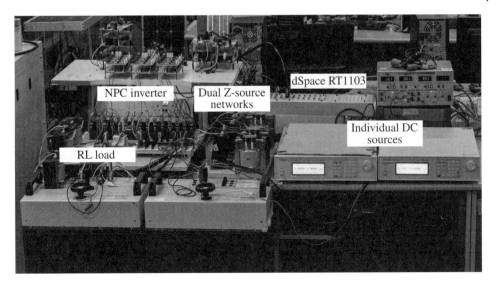

Figure 15.52 Experimental setup of DZSN implementing RD LSPWM.

Table 15.4 System Parameters for Simulation and Experimental Prototypes.

Parameter	Specification
DC supply (V)	40
Z-source inductor (mH)	5
Z-source capacitor (µF)	1000
Switching frequency (kHz)	1
Load resistance (Ω)	30
Load inductor (mH)	6
Fundamental frequency (Hz)	50

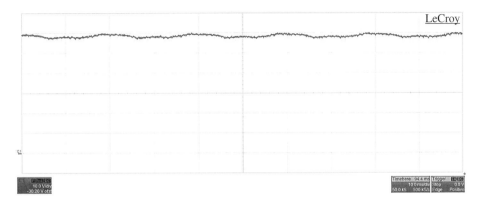

Figure 15.53 Experimental upper capacitor voltage with M = 0.85 (10V/div).

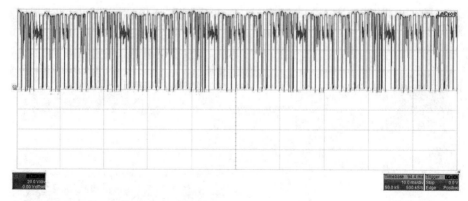

Figure 15.54 Experimental upper dc-link voltage with M = 0.85 (20V/div).

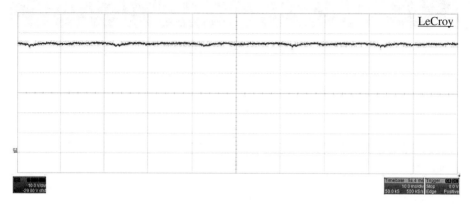

Figure 15.55 Experimental lower capacitor voltage with M = 0.85 (10V/div).

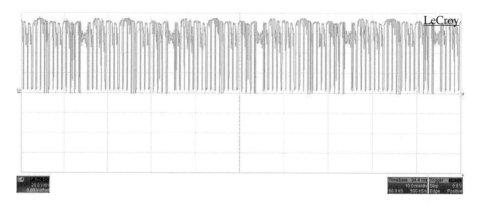

Figure 15.56 Experimental lower dc-link voltage with M = 0.85 (20V/div).

Similarly, the lower network capacitor and dc-link experimental results are illustrated in Figs. 15.55 and 15.56, respectively, which verify the agreement between the theoretical and simulation results. From the dc-link voltage waveforms, it is evident that the output of Z-source network is not constant, unlike the traditional VSI. The voltage reaches zero when the circuit enters the ST state; this is verified with simulations and experiments.

The experimental results of the phase voltages (V_{aN}, V_{bN}, and V_{cN}) and the line to the load neutral voltages for all the three phases of the inverter are illustrated in Figs. 15.57

Figure 15.57 Experimental phase voltages (V_{aN}, V_{bN}, V_{cN}) (75V/div) and their respective FFT spectrum (20 V/div).

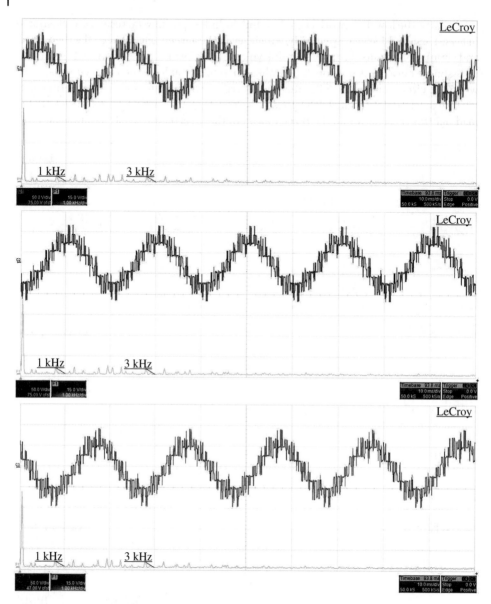

Figure 15.58 Experimental voltages (line to load neutral) (50 V/div) for phases a, b, and c and their respective FFT spectra (15 V/div).

and 15.58, respectively. Furthermore, the corresponding fast fourier transform (FFT) spectrums of the voltages are given in the figures. The line voltages (V_{ab}, V_{bc}, V_{ca}) are shown in Fig. 15.59, and the corresponding FFT spectrum is calculated. The harmonics at the operating switching frequency of 1 kHz are eliminated in the line-to-line voltage. The three-phase output currents are depicted in Fig. 15.60.

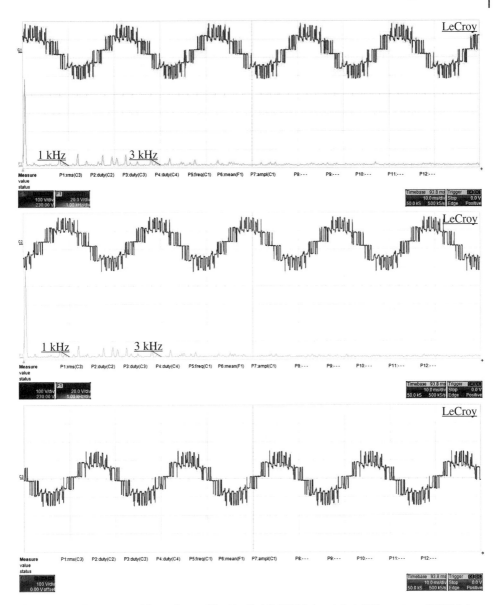

Figure 15.59 Experimental line voltages (V_{ab}, V_{bc}, V_{ca}) (150 V/div) and their FFT spectrum (20 V/div).

15.7 Applications of ZSI

In most applications where energy conversions are required, voltage buck-boost energy conversions are achieved either by two-stage or single-stage conversion methods. The two-stage energy conversion topology includes a dc-dc boost converter added to the front end of a dc-ac VSI or a dc-dc buck converter added to the front end of a dc-ac

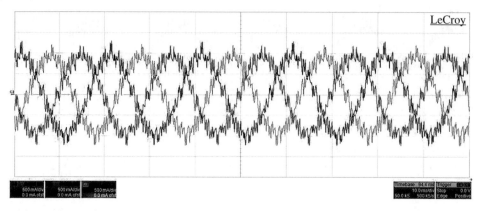

Figure 15.60 Experimental output currents (I_a, I_b, I_c) with RL load (500 mA/div).

CSI. Single-stage conversion uses fewer active or passive components unlike two-stage energy conversion. Some of the widely used converters include the step down or buck converter, step up or boost converter, Cuk, SEPIC, and Z-source converters. ZSI applications have been increasingly investigated by researchers due to their ability to provide efficient energy conversion (dc-dc, dc-ac, ac-dc, and ac-ac) between the source and the load. ZSI can either be fed by a diode or thyristor controlled rectifier, a battery, a fuel-cell stack, or a combination of the dc sources; and the load can be either inductive or capacitive.

In recent years, with the evolution of technology, integration of distributed generation systems into the utility is becoming a popular solution for the future energy demands in many countries. Recognizing the vulnerabilities of the grid dependency, organizations are looking at fuel cells as an attractive option for reliable backup power. Integrating fuel cells in energy neutral buildings along with the grid provides reliable support to the grid.

The main applications of the distributed generation systems are stand-by power, peak shaving, grid support, and stand-alone. These sources have less than few ten kilowatts of power ratings. Therefore, these sources are integrated with energy storage devices such as batteries, ultra-capacitors, and flywheels, when the output is insufficient to meet the load demand. However, direct integration of the distributed energy sources into the load is not advisable, in order to avoid problems that arise due to unexpected fluctuations. Hence, a dc-dc converter is required either to regulate or to boost the voltage to the required level. This increases the complexity, volume, and cost of the system. This drawback can be overcome by using the ZSI.

With the development of the ZSI, the traditional converters are being increasingly replaced for typical research applications. The ZSI possesses advantages like resistance to failure switching and EMI distortions, ride-through capability during voltage sag, low or no in-rush current, and operation as both buck and boost converters. With increase in the demand for power, multilevel inverters have emerged as a satisfactory solution. These inverters have an array of power semiconductors and capacitor voltage sources, and the output voltage generated by these inverters is in the form of stepped waveforms. By synthesizing the ac output voltage from several dc voltages, staircase output waveforms are produced. This allows the multilevel inverter to work efficiently for high-voltage levels. However, few drawbacks are associated with the multilevel inverter. First, its output amplitude is limited to the summation of the dc voltage sources. Second, an additional dc-dc chopper has to be used to boost the output voltage of the dc source. Third, occurrence of short circuit must be prevented to protect the inverter from being damaged. Therefore, the multilevel inverter has to be used with dead-time protection. These problems can be overcome if a ZSI is used. In addition, when cascaded ZSIs are used, additional dc-dc choppers can be cut, thereby reducing the cost and volume. Consequently, the ZSI can be integrated with different renewable sources like wind and solar with the fuel cells to meet the demand.

The topology of the ZSI includes two inductors and capacitors connected in X-shape. By replacing one of the capacitors in the impedance LC network, the ZSI can be used for dc-dc conversion. This configuration eliminates the need for an additional dc-dc converter, and the same impedance network can be used to connect the main inverter circuit to any dc power source. Another advantage is that the input power, the output power, and the state of charge of the battery can be controlled by adjusting the ST duty ratio. Therefore, replacing a capacitor with a battery may increase the system's reliability for applications that require backup, stand-alone, peak shaving, and grid support.

Some of the applications where the ZSI is widely used are electric vehicles and adjustable speed drives. In automotive applications where two-staged energy conversions are required with the battery, the ZSI can be utilized. Using the bidirectional ZSI along with the battery leads to efficient charging of the battery from the grid at night and discharge at times of peak load demand. This process increases the stability of the grid. Numerous studies have been conducted on the integration of the ZSI with fuel-cell vehicles. In motor drive applications where the regenerative braking energy can be recycled, a combination of super-capacitor and bidirectional ZSI can be utilized. The transients and peak load demands are delivered by the super-capacitor during normal operation, and the energy can be stored during regenerative braking. The wide-ranging aforementioned applications of ZSI indicate the indispensability of further research into integration of the ZSI with distributed energy sources.

15.8 Summary

This chapter commenced with an explanation and the operating modes of a two-level ZSI. Subsequent to the explanation of the two-level ZSI, the topologies of three-level Z-source NPC were discussed. The topologies were distinguished by the method used for connecting the neutral point. In the first topology a split capacitor bank (Section 15.3.1), in the second topology a split dc source (Section 15.3.2), and in the third topology two individual Z-source networks (Section 15.3.3) were used. Among the three topologies, the latter two are the most common topologies for the three-level Z-source NPC. The operation of these individual topologies during the ST and N-ST states were explained. Furthermore, this section includes the modulation strategies for careful insertion of the FST, or UST, and LST to the inverter circuitry. Two carrier-based modulation methods, phase opposition disposition modulation, and in-phase disposition modulation, can be used to boost the capability of the inverter by using minimum device commutations per switching cycle. From the analysis, it is deduced that NPC has three ST states when compared to a single ST state in a two-level Z-source. With this feature, different combinations of ST and N-ST states can be achieved in a three-level Z-source

NPC. Compared to other modulation methods for ZSI, MCBC maintains a constant ST duty ratio to achieve maximum voltage gain. Therefore, the MCBC method was modified for integration with the carrier-based disposition modulation methods, thereby resulting in the insertion of ST states and achieving voltage boosting in the three-level Z-source NPC.

This chapter further discusses the newly proposed modulation method for the Z-source NPC inverters. The proposed modulation method utilizes two sets of reference signals, two ST signals, and one carrier signal to provide switching sequences to the inverter. While inserting the ST states, steps were taken to avoid drawbacks like volt-second error. The switching devices at various switching states were tabulated in Table 15.2. As the dual Z-source network is supplied by two individual sources, two different conditions were analyzed with the proposed modulation methods. In general, two dc sources with identical voltage rating are connected with dual ZSI. This results in identical and non-identical conditions of the dc sources. The initial part of the results in Section 15.5.3 describe the performance evaluation of the proposed reference disposition level-shifted PWM. The performance was verified for an ideal circuit. The verification process requires the balancing of the upper and lower capacitor voltages. This balance was confirmed by carrying out simulations in the MATLAB/Simulink and PSIM environments. The results were illustrated in Figs. 15.22 and 15.24. As the capacitors were determined to be balanced, neutral-point balancing was achieved. The results obtained from simulation agree with the theoretical results, validating the proposed modulation method for identical dc voltage sources.

Evaluating the performance of the non-identical dc sources for this modulation method is indispensable, as most natural scenarios do not incorporate identical dc sources. Therefore, for non-identical conditions, achieving the neutral-point balance is an important progression in the analysis of this modulation method. Modification of the ST duty ratio of the upper and lower ZSIs is necessary to achieve this. The results for the non-identical dc sources are presented in the latter part of the results section. In both cases, the implemented modulation scheme achieves maximum voltage boost, low THD, and better output waveform quality with easier implementation. Furthermore, the proposed modulation method can achieve neutral-point balancing for identical and non-identical dc sources.

Following the evaluation of the performance of the new modulation method for an ideal circuit, a similar evaluation was performed for the non-ideal circuit condition. The new modulation method is responsible for the careful insertion of ST for the two individual dc sources supplying the dual ZSIs. Evaluations of the simulation and experimental results were performed, and the findings of the research were presented for the proposed modulation method in practical cases considering the losses. To implement this modulation method for practical cases, the dual Z-source network is considered as non-ideal; i.e. the voltage drops across the circuits. The ST duty ratio and the modulation index for the ST duty ratio were analyzed using the MCBC method. This modulation method uses two level-shifted reference waves that are compared with the carrier to provide the switching. Furthermore, two modified signals were used to control the UST and LST states of the inverter. The full-ST state of the inverter was rejected to avoid the volt-second error. This novel method achieves neutral-point balancing, which eliminates additional voltage stress on the switches. As this method incorporated the phase disposition scheme, it aids in an easier implementation when compared with the space vector for DZSN. The simulation results were presented to validate the performance of the modulation method in a dual impedance network and were subsequently verified with experimental results. From the results, it can be concluded that the dc-link voltages were balanced even during non-ideal conditions. Furthermore, the output line voltage harmonic distortion for the reference-disposition-based modulation method was determined as 41.09%, which is recommended for a three-level operation in industrial and in DER applications. Another important advantage of the proposed modulation method is its operability at relatively low switching frequencies without losing the ability to provide optimal waveforms. Furthermore, it is evident that the boost factor remains unaffected despite the voltage drop across the resistances. Examining the attributes of the proposed modulation methods' performance indicates the utility of the reference disposition level-shifted PWM method in industrial and DER applications.

References

1 F. L. Luo and H. Ye, *Power Electronics: Advanced Conversion Technologies*, Boca Raton: CRC Press/Taylor & Francis, c2010., 2010.

2 F. Z. Peng, "Z-source inverter," *IEEE Transactions on Industry Applications*, vol. 39, no. 2, 504–510, 2003.

3 X. Shi, A. Chen, and C. Zhang, "A neutral-point potential balancing method for Z-source neutral-point-clamped (NPC) inverters by adding the shoot-through offset," in *Power Electronics and Application Conference and Exposition (PEAC), 2014 International*, 2014, pp. 62–67: IEEE.

4 F. Gao, P.C. Loh, F. Blaabjerg, R. Teodorescu, and D. Vilathgamuwa, "Five-level Z-source diode-clamped inverter," *IET Power Electronics*, vol. 3, no. 4, pp. 500–510, 2010.

5 P. C. Loh, F. Gao, and F. Blaabjerg, "Topological and modulation design of three-level Z-source inverters," *IEEE Transactions on Power Electronics*, vol. 23, no. 5, pp. 2268–2277, 2008.

6 M. M. Roomi, A. I. Maswood, H. D. Tafti, and P. H. Raj, "Reference disposition modulation method for non-ideal dual Z-source neutral-point-clamped inverter," *IET Power Electronics*, vol. 10, no. 2, pp. 222–231, 2017.

7 F. Z. Peng, M. Shen, and Z. Qian, "Maximum boost control of the Z-source inverter," *IEEE Transactions on Power Electronics*, vol. 20, no. 4, pp. 833–838, 2005.

8 M. Shen, J. Wang, A. Joseph, F. Z. Peng, L. M. Tolbert, and D. J. Adams, "Constant boost control of the Z-source inverter to minimize current ripple and voltage stress," *IEEE Transactions on Industry Applications*, vol. 42, no. 3, pp. 770–778, 2006.

Part IV

Grid-Integration Applications of Advanced Multilevel Converters

16

Multilevel Converter-Based Photovoltaic Power Conversion

Hossein Dehghani Tafti, Georgios Konstantinou, and Josep Pou

16.1 Introduction

Renewable energy sources are occupying an increasing share in the global power generation market. Among various renewable energy sources, photovoltaic (PV) technology has been in the focus of many governments, and the supported substantial subsidies have resulted a steep increase in the installed capacity. Moreover, with the decreasing trend in PV panel costs from $4.9/W in 1998 to $0.3/W in 2015, the rate of growth of new PV installations is expected to remain at a high level. With the high penetration of installed distributed generation (DG) units, power system operators (PSOs) are facing new challenges in areas such as reliability, availability, and power quality. In order to maintain the power quality and reliability of the power system, medium- and large-scale grid-connected PV power plants (GCPVPPs) should possess low-voltage ride-through (LVRT) capability, as regulated by new standards and grid codes. The extraction of a high amount of power from medium- and large-scale GCPVPPs to the grid can be achieved by advanced multilevel converters. The operation of GCPVPPs with two types of advanced multilevel converters, three-level neutral-point-clamped and cascaded H-bridge, is investigated. The details of the controller design for the operation during voltage sags with LVRT capability are also provided.

16.2 Three-Level Neutral-Point-Clamped Inverter–Based PV Power Plant

16.2.1 Circuit Configuration

The circuit configuration of the multi-string GCPVPP with a three-level neutral-point-clamped (3L-NPC) inverter is presented in Fig. 16.1. Each PV string is connected to a dc–dc boost converter. All of the PV strings are connected to a single dc-bus. This topology improves the maximum power extraction during partial shading, because each PV string is controlled independently from other PV strings, and therefore an MPPT controller can extract the maximum power from the PV string in all irradiation conditions. Furthermore, in the case of failure of some PV strings, all of the other PV strings can continue their normal operation.

Advanced Multilevel Converters and Applications in Grid Integration, First Edition.
Edited by Ali I. Maswood and Hossein Dehghani Tafti.
© 2019 John Wiley & Sons Ltd. Published 2019 by John Wiley & Sons Ltd.

Several types of multi-level inverters are presented in the literature. The three-level flying capacitor inverter utilizes several flying capacitors, which increases the size and cost of the system and reduces the reliability. However, the NPC inverter shows higher reliability by eliminating the flying capacitors and reducing the size. On top of these, there are various commercialized NPC inverter products with different power ratings that can be used for high-power conversion applications.

The grid-tied 3L-NPC inverter consists of two pairs of IGBT switches (S_{a1} and S_{a2}), including the complementary ($S_{a1'}$ and $S_{a2'}$), in each phase. The 3L-NPC operates with three different switching states to synthesize the three output pole voltage levels at $V_{dc}/2$, 0, and $V_{dc}/2$ with respect to the midpoint m of the dc-link. More details about the 3L-NPC inverter can be found in [24].

Despite the increasing demands for renewable energy resources, several regulations (such as those described in [1] and [2]) must be complied with strictly to ensure the power quality of the renewable grid-connected system. Therefore, lower current harmonic distortion can be achieved with the aid of output filters, which are connected between the inverter and the grid, as shown in Fig. 16.1. Due to the intelligent synthesis of three-level output voltage waveforms, the filter size can be much smaller as compared to conventional two-level inverters. The resistance of the inductor filter is usually lesser than 10% of its impedance. Especially for high-current inductor filters, which are used in medium- and large-scale GCPVPPs, the resistance is smaller. Hence, the effect of resistance on the controller is negligible. Accordingly, the resistance of the inductor filter is neglected in the illustrations included in this chapter.

16.2.2 Control Strategy during Grid *Normal* Operation

The connected dc–dc converters to each PV string (see Fig. 16.1) extract the maximum power from PV strings during *Normal* operation. The implemented control algorithm for the operation of the GCPVPP with the 3L-NPC inverter is shown in Fig. 16.2. The dc-link voltage (v_{dc}) is regulated to its reference value, using the proportional-integral (PI) controller, which calculates the d-axis current reference. The outer loop maintains v_{dc} to its reference value by adjusting the inverter output power. The amount of output reactive power is calculated using the q-axis current reference (i^*_q), which is usually set to zero for unity power factor operation. Hereafter, the *dq*-axis current references

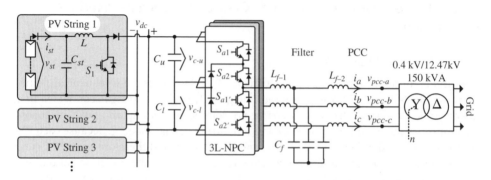

Figure 16.1 Grid-connected multi-string PV power plant with 3L-NPC inverter.

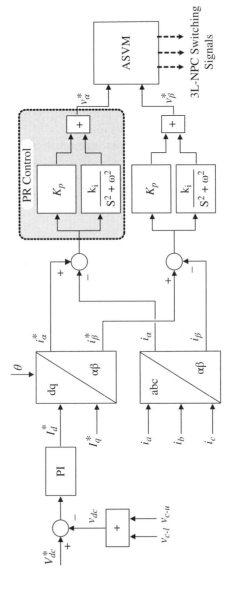

Figure 16.2 Control algorithm of GCPVPP with the 3L–NPC inverter during grid *Normal* operation.

(i^*_d and i^*_q) are converted to obtain their equivalent three stationary-frame current references (i^*_α and i^*_β) using inverse Park transformation.

The respective $\alpha\beta$ current errors are then obtained by subtracting the stationary-frame current references from the instantaneous stationary-frame currents (i_α and i_β), obtained from the measured three-phase inverter currents through inverse Clarke transformation. Finally, the stationary-frame voltage references (v_α^* and v_β^*) are acquired directly from the two independent proportional-resonant (PR) controllers without the need of inverse Park transformation. The transfer function of the PR controller is:

$$G_{PR} = K_p + \frac{K_i}{S^2 + \omega^2} \tag{16.1}$$

Here, G_{PR} is the transfer function of the PR controller; K_p is the proportional gain; K_i is the integral gain; and ω is the resonant frequency.

The resonant term provides an infinite gain at the resonant frequency that minimizes the steady-state error to zero. Therefore, the steady-state current error can be minimized by having the resonant frequency matching with the grid frequency.

It should be noted that the implementation of the ideal PR controller with infinite gain in the experimental setup is not practical, because:

- The grid frequency is not exactly 50 Hz in the practical system, and it varies within the standard margin.
- The voltage measurement contains some errors and does not measure the exact instantaneous value of the voltage.
- A delay exists in the calculation of the processor.

Although the ideal performance cannot be achieved in a practical system, the PR controller is implemented in several research studies due to its fast dynamic performance and zero steady-state error [12, 15, 32, 40].

The output voltage waveforms of the 3L-NPC inverter are synthesized using the adaptive space vector modulation (ASVM) technique. The space vector representation for the 3L-NPC inverter is shown in Fig. 16.3. It can be observed that there are several pairs of vectors (e.g. "100" and "211") in each internal voltage state, which generate the same output voltage level. Conventionally, the classical space vector modulation (SVM) uses all of the vectors during every switching cycle period. Therefore, an increase in the switching loss is expected from the classical SVM, due to the high switching frequency operation. The ASVM technique is implemented in this controller in order to improve the inverter efficiency. The overall switching loss is decreased by reducing the switching frequency, since one of the vector pairs is utilized in each switching cycle [20, 24]. This can be observed from the current flow based on the vector pair "100" and "211," as demonstrated in Fig. 16.4.

Even though both vectors generate same output voltage level, they have different effects on the dc-link capacitor voltages. For instance, when the state vector is "100," the dc-link capacitors C_1 and C_2 are in charging and discharging states, respectively, according to the voltage potentials. Hence, the dc-link capacitor voltage balancing is achieved based on the state vector selection principle as described in equations (16.2) and (16.3). The vector selection principle is applicable for other vector pairs as well.

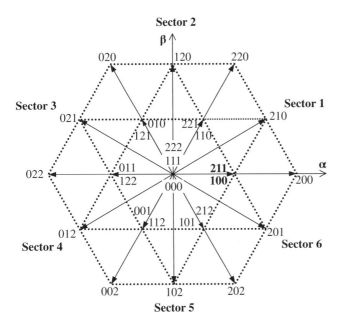

Figure 16.3 Space vector representation of the 3L-NPC inverter.

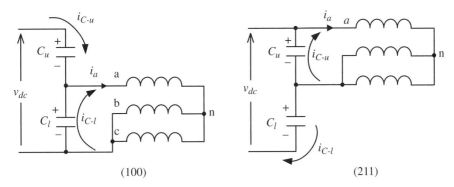

Figure 16.4 Circuit diagram and current path for vectors of one vector pair in the ASVM technique.

$$\left.\begin{array}{ll} v_{c-u} \leq v_{c-l} & \& \quad i_a \geq 0 \\ v_{c-u} > v_{c-l} & \& \quad i_a < 0 \end{array}\right\} \Rightarrow (100) \tag{16.2}$$

$$\left.\begin{array}{ll} v_{c-u} \geq v_{c-l} & \& \quad i_a \geq 0 \\ v_{c-u} < v_{c-l} & \& \quad i_a < 0 \end{array}\right\} \Rightarrow (211) \tag{16.3}$$

The ASVM voltage balancing method is independent of the main controller, and it can balance the capacitor voltages in all operation conditions, including unbalanced voltage sags. On the other hand, the ASVM technique can directly use the stationary frame reference voltages to generate the switching signals, which results in less computational complexity and higher accuracy of the controller.

It can be seen that this conventional controller cannot be implemented for the operation of the GCPVPP during voltage sags, and that it needs some modifications. The following sections present the detailed implementation of the proposed controller for the GCPVPP with the 3L-NPC inverter and the injection of balanced currents during voltage sags.

16.2.3 Control Strategy during Grid Voltage Sags

Two controllers, one for the dc–dc stage and one for the grid-connected 3L-NPC inverter, are required for the operation of the GCPVPP. Each of these controllers require separate control strategies that are specific for the grid operating condition, be it during its normal operation or during network disturbances (voltage sag). Therefore, a fast and precise voltage sag detection strategy is required to determine the operating mode of the controllers. The following section provides the details of the implemented voltage sag detection algorithm.

The extraction of the grid voltage positive sequence is another issue in the execution of the controller, which will be discussed in the following sections, followed by a detailed explanation of the proposed controller.

16.2.4 Voltage Sag Detection

Fast detection of the voltage sag is important for quick operation of the controller at the beginning of the voltage sag [30, 37]. In this study, a second-order generalized integrator (SOGI)–based orthogonal system is applied for the calculation of the amplitude of phase voltages, as illustrated in Fig. 16.5. As output signals, two sine waves ($v_{a-\alpha}$ and $v_{a-\beta}$) with a phase shift of 90° are generated. The component $v_{a-\alpha}$ has the same phase and magnitude as the fundamental of the input signal (v_a). The SOGI block is defined as:

$$GI = \frac{\omega s}{s^2 + \omega^2} \tag{16.4}$$

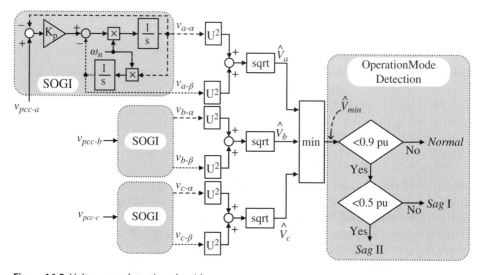

Figure 16.5 Voltage sag detection algorithm.

Here, ω represents the resonance frequency of the SOGI. The closed-loop transfer functions ($H_d = v_{a-\alpha}/v_a$ and $H_q = v_{a-\beta}/v_a$) can be calculated as [10]:

$$H_d(s) = \frac{v_{a-\alpha}}{v_a} = \frac{k\omega s}{s^2 + k\omega s + \omega^2} \tag{16.5}$$

$$H_d(s) = \frac{v_{a-\beta}}{v_a} = \frac{k\omega^2}{s^2 + k\omega s + \omega^2} \tag{16.6}$$

Here, k affects the bandwidth of the closed-loop system. A detailed description of the SOGI block can be found in [10]. The main limitation of the SOGI-based sag detection algorithm is the delay of half of the voltage period.

Since the phase voltage amplitudes are required to detect the single- or two-phase voltage sags, the orthogonal voltages ($v_{x-\alpha}$ and $v_{x-\beta}$, where x is referred to phase a, b, or c) are calculated by independent SOGI blocks for each phase. In this figure, ω_n denotes the fundamental angular frequency of the grid voltage. The amplitude of the phase voltage ($\widehat{V}_x$) is calculated as:

$$\widehat{V}_x = \sqrt{v_{x-\alpha}^2 + v_{x-\beta}^2}. \tag{16.7}$$

Subsequently, the minimum phase voltage amplitude ($\widehat{V_{min}}$) is determined instantaneously, which is utilized for the detection of the voltage sag and operation mode of the controller.

16.2.5 Adaptive Phase-Locked Loop Control

The conventional phase-locked loop (PLL) is suitable for calculation of the positive-sequence voltage angle under balanced voltage conditions. However, it is not able to calculate the positive sequence voltage angle under unbalanced conditions [17]. Therefore, an adaptive PLL is applied for extracting the angle of the positive sequence voltage, as shown in Fig. 16.6. During voltage sags, the grid voltage contains both positive and negative sequences. By using the conventional PLL, the voltage of positive sequence cannot be calculated accurately. With the adaptive PLL [17], first the positive sequence of voltage is extracted using an arbitrary angle that rotates at $2\pi f$ for both the direct and reverse abc-dq transformations. A moving average filter (MAF) is

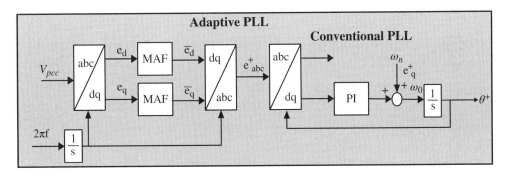

Figure 16.6 Schematic diagram of adaptive phase-locked-loop (PLL) algorithm.

also implemented on the calculated dq voltage, which filters instantaneous oscillations of the dc-link voltages by performing the following operation:

$$\bar{x}(t) = \frac{1}{T_w} \int_{t-T_w}^{t} x(\tau)d\tau \tag{16.8}$$

Here, T_w is the window width of the MAF, and $\bar{x}(t)$ is the moving average value of $x(t)$ calculated over T_w. A window width of $T_w = 1/(2f)$, where f is the grid frequency, can remove all the second-order oscillations of the dc-link voltages [17]. Subsequently, the angle of voltage positive sequence is calculated using the conventional PLL.

16.2.6 Proposed Control Strategy for Inverter and dc–dc Converter during Voltage Sags

A comprehensive structure of the proposed controller of the 3L-NPC inverter during voltage sags is presented in Fig. 16.7.

The main contribution of this study is the proposed method for the calculation of dq-coordinate current references during various operation modes of the GCPVPP. The proposed method ensures that the full current capacity of the grid-connected inverter of the GCPVPP is utilized for injecting active/reactive power during voltage sags.

The q-axis current reference ($I_q{}^*$) can be calculated from the grid code. During *Normal* condition, the inverter should inject purely active power to the grid, and, consequently, $I_q{}^*$ equals to zero. During *Sag I* operation, the amount of the reference reactive current depends on the minimum voltage amplitude ($\widehat{V}_{mn}$), while the inverter should purely inject the reactive current to the grid ($I_q{}^* = -I_{Ndq}$) during *Sag II* condition.

The calculation of d-axis current reference ($I_d{}^*$) is more challenging, because it relates to the injected active power to the grid (p) and extracted power from PV strings. This chapter proposes the following algorithm for calculating $I_d{}^*$ during various operation conditions in order to utilize the maximum capability of the 3L-NPC inverter in injecting

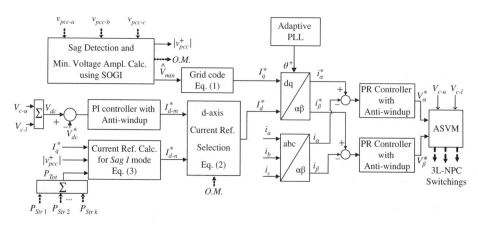

Figure 16.7 Proposed current controller for multi-string GCPVPP with 3L-NPC inverter during voltage sags.

active/reactive power to the grid during voltage sags.

$$I_d^* = \begin{cases} I_{d-m}^*, & \text{Grid}: \textit{Normal} \\ I_{d-n}^*, & \text{Grid}: \textit{SagI} \\ I_{d-m}^*, & \text{Grid}: \textit{SagII} \end{cases} \tag{16.9}$$

Here, the proposed algorithms for calculating I_{d-m}^* and I_{d-n}^* under three different operation modes are as follows.

Normal. During *Normal* operation, the dc–dc converters extract the maximum power from PV strings, while the 3L-NPC inverter controller is used to maintain the dc-link voltage (v_{dc}) by controlling the injected active power into the grid. It is realized by manipulating the active current reference I_{d-m}^*, as shown in Fig. 16.7. Due to the objective of the dc-d-m link voltage controller, the injected active power to the grid is adjusted to the extracted power from PV strings. The PI dc-link voltage controller with anti-windup is implemented to calculate I_{d-m}^*. The implementation of the anti-windup prevents the saturation of the d-m integrator part of the PI controller. If the output of the PI controller is beyond its operation range, the difference is reduced from the input error of the integrator. This reduces the input error to the integrator and prevents its saturation. A detailed description of the anti-windup scheme can be found in [33].

Sag I. In this operation mode, I_q^* is a fraction of the inverter nominal current. Therefore, the inverter is also able to simultaneously inject active power into the grid. Since priority should be given to the reactive power, I_q^* is first calculated from the grid code, and then the amount of d-axis current reference I_{d-n}^* should be calculated as follows:

$$I_{d-n}^* = \min\left[\sqrt{I_{N_{dq}}^2 - I_q^{*2}}, \; \frac{P_{Tot}}{|v_{pcc}^+|}\right] \tag{16.10}$$

Here, P_{Tot} is the instantaneous total extracted power from all PV strings ($P_{Str1}, P_{Str2}, \ldots$), as shown in Fig. 16.7. Also, $|v_{pcc}^+|$ is the amplitude of the positive-sequence of the voltage of point of common coupling (PCC), which is calculated dynamically through the controller.

$I_{N_{dq}}$ is related to the inverter nominal current, which usually is chosen based on the nominal maximum power of PV strings. However, during voltage sags, the nominal maximum power of PV strings might not be available. Therefore, by injection of the d-axis current of $\sqrt{I_{N_{dq}}^2 - I_q^{*2}}$ to the grid, the injected active power to the grid is larger than the extracted power from PV strings, which results in the discharging of dc-link capacitors and the decreasing of the dc-link voltage to lower than the nominal range. In order to avoid this problem, the instantaneous amount of the extracted power from PV strings should be considered in the calculation of I_d^*, as in equation (16.10). The amount of $P_{Tot} / |v_{pcc}^+|$ refers to the available d-axis current that can be injected to the grid based on the instantaneous extracted power from PV strings.

On the other hand, if the total extracted power from PV strings is higher than the injected active power to the grid, the excessive energy between the extracted power from PV strings and the injected active power to the grid ($P_{Tot} - p$, where p is the instantaneous injected active power to the grid) will be stored in the dc-link capacitors, which may lead to overvoltage. Hence, the extracted power from PV strings should be reduced during *Sag I*. Thus, a power reduction algorithm is implemented by controlling the dc-link voltage through the dc–dc converters, as shown in Fig. 16.8. In order to

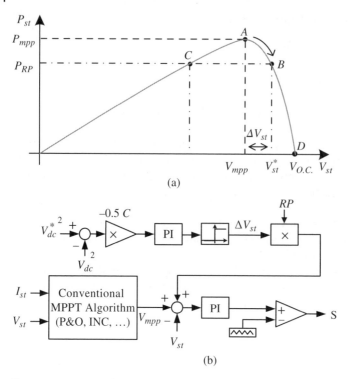

Figure 16.8 Power reduction strategy from PV strings during *Sag I*: (a) power–voltage curve of the PV string; and (b) dc–dc converter controller.

reduce the extracted power from the PV string, a reduced power (*RP*) signal is defined. This signal is set to 1 for all PV strings during *Sag I*, while it is set to 0 for all PV strings during *Normal* condition. As illustrated in Fig. 16.8(a), while $RP = 1$, the voltage (ΔV_{st}) is added to V_{mpp}, which results in moving the operation point from *A* to *B*.

It should be mentioned that, according to Fig. 16.8(a), moving the operation point to points *B* or *C* can result in the reduction of the extracted power from the PV string. In this chapter, the operation point of the PV string during *Sag I* is moved to point *B* (in the right-side of the *MPP*), because:

- The difference between v_B and v_{mpp} ($|\, v_{mpp} - v_B \,|$) is smaller than $|\, v_{mpp} - v_C \,|$, and therefore moving the operation from point *A* to point *B* can be executed faster than moving to point *C*, which results in a faster response.
- For small power references, the operation of the PV string in the left-side of *MPP* (point *C*) necessitates the PV string voltage to be close to the short-circuit voltage. This imposes the dc–dc converter duty cycle to be close to one. As a result, controlling the dc–dc converter becomes more difficult, and large oscillations may appear in the input/output voltages. This issue is avoided by moving the operation point to the right-side of the *MPP* (*B*), where the PV string voltage becomes closer to its open-circuit voltage, resulting in a decrease of the duty cycle.

In order to control the dc-link voltage during *Sag I*, the stored energy of the dc-link ($0.5\ Cv^2_{dc}$, where *C* is the equivalent value of all dc-link capacitors) is manipulated.

Therefore, ΔV_{st} is calculated with the aim of controlling the dc-link energy. Accordingly, the square of the dc-link voltage reference (v^{*2}_{dc}) is compared with the instantaneous dc-link voltage square (v^2_{dc}), and is multiplied with 0.5 C. This error is fed into the PI controller in order to calculate the required ΔV_{st}. The duration of the voltage sag is short; hence, it is assumed that irradiation and temperature do not change during this condition, and that V_{mpp} is constant. Accordingly, the MPPT algorithm stops its operation during the voltage sag, and the calculated V_{mpp} before the occurrence of the voltage sag is considered during the voltage sag.

As shown in Fig. 16.8(b), various types of MPPT algorithms, such as perturb and observe (P&O) or incremental conductance (INC), can be implemented in the proposed controller. Partial shading has been extensively addressed in several research articles and can also be implemented in the MPPT algorithm. However, since the main focus of this chapter is to study the LVRT of the grid-connected GCPVPP, partial shading is not investigated in this study. Accordingly, the well-known P&O algorithm, as introduced in [26], is implemented. As shown in Fig. 16.8(b), the calculated reference voltage is then compared to the instantaneous voltage of the PV string (v_{st}), and the difference is fed into a PI controller in order to generate the duty cycle of the boost converter.

Sag II. During *Sag II*, grid codes require GCPVPPs to inject the maximum possible reactive current of the inverter into the grid. Therefore, due to the current limitation of the grid-connected inverter, the active power injection is maintained at zero in order to inject the maximum possible reactive current to compensate for the voltage sag. Accordingly, PV strings are open-circuited (OC) in order to reduce the extracted power from all of the PV strings to zero. The operation point of the PV sting is moved to point *D* in Fig. 16.8(b), which results in zero power extraction from the PV string. During this operation mode, $I_d{}^*$ is considered to be $I_{d\text{-}m}$, which is calculated from the dc-link voltage controller, as shown in Fig. 16.7. Since, no power is injected to the dc-link from PV strings during *Sag II*, the amount of $I_{d\text{-}m}$ is almost zero. However, it can help to keep the dc-link voltage at its reference value, especially during unbalanced voltage sags that result in the oscillation of the dc-link voltage.

As shown in Fig. 16.7, the computed *dq*-axis current references ($I_d{}^*$ and $I_q{}^*$) are converted to their equivalent reference stationary frame currents (i^*_α and i^*_β) using the inverse Park transformation. An adaptive PLL is applied for extracting the angle of the voltage positive sequence, as shown in Fig. 16.7. The respective $\alpha\beta$ current errors are then obtained by subtracting the stationary frame current references from the stationary frame currents (i_α and i_β) obtained from the measured three-phase output inverter currents through inverse Clarke transformation. The errors between the calculated current references and instantaneous phase currents are fed into the PR with anti-windup controller, which is implemented in αβ-coordinate. Although, the implementation of the conventional PI controller for balanced current injection is simple, it requires multiple frame transformation, and results in slow performance. The PR controller with anti-windup shows faster dynamic response and zero steady-state error [18]. Finally, an ASVM technique is applied for generating the switching signals of the three-wire 3L-NPC inverter by considering dc capacitor voltage balancing.

16.2.7 Validation

The multi-string GCPVPP (Fig. 16.1) is modelled and developed using Matlab/ Simulink© and PLECS toolbox. Three parallel PV strings are connected to the dc-link

Table 16.1 Simulation parameters of the GCPVPP with the NPC inverter.

Parameter	Symbol	Value
Grid line–line voltage	v_{pcc-ab}	12.47 kV$_{rms}$
PCC phase–neutral voltage	v_{pcc-an}	230 V$_{rms}$
dc-link voltage	v_{dc}	700 V$_{dc}$
dc-link capacitor	C_u/C_l	2.2 mF
Fundamental frequency	f_0	50 Hz
3L-NPC switching frequency	f_s	3.6 kHz
dc–dc converter switching frequency	$f_{s-dc-dc}$	10 kHz
dc–dc converter inductor	L	1 mH

of the grid-connected 3L-NPC inverter. Each PV string is constructed using 15 parallel PV branches, with each branch having 14 series-connected PV panels.

Consequently, the maximum output power of each PV string is 50 kW, and the maximum total power of the GCPVPP equals 150 kW. The detailed parameters of the simulated GCPVPP are presented in Table 16.1. In order to validate the performance of the proposed controller under various irradiation and temperature values, different amounts of irradiance and temperature are induced for the PV strings – String I ($I_r = 1$ kW/m², $T = 25°$C); String II ($I_r = 0.5$ kW/m², $T = 25°$C); and String III ($I_r = 1.1$ kW/m², $T = 35°$C).

It is necessary to perform a comprehensive power system test case simulation in order to investigate the performance of the designed GCPVPPs in terms of the control and dynamic performance. Accordingly, a medium-voltage grid test case, based on North American Network, depicted in Fig. 16.9, is introduced in [39]. The grid voltage is 12.47 kV, and the GCPVPP is connected to Bus 3 (PCC) using a step-up transformer (0.4 kV/12.47 kV). The test case system consists of both single-phase and three-phase loads. Single-phase loads are connected to Bus 2 and Bus 5. The X/R ratio of lines equals to 6, and the medium-voltage grid is connected to the 115 kV line through a 12.47 kV/115 kV transformer. The total load of Sub-network 1, shown in Fig. 16.9, is 200 kVA, while 150 kVA of this capacity can be supplied by the only DG unit in this network, which is the GCPVPP connected to Bus 3.

Two different study sets are performed on the modelled grid-connected GCPVPP, and the results are presented in the following subsections. First, the dynamic performance of the proposed controller is investigated during the different voltage sag conditions. Subsequently, the effect of the proposed active/reactive power injection method is evaluated in enhancing the PCC voltage during various fault conditions at different buses of the system.

In order to validate the applicability of the proposed controller, the operation of GCPVPP is examined under two different voltage sags:

- *Sag I*: It includes a single phase to ground fault at Bus 4, which results in a 27% two-phase voltage sag between $t = 0.15$ s and $t = 0.3$ s in the PCC, as shown in Fig. 16.10(a).
- *Sag II*: It consist of a two-phase fault at Bus 4, which leads to a 64% three-phase voltage sag between $t = 0.45$ s and $t = 0.6$ s in the PCC, as shown in Fig. 16.10(a).

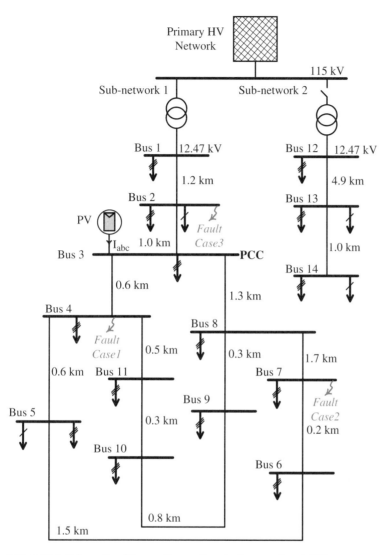

Figure 16.9 Schematic of the medium-voltage test-case network [39].

The amounts of injected active and reactive power to the grid are shown in Fig. 16.10(b). It can be observed that the 3L-NPC inverter injects maximum active power to the grid during *Normal* operation. However, the amount of active power is reduced during *Sag I*, owing to the injection of the reactive power in this period. After $t = 0.3$ s, the grid recovers to *Normal* condition, and hence the inverter output active power is increased while the injected reactive power becomes zero. The injection of reactive power increases to its maximum possible value at $t = 0.45$ s due to *Sag II*, while the injected active power is zero. It should be mentioned that the oscillation of injected active and reactive power during *Sag I* and *Sag II* is because of the injection of balanced currents to the grid during unbalanced grid voltages. The oscillation of the injected active power results in the oscillation of the dc-link voltages, as illustrated in

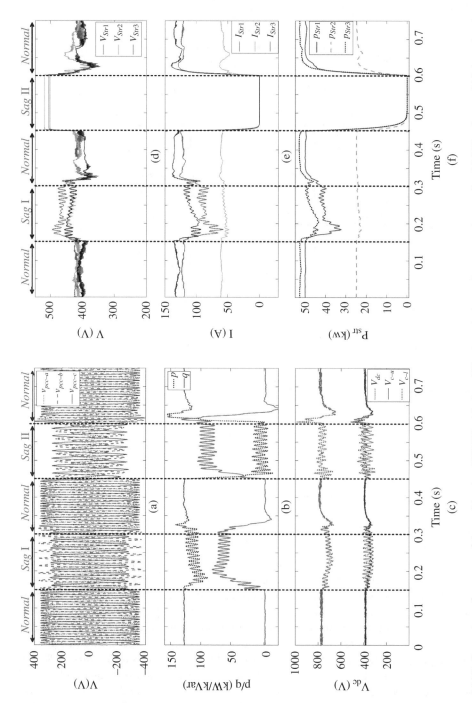

Figure 16.10 Dynamic performance of the multi-string GCPVPP with three PV strings during various operation modes: (a) PCC voltages; (b) injected active and reactive power to the grid; (c) dc-link and dc capacitor voltages; (d) PV string voltages; (e) PV string currents; and (f) extracted power from PV strings.

Fig. 16.10(c). Besides, the proposed control method maintains v_{dc} at its reference value during all the operation modes. The dc-link capacitor voltages remain balanced due to the implementation of the ASVM with voltage balancing capability.

The dynamic performance of PV strings is presented in Fig. 16.10(d). Since the PV strings are exposed to different amounts of irradiation and temperature, they have different maximum power values, voltages, and currents during *Normal* condition. During *Sag I*, the extracted power from PV strings reduces to a certain value that is calculated by the dc-link voltage controller. Accordingly, the voltage of PV strings increases. This moves the PV strings operation point to the right-side of the *MPP* point. Moreover, the current of PV strings reduces during *Sag I*, as depicted in Fig. 16.10(e). The total extracted PV power is decreased to zero in order to deliver pure reactive power to the grid during *Sag II*. During this condition, the current reduces to zero, and PV string voltages increase to their open-circuit value.

The dynamic performances of the current controller of the 3L-NPC inverter at the beginning and end of the voltage sag are depicted in Fig. 16.11. During *Normal* operation, the 3L-NPC inverter delivers only active power to the grid under unity power factor operation, and no phase-shift exists between the PCC phase voltages and the injected currents, as shown in Fig. 16.11(b). However, there is a phase-shift (φ_1) between the voltage and currents during *Sag I*, which shows the simultaneous injection of active and reactive powers to the grid. Fig. 16.11(d) shows the injected currents during *Sag II*. A phase-shift of $\varphi_2 = \pi/2$ exists between the PCC voltage and injected currents,

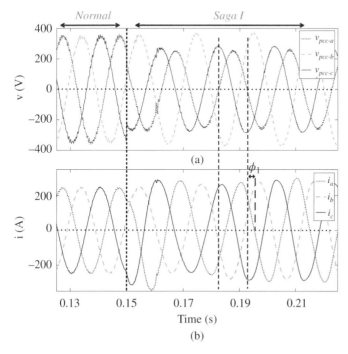

Figure 16.11 Dynamic current controller performance of the grid-connected 3L-NPC inverter: (a) starting of *Sag I* – PCC voltages; (b) starting of *Sag I* – injected currents; (c) ending of *Sag II* – PCC voltages; and (d) ending of *Sag II* – injected currents.

which implies pure reactive power injection during *Sag II*. Moreover, the currents of the inverter recover to *Normal* mode with unity power factor operation quickly, owing to the fast performance of the PR current controller. It should be noticed that the amplitude of the inverter current remains at its nominal current amplitude during all sag conditions, which shows the effectiveness of the proposed *dq*-axis current reference calculation in using the full current capacity of the inverter when injecting active/reactive power to the grid during voltage sags.

16.2.8 Experimental Verification

A three-phase 3L-NPC converter setup (Fig. 16.12) has been utilized in order to experimentally validate the effectiveness of the proposed controller under unbalanced voltage sags. The 3L-NPC setup consists of Semikron SKM145GB176D 1700V IGBT modules. The grid is generated using a TopCon TC.ACS four-quadrant grid simulator, while the PV side is emulated using an ETS600/8 Terra SAS PV simulator. A dc–dc boost converter is connected between the PV panel and dc-link, which extracts the maximum power for the PV panel during normal operation mode. The proposed controller and protection functions of the converter are implemented in a dSPACE 1006 platform. The parameters of the experimental setup are given in Table 16.2. The proposed controller has been evaluated under two different cases corresponding to *Sag I* and *Sag II*.

Sag I. This case study includes a 36% two-phase voltage sag for a duration of 150 ms. The three-phase grid voltages (v_{pcc}) and injected currents (i_{abc}) are depicted in Fig. 16.13(a). Due to the implementation of the proposed control algorithm, the injected three-phase currents are balanced during such unbalanced voltage sags. Additionally, because the injected currents are equal to the inverter nominal current, maximum current capacity of the inverter is applied for enhancing the grid voltages during the voltage sag. A delay of maximum $T/4$, where T is the grid voltage period, is introduced by the SOGI due to the single-phase quadrature signal generation. It should be noted that this delay also depends on the amplitude of the voltage sag. During voltage sags with smaller amplitudes (e.g. 10–30%), the sag can be detected faster as compared to voltage sags with larger amplitudes (e.g. 70–90%).

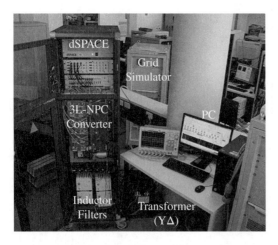

Figure 16.12 Experimental setup of the grid-connected 3L-NPC inverter.

Table 16.2 Experimental parameters of the GCPVPP with the NPC inverter.

Parameter	Symbol	Value
PV panel and dc–dc converter parameters		
PV panel maximum power	p_{mpp}	3.3 kW
PV panel maximum power point voltage	v_{mpp}	480 V
PV panel maximum power point current	i_{mpp}	7 A
PV panel filling factor	FF	0.8
PV panel capacitor	C_{pv}	200 µF
dc–dc converter switching frequency	f_{sw}	10 kHz
3L-NPC inverter parameters		
Apparent power	S	3.3 kVA
PCC phase-neutral voltage	$v_{pcc\text{-}abc}$	320 V_{rms}
dc-link voltage	v_{dc}	560 v_{dc}
dc-link capacitor	Cu / Cl	4.9 mF
Fundamental frequency	f_0	50 Hz
3L-NPC switching frequency	f_s	1.5 kHz
Inductor filter	L_f	4 mH

The dc-link voltage and capacitor voltages are depicted in Fig. 16.13(b). The dc-link voltage during *Normal* operation is 560 V_{dc} and remains constant during *Sag I*. Additionally, dc-link capacitors remain balanced during all operation modes. The PV voltage (v_{st}) is equal to 480 V_{dc} during *Normal* mode, which shows the operation of the PV string at MPPT. During *Sag I*, v_{st} increases to 555 V_{dc} with the proposed controller in Fig. 16.8 in order to decrease the extracted power. The extracted power from PV string (p_{st}) and output current of the PV string (i_{st}) are reduced during *Sag I*.

The quantities of the injected active and reactive powers (p and q) during *Sag I* are illustrated in Fig. 16.13(c). During *Normal* operation, the 3L-NPC inverter injects purely active power to the grid, equal to 3.1 kW. The average active power is reduced to 1.3 kW for the duration of *Sag I*. As expected, a second harmonic oscillation exists in the injected active power, due to the injection of balanced currents during the unbalanced voltage sag. On the other hand, the average of reactive power Q is increased to 2 kVAR. Consequently, the average injected apparent power is equal to 2.4 kVA ($S = \sqrt{1.3^2 + 2^2}$), which is 77% of the apparent power during *Normal* operation ($S_N = 3.1$ KVA). The reduction of the injected apparent power is due to the decrease of the grid voltages during the voltage sag, while the injected currents remain at a nominal value.

Sag II. This case study consists of a three-phase voltage sag of 70%, as shown in Fig. 16.14(a).

Phase *a* experiences the smallest voltage amplitude at 30% of the nominal value. However, due to the implementation of the proposed controller, the injected currents remain balanced, and the maximum current capacity of the inverter is utilized during the voltage sag.

The dc-link voltage, shown in Fig. 16.14(b), remains at its nominal value during *Sag II*, while v_{st} increases to 165 V_{dc}, which is equal to the PV panel open-circuit voltage.

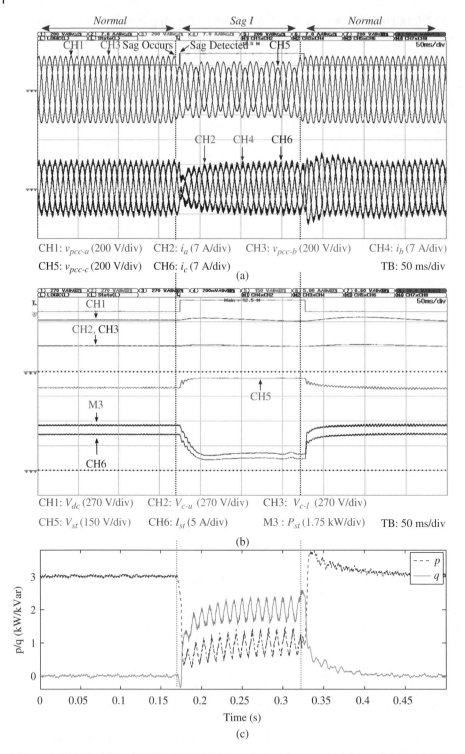

Figure 16.13 Experimental results: *Sag I* – 36% two-phase voltage sag: (a) three-phase voltages and currents; (b) dc-link voltage, PV string voltage, current, and power; and (c) injected active/re-active power.

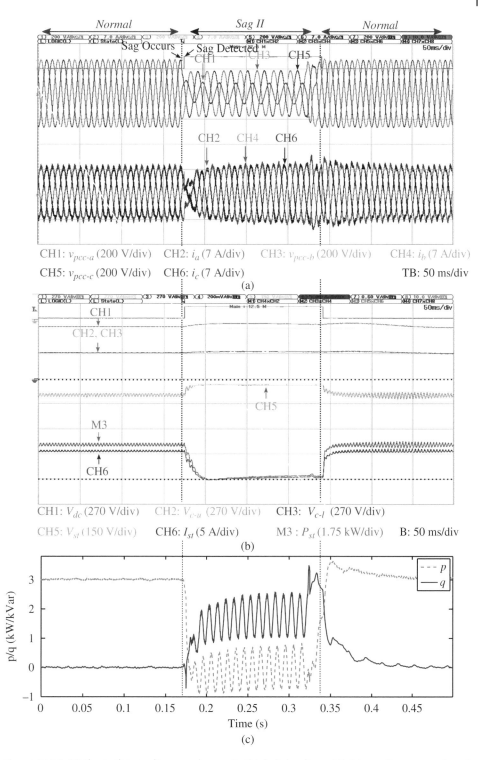

CH1: v_{pcc-a} (200 V/div)   CH2: i_a (7 A/div)   CH3: v_{pcc-b} (200 V/div)   CH4: i_b (7 A/div)

CH5: v_{pcc-c} (200 V/div)   CH6: i_c (7 A/div)   TB: 50 ms/div

(a)

CH1: V_{dc} (270 V/div)   CH2: V_{c-u} (270 V/div)   CH3: V_{c-l} (270 V/div)

CH5: V_{st} (150 V/div)   CH6: I_{st} (5 A/div)   M 3 : P_{st} (1.75 kW/div)   B: 50 ms/div

(b)

(c)

Figure 16.14 (a) Three-phase voltages and currents; (b) dc-link voltage, PV string voltage, current, and power; and (c) injected active/reactive power.

Accordingly, p_{st} and i_{st} are also reduced to zero in this condition. The injected active and reactive powers are depicted in Fig. 16.14(c). It can be seen that the average active power is reduced to zero. The average reactive power (q) is increased to 1.3 kVAR during the voltage sag – that is, 45% of S_N. It should be noted that, in this case study, the delay of the voltage sag detection algorithm is smaller than *Sag I*, because of the smaller voltage amplitude of the voltage sag.

16.3 Seven-Level Cascaded H-Bridge Inverter–Based PV Power Plant

Among various multilevel converter topologies, the multilevel cascaded H-bridge (CHB) converter, shown in Fig. 16.15, was introduced as a potential candidate for medium- and large-scale GCPVPPs in [4, 11, 25, 27, 35, 42–44] because of the following reasons:

- The low switching frequency of each H-bridge semiconductor results in the reduction of semi-conductor losses.
- The total power losses of the GCPVPP with CHB converter are smaller than those from other topologies [15].
- Direct connection to medium-voltage grids eliminates the need for bulky and expensive line-frequency transformers.
- Multiple dc-links enable independent maximum power point tracking (MPPT) of the PV strings, which maximizes the total extracted power.

A grid-connected PV CHB inverter may experience inter-phase imbalance, which occurs when each phase generates a different amount of power as a result of the unbalanced phase currents or partial shading; and an additional inter-bridge imbalance

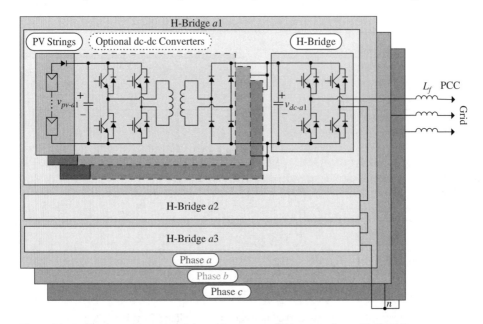

Figure 16.15 Circuit configuration of the grid-connected PV power plant with 7L-CHB converter.

that happens when each bridge in the same phase leg generates a different amount of power as a consequence of unequal power generation from the PV strings [42, 43]. These imbalances result in several issues: (a) the output voltage quality of the converter deteriorates; (b) the voltage of one dc-link may increase beyond its normal range, which can activate the dc-link over-voltage protection relay, which in turn disconnects the converter from grid; and (c) in the single-stage power conversion topology (without dc–dc converters), the deviation of dc-link voltages from their average value ($\overline{V_{dc}}$), which is set to the voltage of the PV strings' maximum power point (v_{mpp}), decreases the extracted power from PV strings [36].

With respect to the above-mentioned issues, several algorithms are introduced in the literature to provide a balanced current injection to the grid under different conditions. Various zero-sequence voltage injection algorithms are proposed in [4, 42, 43] to provide balanced current injection to the grid under dc-side imbalances, including partial shading or mismatch between PV strings. The issue of balanced current injection under the failure of some of the bridges of the CHB inverter is addressed in [41]. The main goal of these studies is the zero-sequence voltage injection for achieving inter-phase balancing and injecting active power (p) with balanced currents to the grid during grid normal operation. A new GCPVPP topology with part-row and full-row connection of dc–dc converters to the CHB inverter is proposed in [35], which reduces the second harmonic oscillation of the dc-link voltage and provides balanced current injection during partial shading. An inter-bridge balancing through the modification of bridge voltage references is also presented in [4]. The operation of the CHB inverter during grid normal operation with balanced voltages and active power injection is addressed in these studies.

During unbalanced voltage sags, a grid-connected PV CHB inverter experiences inter-phase imbalance as a result of unbalanced phase currents. Further inter-phase and inter-bridge imbalance can simultaneously occur as a result of unequal power generation from the PV strings (depending on irradiance conditions). Accordingly, various strategies are presented in [22, 23, 25, 28], which achieve low-voltage ride-through capability for the CHB-based static synchronous compensator (STATCOM). Moreover, zero-sequence and negative-sequence current injection of the CHB-based STATCOM during grid disturbances are also taken into consideration in [6, 9]. The main disadvantage is that only reactive power injection is implemented on the grid-connected inverter in these studies. However, modern grid codes also require simultaneous active power injection with either balanced or unbalanced currents from grid-connected PV inverters [29, 38]. On top of that, a deep consideration with respect to the inter-phase and inter-bridge balancing is required. Therefore, there is need for a flexible control strategy, which can ensure dc-link voltage balancing for the grid-connected PV CHB inverter during voltage sags. However, this is not investigated in the literature [4, 6, 9, 11, 22, 23, 25, 27, 28, 35, 36, 41–43].

Motivated by the methods discussed in the preceding text, this chapter introduces a flexible control strategy for the operation of grid-connected PV CHB inverters during unbalanced voltage sags. The proposed strategy is capable of balancing dc-link voltages while injecting both active and reactive powers with either balanced or unbalanced currents. A zero-sequence voltage injection algorithm is proposed for inter-phase balancing, which allows for energy exchanges between various phases. For inter-bridge balancing, the extracted power from different bridges of one phase is changed by modifying bridge voltage references. A feedforward voltage compensation is also applied to decrease the inverter transient current at the beginning of the voltage sag. The performance of the proposed control strategy is evaluated on the 9 kVA grid-connected 7L-CHB inverter simulation, and experimental setups under various voltage sags, to validate the capability of the proposed algorithm for the operation of the grid-connected CHB inverter during voltage sags. Unbalanced voltage sags are more common as compared to balanced voltage sags, and the operation of the grid-connected PV CHB inverter during balanced voltage sags or normal grid condition is also easier to deal with. Therefore, unbalanced voltage sags are taken into consideration in all of the investigated case studies in this chapter, while the proposed control strategy can also be used during grid normal operations or balanced voltage sags. Notice that, for those operations, MPPT can always be implemented; however, it is not a mandatory objective of this manuscript.

16.3.1 Circuit Configuration

Various topologies of isolated dc–dc converters are introduced in the literature for medium- and large-scale GCPVPPs [45]. Each dc–dc converter reduces the oscillation of the extracted power from PV strings, which exists in the direct connection of PV strings to dc-links (without dc–dc converters) as a consequence of the existence of second harmonic oscillation in dc-link voltages. Moreover, the isolated dc–dc converters eliminate the leakage current of PV modules, which is the result of the common mode voltage across the parasitic capacitance between PV modules and the ground.

Inductor filters are connected in the output of the inverter to reduce the current harmonics and to comply with grid standards. This topology is designed for the direct connection of the GCPVPP to the medium- or high-voltage grid. The inductor filter is usually used for the connection of power converters to medium- or high-voltage grids, due to challenges in the design of high-voltage capacitors and their low reliability. Hence, the inductor filter is considered for the connection of the CHB inverter to the grid. The tuned arm filter is designed to eliminate the high-frequency current harmonic from the injected current to the grid. The design of the cut-off frequency for this type of filter is normally 10 times larger than the fundamental frequency. The frequency bandwidth of voltage sags is in the range of the fundamental frequency. Hence, the performance of the proposed control algorithm does not change while the LCL filter is connected between the CHB inverter and the grid.

The implementation of a high-power isolated dc–dc converter is very challenging. However, in the implemented GCPVPP, parallel dc–dc converters are applied to transfer the large amount of power from PV strings to the dc-link of each H-bridge. In this case, the available high-power isolated dc–dc converters, as mentioned in the following references, can be used in the implementation of the GCPVPP. For example, a 166 kW/20 kHz prototype of isolated dc–dc converter has been implemented in [21]. Several other designs are also proposed in [5, 7, 8] for high-power dc–dc converters.

16.3.2 Control Algorithm

A circuit configuration of the GCPVPP with the 7L-CHB converter is illustrated in Fig. 16.15. A detailed description of the GCPVPP with grid-connected CHB inverter can be found in [42, 43].

In order to generalize the proposed strategy, an N-level CHB is considered in this section. The structure of the proposed control strategy, divided into five different parts, is depicted in Fig. 16.16. These parts will be described in detail in the following subsections.

The controller of the CHB inverter is used to maintain the average dc-link voltage $(\overline{V_{dc}})$ by controlling the injected active power into the grid. $\overline{V_{dc}}$ is calculated as follows:

$$\overline{V_{dc}} \triangleq \frac{\sum_{x=a}^{c} \sum_{j=1}^{N} v_{dc-xj}}{N \times 3} \tag{16.11}$$

Here, $x \in \{a, b, c\}$. The bridge number is denoted by j, while v_{dc-xj} is the dc-link voltage of bridge xj. As illustrated in Fig. 16.16, the error between $\overline{V_{dc}}$ and V^*_{dc}, which is the dc bridge voltage reference, is fed into a PI controller that calculates the d-axis current reference (i^*_d). In this study, a sag detection SOGI-based orthogonal system is applied for calculating the amplitude of phase voltages and for determining the minimum phase voltage amplitude. Detailed descriptions of the SOGI implementation and voltage sag detection can be found in [38]. Based on the amplitude of the voltage sag, the q-axis current reference (i^*_q) is calculated from the grid codes, as implemented in [6].

In order to show the flexibility of the proposed strategy, two different current reference calculation strategies are demonstrated in this study: (1) balanced currents during balanced/unbalanced voltage sags, as in [31], which uses the full current capacity of the inverter to enhance the PCC during voltage sags; and (2) unbalanced currents during unbalanced voltage sags in order to achieve zero active power oscillation, as in [19].

The errors between the calculated current references and instantaneous phase currents, transformed to the $\alpha\beta$ coordinate (i_α and i_β), are fed into the PR with anti-windup controllers. Although the implementation of the conventional PI controller for balanced current injection is simpler, it requires multiple frame transformations, and results in reduced dynamic performance as compared to abc-framework-based controllers [3, 14, 34]. The anti-windup control loop prevents the saturation of the output voltage of the PR controller and maintains the sinusoidal shape of the output voltage reference. Furthermore, for the injection of accurate unbalanced currents under unbalanced voltage sags, the calculation of both positive- and negative-sequence voltages in the dq frame is required, which increases the computational complexity [14]. Furthermore, the PR controller with anti-windup shows faster dynamic response and zero steady-state error [18].

The voltage references ($v^*_{\alpha-1}$ and $v^*_{\beta-1}$) are calculated through PR controllers. Subsequently, feedforward compensators and voltage balancing controllers are implemented. Their detailed descriptions are provided in the following subsections. Finally, the conventional phase-shifted pulse-width modulation algorithm is implemented to generate the switching signals. The detailed explanation of this algorithm can be found in [13, 16].

16.3.3 Effects of Feedforward Voltage Compensation

Fast detection of voltage sags is important for the quick operation of the controller [30]. A maximum delay of $T/4$, where T is the grid voltage period, is introduced by the SOGI

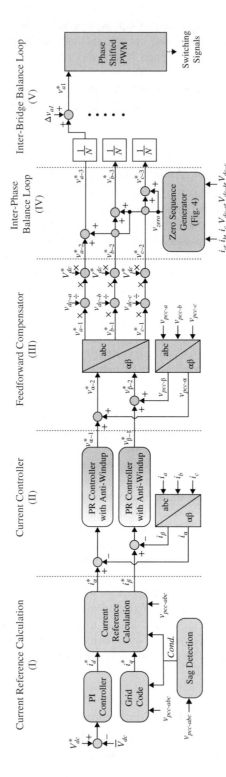

Figure 16.16 Proposed control strategy for grid-connected CHB inverter during voltage sags.

due to the single-phase quadrature signal generation. It should be noted that this delay also depends on the amplitude of the voltage sag.

The delay of the voltage sag detection algorithm leads to a high transient current at the beginning of voltage sags, because the controller continues to inject maximum power to the grid in the period between the occurrence of the voltage sag and its detection. Thus, the use of a feedforward voltage compensator is necessary to decrease the transient current during this period. A feedforward voltage compensator is implemented based on [18], as depicted in Fig. 16.16. The instantaneous PCC voltages ($v_{pcc\text{-}\alpha}$ and $v_{pcc\text{-}\beta}$), which are transformed to the $\alpha\beta$ coordinate using Clark transformation, are added to $v^{*}_{\alpha\text{-}1}$ and $v^{*}_{\beta\text{-}1}$, respectively, as follows:

$$v^{*}_{\alpha\text{-}2} = v^{*}_{\alpha\text{-}1} + v_{pcc\text{-}\alpha}$$
$$v^{*}_{\beta\text{-}2} = v^{*}_{\beta\text{-}1} + v_{pcc\text{-}\beta} \tag{16.12}$$

The vector diagram during the grid normal operation is depicted in Fig. 16.17(a). During this condition, the inverter operates under unity power factor, and the output voltage of the inverter has a similar phase and amplitude as v_{pcc}. Accordingly, a large portion of $v^{*}_{\alpha\text{-}2}$ and $v^{*}_{\beta\text{-}2}$ are provided by $v_{pcc\text{-}\alpha}$ and $v_{pcc\text{-}\beta}$, respectively, while the current controller only regulates the required voltage drop on the filter inductors for the intended current references. Therefore, the amplitudes of $v^{*}_{\alpha\text{-}1}$ and $v^{*}_{\beta\text{-}1}$ are small during the grid normal operation.

At the beginning of the voltage sag, as shown in Fig. 16.17(b), the instantaneous values of $v_{pcc\text{-}\alpha}$ and $v_{pcc\text{-}\beta}$ are reduced immediately, following the voltage sag amplitude. At the same time, the values of $v^{*}_{\alpha\text{-}1}$ and $v^{*}_{\beta\text{-}1}$ cannot change immediately, because they depend on the voltage sag detection algorithm, current reference calculation method, and current controller. Consequently, $v^{*}_{\alpha\text{-}2}$ and $v^{*}_{\beta\text{-}2}$ decrease instantaneously at the beginning of the voltage sag. This reduces the output voltage of the inverter at the beginning of the voltage sag and decreases its transient current. It should be noted that, without this feedforward compensator, the voltage references are only regulated through the current controller, which cannot react instantaneously to the voltage sag, and hence the transient current becomes large at the beginning of the voltage sag. Finally, a feed-forward voltage controller for reducing the effect of the variation of dc-link voltage is also implemented by multiplying the voltage references by $V^{*}_{dc}/v_{dc\text{-}xj}$.

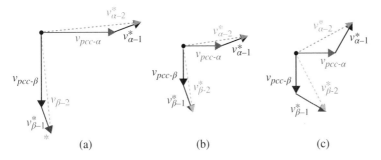

Figure 16.17 Feedforward voltage compensation principle – voltage references: (a) during the grid normal operation; (b) at the beginning of the voltage sag; and (c) during the steady-state period of the voltage sag.

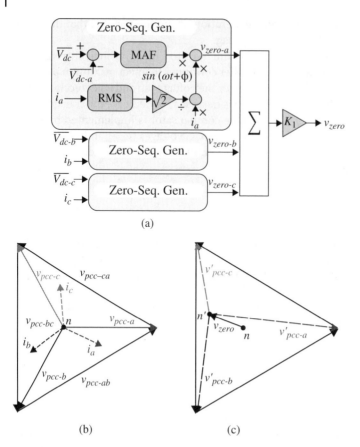

(a)

(b)   (c)

Figure 16.18 Proposed inter-phase balancing algorithm: (a) proposed zero-sequence generation strategy; (b) vector diagram during balanced operation; and (c) vector diagram during zero-sequence voltage injection.

16.3.4   Proposed Inter-Phase Balancing Strategy

An inter-phase balancing strategy is implemented by adding a zero-sequence voltage (v_{zero}) to the voltage references (v^*_{a-2}, v^*_{c-2}, and v^*_{c-2}), as shown in Part IV of Fig. 16.16. The proposed algorithm for the calculation of v_{zero} is presented in Fig. 16.18(a). The error between $\overline{V_{dc}}$ and the average of the bridge voltages of phase a ($\overline{V_{dc-a}} = (v_{dc-a1} + v_{dc-a2} + \ldots + v_{dc-aN})/N$) is fed into a moving average filter (MAF), which filters instantaneous oscillations of the dc-link voltages by performing the following operation:

$$\bar{x}(t) = \frac{1}{T_w} \int_{t-Tw}^{t} x(\tau)d\tau \tag{16.13}$$

Here, T_w is the window width of the MAF, and $\bar{x}(t)$ is the moving average value of $x(t)$ calculated over T_w. A window width of $T_w = 1/(2f)$, where f is the grid frequency, can remove all the second-order oscillations of the dc-link voltages. The proposed inter-phase voltage balancing algorithm is designed to balance the average dc component of the capacitor voltages with a bandwidth of 20 Hz. Therefore, the effect of delay of the MAF block ($1/(2f)$), with a bandwidth of 100 Hz, can be neglected on the performance of the proposed inter-phase voltage balancing algorithms [17, 18].

The calculated average error from the MAF is then multiplied by $sin(\omega t + \varphi)$, while $i_a = I_m sin(\omega t + \varphi)$. As shown in Fig. 16.18(a), the instantaneous value of $sin(\omega t + \varphi)$ is approximated as:

$$sin(\omega t + \phi) \approx \frac{i_a}{RMS(i_a) \times \sqrt{2}} \tag{16.14}$$

Here, $RMS(i_a)$ is the RMS value of the current, calculated dynamically in the controller. Subsequently, the calculated zero-sequence voltage from phase a (v_{zero-a}) is added to v_{zero-b} and v_{zero-c}. The total zero voltage (v_{zero}) is adjusted using a proportional controller with a proportional gain of K_1, which ensures the inter-phase balancing of the grid-connected CHB inverter during unbalanced voltage sags.

The vector diagram of the CHB inverter, during a balanced operation, is depicted in Fig. 16.18(b). In this condition, the injected zero-sequence voltage is zero. The operation of the proposed inter-phase balancing strategy is investigated, considering $v_{dc-a} > \overline{V_{dc}}$ and v_{dc-b}, $v_{dc-c} < \overline{V_{dc}}$. The corresponding voltage and current vector diagrams after the injection of the proposed zero-sequence voltage are presented in Fig. 16.18(c). Since $v_{dc-a} - \overline{u_{dc}} < 0$, v_{zero-a} is in the opposite direction of i_a. Similarly, v_{zero-b} and v_{zero-c} are in the same direction as i_b and i_c, respectively. Thus, the neutral point of the CHB inverter is moved from n to n', by a vector of v_{zero}. The amplitude of the new voltage vector of phase a (v_a') is larger than those of phases b and c (v_b' and v_c'). Therefore, the extracted power from phase a ($p'_a = v_a'.i_a$) is larger than p'_b and p'_c, which means that the extracted power from the capacitors of phase a is larger than those from phases b and c. As a consequence, v_{dc-a} reduces, and dc-link capacitors become balanced. It should be noted that, because the proposed method of zero-sequence voltage injection depends on the current direction, it can achieve inter-phase balancing under all different types of current injection, including active/reactive power injection with balanced/unbalanced currents.

16.3.5 Inter-Bridge Balancing Strategy

It is important that all of the dc-link voltages of each phase remain equal to achieve high-quality output voltages. Since the current of all the bridges of one phase are equal, their voltage varies if their power is different. An inter-bridge balancing mechanism, shown in Part V of Fig. 16.16, is implemented in this study as in [4]. The voltage reference of each bridge is calculated by dividing the voltage reference of the phase (v^*_{a-3}) by N and, subsequently, is added to Δv_{xj}, which is the required deviation of the voltage reference of each bridge for inter-bridge balancing. The proposed algorithm for the calculation of Δv_{xj} is presented in Fig. 16.19(a). The difference between v_{dc-xj} and $\overline{V_{dc}}$ is fed

Figure 16.19 Inter-bridge balancing algorithm.

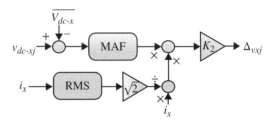

into the MAF, as defined in (16.13). The filtered voltage difference is then multiplied by $sin(\omega t + \varphi)$ (16.14) and fed into a proportional controller with a proportional gain of K_2 to calculate Δv_{xj}.

16.3.6 Validation

The GCPVPP with a 7L-CHB inverter (Fig. 16.15(a)) is modeled and developed using Matlab/Simulink© and PLECS toolbox. The main system parameters are listed in Table 16.3. The aim of the chapter is to analyze the performance of the proposed inter-phase and inter-bridge voltage balancing algorithms during grid voltage sags. Although the MPPT stage is not implemented, the dc-link voltage is still regulated by the current reference generation. This does not affect the purpose of this chapter to demonstrate the control and operation of the CHB inverter [41–43], because: (i) the extracted power from PV strings is close to their maximum power by setting the dc-link voltages to the voltage of the maximum power; (ii) the oscillation of the extracted power from PV strings, due to the oscillation of the dc-link voltage, is small as compared to its average value; and (iii) during voltage sags, the injection of reactive power to the grid is required by the grid codes. The extracted power from PV strings is reduced based on the amount of injected reactive current and the inverter nominal current.

Table 16.3 Simulation and experimental parameters of the GCPVPP with 7L-CHB inverter.

Parameter	Symbol	Value
PV panel parameters		
PV string maximum power	P_{mpp}	1 kW
PV panel MPP voltage	v_{mpp}	145 V
PV panel MPP current	i_{mpp}	7 A
Number of series-connected PV panels in the PV string	N_s	1
Number of parallel-connected PV panels in the PV string	N_p	1
7L-CHB inverter parameters		
Apparent power	S	9 kVA (1 pu)
PCC voltage	$v_{pcc\text{-}ll}$	430 V_{rms} (1 pu)
Inductor filter	L_f	8 mH (0.15 pu)
dc-link voltage	V_{dc}	145 V
dc-link capacitor	$C_{dc\text{-}xj}$	4.5 mF
Carrier frequency	F_{carr}	600 Hz
Apparent switching frequency	f_{sw}	3.6 kHz
Control parameters		
PR current controller	k_{p1}, k_r	0.003, 0.1
Inter-phase/inter-bridge voltage balancing	k_1, k_2	0.005, 0.005
dc-link voltage controller	k_{p2}, k_i	0.15, 0.6
Feed-forward filter time-constant	τ	0.00001 s

Since three H-bridges are cascaded in each phase leg, the converter output phase voltages feature seven-level waveforms with an equivalent switching frequency of 3.6 kHz. In order to evaluate the flexibility of the proposed strategy, two different voltage sag conditions are investigated: 30% single-phase voltage sag with simultaneous injection of active and reactive powers with unbalanced currents (*Case I*); and 70% unbalanced three-phase voltage sag with reactive power injection and balanced currents (*Case II*).

The simulation results on a 9 kVA GCPVPP are presented in this chapter. These case studies have also been tested on a 10 MVA GCPVPP, connected to the 6.6 kV distribution network, and similar performances were achieved.

Case I

This case demonstrates the operation of the CHB inverter during a 30% single-phase voltage sag, which can occur because of a single line to ground fault in the power system.

This voltage sag requires simultaneous injection of active and reactive power, according to the grid codes [37]. The unbalanced current injection method is chosen in this case, which results in zero active power oscillation from the inverter. The voltage sag occurs at phase b with a duration of 150 ms. The three-phase grid voltages (v_{pcc}) and injected currents (i_{abc}) are depicted in Fig. 16.20(a). Before $t = 0.05$ ms, the grid is at normal condition, and the inverter injects active power to the grid with balanced currents (Fig. 16.20(c)), while no phase displacement exists between the phase currents and voltages. Since the minimum amplitude of the voltage sag in this test study is larger than 0.5 pu, the inverter injects both active and reactive powers to the grid between $t = 0.05$ ms and $t = 0.2$ ms. A second harmonic voltage oscillation exists in both active and reactive powers, due to the injection of balanced currents to the PCC with unbalanced voltages.

It can be seen from Fig. 16.20(b) that the transient current of the inverter at the beginning of the voltage sag is reduced through the feedforward voltage compensator. The current of phase b, the phase with the voltage sag, remains equal to the nominal current of the inverter (11 A$_{rms}$), while the currents of nonfaulty phases reduce to 9.2 A$_{rms}$. The bridge capacitor voltages (v_{dc-xj}), presented in Fig. 16.20(d), are balanced, and their average remains close to 165 V during this unbalanced voltage sag condition. The extracted powers from PV strings are decreased to 6 kW, due to the current limitation of the inverter as a consequence of reactive power injection during the voltage sag. After the clearance of the voltage sag, the output power of the inverter recovers to 9 kW and the reactive power reduces to zero, while capacitor voltages stay balanced.

Case II

The performance of the proposed control strategy during a deep unbalanced voltage sag with reactive power injection and balanced currents is examined in this test, considering an unbalanced three-phase 70% voltage sag (phase a experiences deeper voltage sag as compared to phases b and c), as shown in Fig. 16.21(a). The inverter injects 3.8 kVAR average reactive power to the grid, while the injected active power decreases to zero. The dc-link voltages remain balanced. The presented simulation results verify the flexibility of the proposed strategy under various voltage sags and different power/current injection strategies.

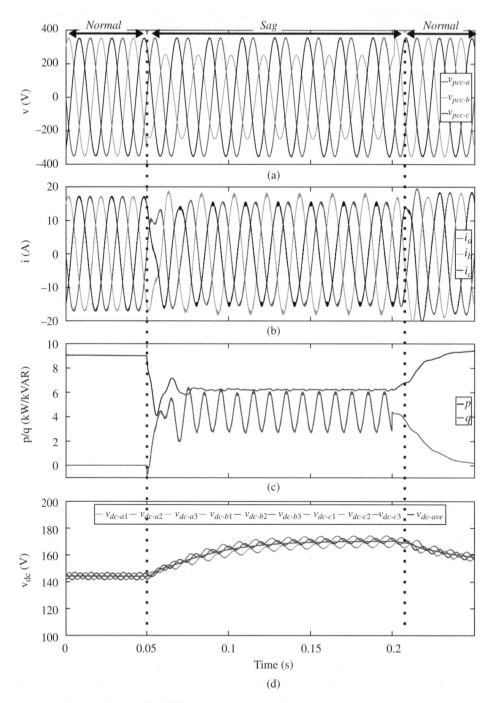

Figure 16.20 Simulation results of 9 kVA grid-connected 7L-CHB inverter. *Case I – 30%* single-phase voltage sag with unbalanced current injection: (a) PCC voltages; (b) inverter output currents; (c) injected active/reactive power to the grid; and (d) dc-link voltages.

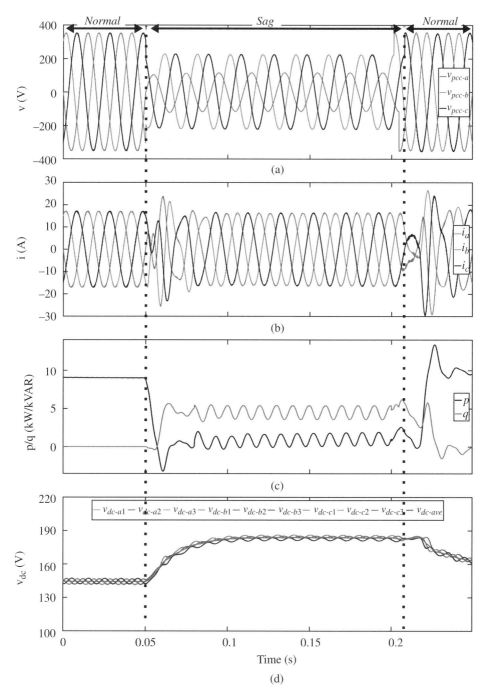

Figure 16.21 Simulation results of a 9 kVA grid-connected 7L-CHB inverter. *Case II* – 70% three-phase voltage sag with balanced current injection: (a) PCC voltages; (b) inverter output currents; (c) injected active/reactive power to the grid; and (d) dc-link voltages.

16.3.7 Experimental Verification

Fig. 16.22(a) and (b) show the circuit diagram and hardware setup, respectively, for a 9 kVA three-phase 7L-CHB converter. The parameters of the CHB prototype are listed in Table 16.3. Due to safety issues and the limited testing capabilities of the university laboratory, the experimental validation was performed on a scaled-down GCPVPP. The grid interfacing inductor for the simulation was selected to have a close per-unit value as the experimental setup. Therefore, these two systems were equivalent and showed similar performances. The converter was fed by nine isolated Elgar TerraSAS PV simulators from Ametek, each of which was programmed to simulate the electrical behavior of a PV string, with a solar irradiance of 1000 W/m² and a module temperature of 25°C. The control strategy was implemented on a dSPACE DS1006 platform with onboard Xilinx FPGA modules operating at 100 MHz. A programmable 20 kVA ac power source, model TopCon TC.ACS 4 quadrant, with the capability of simulating various voltage

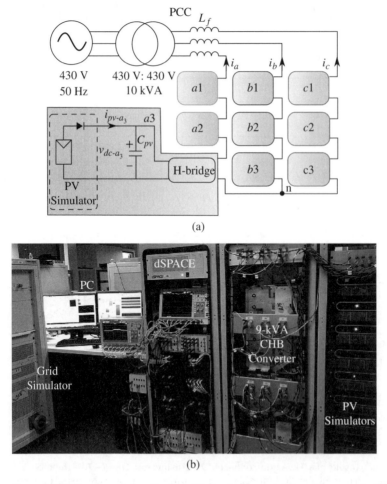

(a)

(b)

Figure 16.22 Experimental verification: (a) circuit diagram of the grid-connected CHB inverter test system; and (b) setup of the experiment.

sag conditions, was used as a grid. The proposed controller was evaluated under three different voltage sag conditions and power and current injection strategies.

Case I

This case experimentally examines the performance of the proposed control strategy during a 30% single-phase voltage sag, which requires simultaneous injection of active and reactive power to the grid. Unbalanced current injection is also taken into consideration. The sag occurs in phase c for a duration of 150 ms.

The three-phase grid voltages (v_{pcc}) and injected currents (i_{abc}) are depicted in Fig. 16.23(a). The delay between the beginning of the voltage sag and its detection is also presented in this figure. The transient currents during this time are not large, as a result of the proposed feedforward voltage compensator. The current of phase c, the phase with the voltage sag, remains equal to the nominal current of the inverter ($11\ A_{rms}$), while the currents of nonfaulty phases reduce to $9.2\ A_{rms}$. This leads to zero active power oscillation to the grid during such unbalanced voltage sags, as depicted in Fig. 16.23(c). The average active power reduces to 5.3 kW, while the average reactive power increases to 3.1 kVAR.

The bridge capacitor voltages are depicted in Fig. 16.23(b). The grid is in normal operation mode before $t = 0.175$ s. During this period, the capacitor voltages are balanced, and their average voltage is equal to $145\ V_{dc}$, which is equal to v_{mpp} of the simulated PV string. This results in the extraction of maximum power from all of the PV strings, which in total is equal to 9 kW, as presented in Fig. 16.23(c). The implementation of the proposed inter-phase and inter-bridge balancing schemes leads to balanced capacitor voltages during the voltage sag. The reduction of the power from PV strings during the voltage sag is performed by increasing the voltage of PV panels to 168 V.

Case II

This case study demonstrates a three-phase 70% unbalanced voltage sag (phase a experiences deeper voltage sag as compared to phases b and c) in order to verify the performance of the proposed control strategy during reactive power injection with balanced currents, and the results are displayed in Fig. 16.24. A three-phase to ground fault can be the reason of such voltage sag at PCC. It can easily be seen that the transient current of the inverter at the beginning of such a deep voltage sag is not large. During the voltage sag, the amplitude of currents remains nominal, while a phase displacement of $\pi/2$ exists between the phase current and voltage. This indicates the injection of reactive power to the grid with an average value of 3.4 kVAR, while the average injected active power is zero. A second harmonic oscillation exists in both active and reactive powers due to the injection of balanced currents to the PCC with unbalanced voltages. The dc-capacitor voltages are balanced during the voltage sag, and their average is increased to 177 V (equal to the open-circuit voltage of the PV panel) in order to reduce the extracted power from the PV panels to zero.

Case III

The performance of the proposed controller in balancing the capacitor voltages under an unbalanced voltage sag with a long duration is evaluated in this case study. The simulated voltage sag is similar to that in *Case II*, although its duration is increased to 1 s. Besides, the unbalanced current injection strategy is taken into consideration in order to verify the flexibility of the proposed strategy, and the results are depicted in Fig. 16.25.

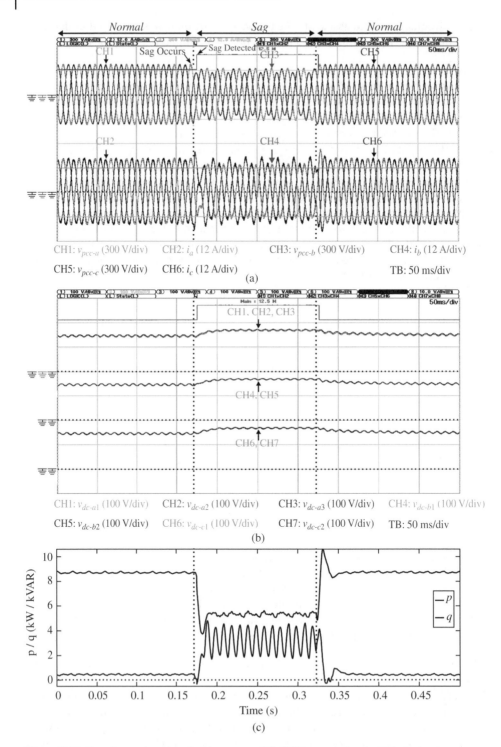

CH1: $v_{pcc\text{-}a}$ (300 V/div)   CH2: i_a (12 A/div)   CH3: $v_{pcc\text{-}b}$ (300 V/div)   CH4: i_b (12 A/div)

CH5: $v_{pcc\text{-}c}$ (300 V/div)   CH6: i_c (12 A/div)   TB: 50 ms/div

(a)

CH1: $v_{dc\text{-}a1}$ (100 V/div)   CH2: $v_{dc\text{-}a2}$ (100 V/div)   CH3: $v_{dc\text{-}a3}$ (100 V/div)   CH4: $v_{dc\text{-}b1}$ (100 V/div)

CH5: $v_{dc\text{-}b2}$ (100 V/div)   CH6: $v_{dc\text{-}c1}$ (100 V/div)   CH7: $v_{dc\text{-}c2}$ (100 V/div)   TB: 50 ms/div

(b)

(c)

Figure 16.23 Experimental results of grid-connected 7L-CHB inverter. *Case I* – 30% single-phase voltage sag with unbalanced current injection: (a) PCC voltages and currents; (b) dc-link voltages; and (c) injected active/reactive power to the grid.

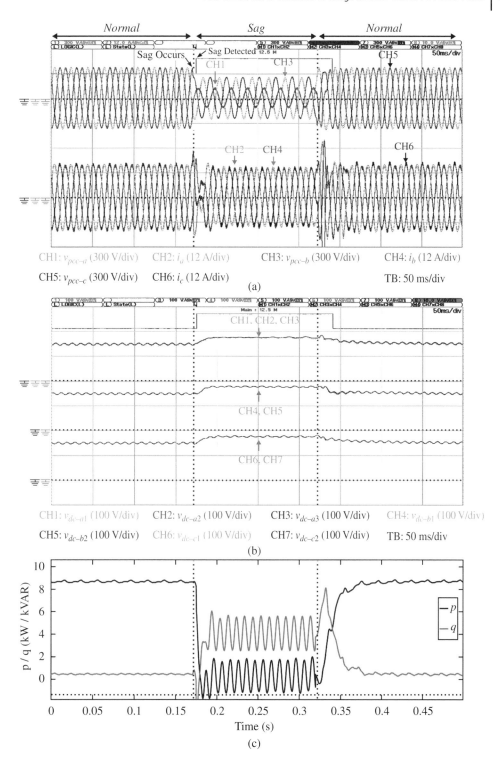

CH1: v_{pcc-a} (300 V/div) CH2: i_a (12 A/div) CH3: v_{pcc-b} (300 V/div) CH4: i_b (12 A/div)

CH5: v_{pcc-c} (300 V/div) CH6: i_c (12 A/div) TB: 50 ms/div

(a)

CH1: v_{dc-a1} (100 V/div) CH2: v_{dc-a2} (100 V/div) CH3: v_{dc-a3} (100 V/div) CH4: v_{dc-b1} (100 V/div)

CH5: v_{dc-b2} (100 V/div) CH6: v_{dc-c1} (100 V/div) CH7: v_{dc-c2} (100 V/div) TB: 50 ms/div

(b)

(c)

Figure 16.24 Experimental results of grid-connected 7L-CHB inverter. *Case II* – 70% three-phase voltage sag with balanced current injection: (a) PCC voltages and currents; (b) dc-link voltages; and (c) injected active/reactive power to the grid.

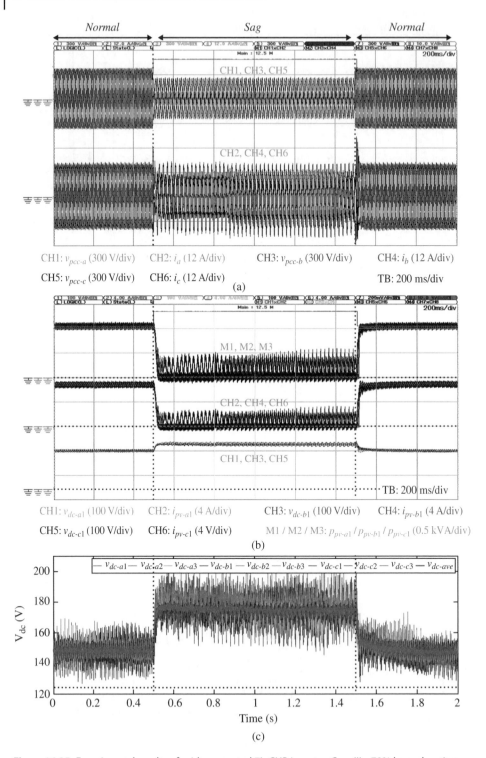

Figure 16.25 Experimental results of grid-connected 7L-CHB inverter. *Case III* – 70% long-duration three-phase voltage sag with unbalanced current injection: (a) PCC voltages and currents; (b) PV panel voltages, powers; and (c) dc-link voltages.

The extracted power and current from the PV modules decrease to zero during the voltage sag, as shown in Fig. 16.25(b). The dc-link voltages of all the nine bridges and their average are depicted in Fig. 16.25(c). V_{dc} is equal to 145 V (v_{mpp}) during the grid normal operation, while it increases to 177 V during the voltage sag. It can be seen that dc-link voltages remain balanced during such an unbalanced voltage sag with a long duration, which shows the stable operation of the CHB inverter during unbalanced voltage sags.

It can be seen that the proposed control algorithm is able to provide voltage balancing between all of the capacitors of the CHB converter during all operation conditions, including balanced or unbalanced current injection to the grid.

16.4 Summary

Active/reactive power injection is required for medium- and large-scale GCPVPPs during voltage sags. However, there is a lack of detailed studies in the literature on the operation of the grid-connected CHB inverter, which is a promising candidate for these types of GCPVPPs. Accordingly, a flexible control strategy for the operation of the PV grid–connected CHB inverter during unbalanced voltages sags has been introduced in this chapter. A zero-sequence voltage injection method has been introduced in order to achieve inter-phase balancing, while inter-bridge balancing has been accomplished through the modification of bridge voltage references. Transient currents of the inverter, at the beginning of the voltage sag, have been greatly reduced by implementing a feed-forward voltage compensator.

Detailed implementations of the proposed control strategies have been presented, and their effectiveness has been demonstrated through simulation and experimental results on a 9 kVA GCPVPP during different unbalanced voltage sag conditions. The CHB inverter is capable of achieving low-voltage ride-through and is able to inject the required amount of active and reactive power to the grid according to the grid codes. The flexibility of the proposed controller has also been verified under two different current injection strategies: (a) balanced current injection with the advantage of using the full current capacity of the inverter in enhancing the PCC voltage; and (b) unbalanced current injection with zero active power oscillation. The evaluation results verify the applicability of the proposed control strategy for medium- and large-scale GCPVPPs during voltage sags.

References

1 IEEE recommended practice for interconnecting distributed resources with electric power systems. *IEEE Std* 1547.6.

2 IEEE recommended practice for utility interface of photovoltaic (PV) systems. *IEEE Std.* 929, 2000.

3 R. P. Aguilera, P. Acuna, Y. Yu, G. Konstantinou, C. D. Townsend, B. Wu, and V. G. Agelidis. "Predictive control of cascaded H-bridge converters under unbalanced power generation," *IEEE Transactions Industrial Electronics*, vol. 64, no. 1, pp. 4–13, Jan 2017.

4 H. Akagi, S. Inoue, and T. Yoshii. "Control and performance of a transformerless cascade PWM STATCOM with star configuration," *IEEE Transactions on Industrial Applications*, vol. 43, no. 4, pp. 1041–1049, Jul 2007.

5 N. H. Baars, J. Everts, H. Huisman, J. L. Duarte, and E. A. Lomonova. "A 80-kw isolated dc–dc converter for railway applications," *IEEE Transactions on Power Electronics*, vol. 30, no. 12, pp. 6639–6647, Dec 2015.

6 E. Behrouzian and M. Bongiorno. "Investigation of negative-sequence injection capability of cascaded H-bridge converters in star and delta configuration," *IEEE Transactions on Power Electronics*, vol. 32, no. 2, pp. 1675–1683, Feb 2017.

7 R. Bosshard, U. Iruretagoyena, and J. W. Kolar. "Comprehensive evaluation of rectangular and double-d coil geometry for 50 kw/85 khz IPT system," *IEEE Journal of Emerging and Selected Topics in Power Electronics*, vol. 4, no. 4, pp. 1406–1415, Dec 2016.

8 H. Cha, R. Ding, Q. Tang, and F. Z. Peng. "Design and development of high-power dc–dc converter for metro vehicle system," *IEEE Transactions on Industry Applications*, vol. 44, no. 6, pp. 1795–1804, Nov 2008.

9 H. C. Chen, P. H. Wu, C. T. Lee, C. W. Wang, C. H. Yang, and P. T. Cheng. "Zero-sequence voltage injection for DC capacitor voltage balancing control of the star-connected cascaded H-bridge PWM converter under unbalanced grid," *IEEE Transactions on Industry Applications*, vol. 51, no. 6, pp. 4584–4594, Nov 2015.

10 M. Ciobotaru, R. Teodorescu, and F. Blaabjerg. "A new single-phase PLL structure based on second order generalized integrator," in *Proceedings of IEEE 37th Power Electronics Specialists Conference*, pp. 1–6, Jun 2006.

11 C. D. Fuentes, C. A. Rojas, H. Renaudineau, S. Kouro, M. A. Perez, and M. Thierry. "Experimental validation of a single dc bus cascaded H-bridge multilevel inverter for multistring photovoltaic systems," *IEEE Transactions on Industrial Electronics*, vol. 64, no. 2, pp. 930–934, Oct 2016.

12 J. Gao, X. Wu, S. Huang, W. Zhang, and L. Xiao. "Torque ripple minimisation of permanent magnet synchronous motor using a new proportional-resonant controller," *IET Power Electronics*, vol. 10, no. 2, pp. 208–214, 2017.

13 D. Grahame Holmes and Thomas A. Lipo. *Carrier-Based PWM of Multilevel Inverters*, pp. 453–530. Wiley-IEEE Press, 2003.

14 R. Kabiri, D. G. Holmes, and B. P. McGrath. "Control of active and reactive power ripple to mitigate unbalanced grid voltages," *IEEE Transactions on Industry Applications*, vol. 52, no. 2, pp. 1660–1668, Mar 2016. ISSN 0093-9994.

15 A. Kuperman. "Proportional-resonant current controllers design based on desired transient performance," *IEEE Transactions on Power Electronics*, vol. 30, no. 10, pp. 5341–5345, Oct 2015.

16 Y. Li, Y. Wang, and B. Q. Li. "Generalized theory of phase-shifted carrier PWM for cascaded H-bridge converters and modular multilevel converters," *IEEE Journal of Emerging and Selected Topics in Power Electronics*, vol. 4, no. 2, pp. 589–605, Jun 2016.

17 M. Mirhosseini, J. Pou, V. G. Agelidis, E. Robles, and S. Ceballos. "A three-phase frequency-adaptive phase-locked loop for independent single-phase operation," *IEEE Transactions on Power Electronics*, vol. 29, no. 12, pp. 6255–6259, Dec 2014.

18 M. Mirhosseini, J. Pou, B. Karanayil, and V. G. Agelidis. "Resonant versus conventional controllers in grid-connected photovoltaic power plants under unbalanced grid voltages," *IEEE Transactions on Sustainable Energy*, no. 99, pp. 1–9, March 2016. ISSN 1949-3029.

19 F. Nejabatkhah, Y. W. Li, and B. Wu. "Control strategies of three-phase distributed generation inverters for grid unbalanced voltage compensation," *IEEE Transactions on Power Electronics*, vol. 31, no. 7, pp. 5228–5241, July 2016.

20 Teresa Orlowska-Kowalska, Frede Blaabjerg, and Jose Rodriguez. *Advanced and Intelligent Control in Power Electronics and Drives*. Springer, 2014.

21 G. Ortiz, M. G. Leibl, J. E. Huber, and J. W. Kolar. "Design and experimental testing of a resonant dc–dc converter for solid-state transformers," *IEEE Transactions on Power Electronics*, vol. 32, no. 10, pp. 7534–7542, Oct 2017.

22 J. I. Y. Ota, Y. Shibano, and H. Akagi. "Low-voltage-ride-through (LVRT) capability of a phase-shifted-PWM STATCOM using the modular multilevel cascade converter based on single-star bridge-cells (MMCC-SSBC)," in *Proceedings of the IEEE Energy Conversion Congress and Exposition*, pp. 3062–3069, Sept 2013.

23 J. I. Y. Ota, Y. Shibano, and H. Akagi. "A phase-shifted PWM D-STATCOM using a modular multilevel cascade converter (SSBC)-part II: Zero-voltage-ride-through capability," *IEEE Transactions on Industry Applications*, vol. 51, no. 1, pp. 289–296, Jan 2015.

24 H. R. Pinkymol, A. I. Maswood, O. H. P. Gabriel, and L. Ziyou. "Analysis of 3-level inverter scheme with DC-link voltage balancing using LS-PWM and SVM techniques," in *Proceedings of the IEEE International Conference on Renewable Energy Research and Applications (ICRERA)*, pp. 1036–1041, Oct 2013.

25 J. Sastry, P. Bakas, H. Kim, L. Wang, and A. Marinopoulos. "Evaluation of cascaded H-bridge inverter for utility-scale photovoltaic systems," *Renewable Energy*, vol. 69, pp. 208–218, Sep 2014.

26 D. Sera, L. Mathe, T. Kerekes, S. V. Spataru, and R. Teodorescu. "On the perturb-and-observe and incremental conductance MPPT methods for PV systems," *IEEE Journal of Photovoltaics*, vol. 3, no. 3, pp. 1070–1078, Jul 2013.

27 P. Sochor and H. Akagi. "Theoretical comparison in energy-balancing capability between star-and delta-configured modular multilevel cascade inverters for utility-scale photovoltaic systems," *IEEE Transactions on Power Electronics*, vol. 31, no. 3, pp. 1980–1992, Mar 2016.

28 Q. Song and W. Liu. "Control of a cascade STATCOM with star configuration under unbalanced conditions," *IEEE Transactions on Power Electronics*, vol. 24, no. 1, pp. 45–58, Jan 2009.

29 H. D. Tafti, A. I. Maswood, G. Konstantinou, J. Pou, and F. Blaabjerg. "A general constant power generation algorithm for photovoltaic systems," *IEEE Transactions on Power Electronics*, vol. 33, no. 5, pp. 4088–4101, May 2018.

30 H. Dehghani Tafti, B. Vahidi, R.A. Naghizadeh, and S.H. Hosseinian. "Power quality disturbance classification using a statistical and wavelet-based hidden Markov model with Dempster–Shafer algorithm," *International Journal of Electrical Power and Energy Systems*, no. 47, pp. 368–377, Dec 2013.

31 H. Dehghani Tafti, A. I. Maswood, G. Konstantinou, J. Pou, K. Kandasamy, Z. Lim, and G. H. P. Ooi. "Low-voltage ride-thorough capability of photovoltaic grid-connected neutral-point-clamped inverters with active/reactive power injection," *IET Renewable Power Generation*, vol. 11, no. 8, pp. 1182–1190, Jul 2017b.

32 R. Teodorescu, F. Blaabjerg, M. Liserre, and P. C. Loh. "Proportional-resonant controllers and filters for grid-connected voltage-source converters," *in Proceedings of the IEEE Electric Power Applications*, vol. 153, no. 5, pp. 750–762, Sep 2006.

33 E. Tomaszewski and J. Jiangy. "An anti-windup scheme for proportional resonant controllers with tuneable phase-shift in voltage source converters," in *Proceedings of the IEEE Power and Energy Society General Meeting (PESGM)*, pp. 1–5, Jul 2016a.

34 E. Tomaszewski and J. Jiangy. "An anti-windup scheme for proportional resonant controllers with tuneable phase-shift in voltage source converters," in *Proceedings IEEE Power and Energy Society General Meeting (PESGM)*, pp. 1–5, Jul 2016b.

35 C. D. Townsend, Y. Yu, G. Konstantinou, and V. G. Agelidis. "Cascaded H-bridge multi-level PV topology for alleviation of per-phase power imbalances and reduction of second harmonic voltage ripple," vol. 31, no. 8, pp. 5574–5586, Aug 2016. ISSN 0885-8993.

36 B. Xiao, L. Hang, J. Mei, C. Riley, L. M. Tolbert, and B. Ozpineci. "Modular cascaded H-bridge multilevel PV inverter with distributed MPPT for grid-connected applications," vol. 51, no. 2, pp. 1722–1731, Mar 2015. ISSN 0093-9994.

37 Y. Yang and F. Blaabjerg. "Low-voltage ride-through capability of a single-stage single-phase photovoltaic system connected to the low voltage grid," *International Journal of Photoenergy*, pp. 1–9, Dec 2013.

38 Y. Yang, P. Enjeti, F. Blaabjerg, and H. Wang. "Wide-scale adoption of photovoltaic energy: Grid code modifications are explored in the distribution grid," vol. 21, no. 5, pp. 21–31, Sep 2015. ISSN 1077-2618.

39 A. Yazdani, A. R. Di Fazio, H. Ghoddami, M. Russo, M. Kazerani, J. Jatskevich, K. Strunz, S. Leva, and J. A. Martinez. "Modeling guidelines and a benchmark for power system simulation studies of three-phase single-stage photovoltaic systems," *IEEE Transactions on Power Delivery*, vol. 26, no. 2, pp. 1247–1264, Apr 2011.

40 T. Ye, N. Dai, C. S. Lam, M. C. Wong, and J. M. Guerrero. "Analysis, design, and implementation of a quasi-proportional-resonant controller for a multifunctional capacitive-coupling grid-connected inverter," *IEEE Transactions on Industry Applications*, vol. 52. no. 5, pp. 4269–4280, Sep 2016.

41 Y. Yu, G. Konstantinou, B. Hredzak, and V. G. Agelidis. "Operation of cascaded H-bridge multilevel converters for large-scale photovoltaic power plants under bridge failures," *IEEE Transactions on Industrial Electronics*, vol. 62, no. 11, pp. 7228–7236, Nov 2015. ISSN 0278-0046.

42 Y. Yu, G. Konstantinou, B. Hredzak, and V. G. Agelidis. "Power balance optimization of cascaded H-bridge multilevel converters for large-scale photovoltaic integration," *IEEE Transactions on Power Electronics*, vol. 31, no. 2, pp. 1108–1120, Feb 2016a. ISSN 0885-8993.

43 Y. Yu, G. Konstantinou, B. Hredzak, and V. G. Agelidis. "Power balance of cascaded H-bridge multilevel converters for large-scale photovoltaic integration," *IEEE Transactions on Power Electronics*, vol. 31, no. 1, pp. 292–303, Jan 2016b. ISSN 0885-8993.

44 Y. Yu, G. Konstantinou, C. D. Townsend, R. P. Aguilera, and V. G. Agelidis. "Delta-connected cascaded H-bridge multilevel converters for large-scale PV

grid integration," *IEEE Transactions on Industrial Electronics*, vol. 64, no. 11, pp. 8877–8886, Nov 2017.

45 Z. Zhang, X. F. He, and Y. F. Liu. "An optimal control method for photovoltaic grid-tied-interleaved flyback microinverters to achieve high efficiency in wide load range," *IEEE Transactions on Industrial Electronics*, vol. 28, no. 11, pp. 5074–5087, Nov 2013. ISSN 0885-8993.

17

Multilevel Converter–based Wind Power Conversion
Md Shafquat Ullah Khan

17.1 Introduction

The limited availability of fossil fuels and the environmental impacts associated with them are making renewable energy sources increasingly important. Wind turbine systems (WTSs) and wind parks have large power-capturing abilities. Wind power production has escalated from 6 MW in 1996 to projected estimations of up to 800 GW by 2021 [1]. The scale and availability of wind power has made WTS one of the most reliable renewable energy resources. Wind turbines are getting bigger in size, as are the demands of capacity and performance from power converters. Utilities are now able to install wind turbine generators of around 10 MW power rating [2], and it is believed that this figure will double in the coming decade. Hefty wind generators are making advanced multilevel converter topologies a popular choice to accommodate higher power. Multilevel power converters are used both as generator-side rectifiers and grid-side inverters. With the stochastic behavior of wind, grid integration is one of the major challenges of wind power conversion.

17.2 Wind Power Conversion Principles

Basically, a WTS has a wind turbine for converting the kinetic energy of incident wind into rotational mechanical energy for generator input. The two main types of wind turbine configurations *vertical axis wind turbines* (VAWTs) and *horizontal axis wind turbines* (HAWTs), the latter being more popular. In this configuration, the blades are connected to a central hub, which is attached to the top section of the tower. The nacelle contains the drive train, generator, and the electronic devices necessary for power conversion. It is expected that, in the future, power converters for high-power wind turbines too will become smaller and produce less heat, considering the space constrains in the nacelle.

17.2.1 Wind Turbine Modeling Principles

The mechanical power obtained from the rotating input of the wind turbine is converted into electrical energy by the electrical machine or generator. The generated electrical

Advanced Multilevel Converters and Applications in Grid Integration, First Edition.
Edited by Ali I. Maswood and Hossein Dehghani Tafti.

power is conditioned by the power electronic interfaces, with the aim of maintaining control over the different parameters. The potential power obtained from a wind turbine is represented by the equation,

$$P_{mech} = \frac{1}{2}\rho_w A_T C_p(\lambda, \beta) v_{wind}^2 \tag{17.1}$$

Here, ρ_w is the air density (kg/m³); v_{wind} represents wind speed (m/s); A_T is the area swept by the rotating turbine blades with radius R_T ($A_T = \pi R_T^2$ m²); C_p represents the coefficient of performance; and λ is a function of the blade pitch angle (β) and tip speed ratio ($\lambda = \omega r/v_{wind}$). C_p is called the *turbine power coefficient*; it illustrates the overall turbine efficiency, and can be modeled based on the following equations (equations 17.2 and 17.3) [3]:

$$C_p(\lambda, \beta) = C_1\left(\left(\frac{C_2}{\lambda_i}\right) - C_3\beta - C_4\right) e^{\left(\frac{-C_5}{\lambda_i}\right)} + C_6\lambda \tag{17.2}$$

The system is designed using the values presented in the literature [4, 5]: $C_1 = 0.5176$, $C_2 = 116$, $C_3 = 0.4$, $C_4 = 5$, $C_5 = 21$, and $C_6 = 0.0068$.

$$\frac{1}{\lambda_i} = \frac{1}{\lambda + 0.08\beta} - \frac{0.035}{\beta^3 + 1} \tag{17.3}$$

Similar to the PV systems, WTSs too are designed to extract the maximum power. WTSs were previously designed to operate at a constant wind speed. Advancements in the technologies associated with power conversion have now made it possible to operate the system in variable wind speed conditions. WTSs are expected to operate close to optimum tip speed ratio (TSR) to attain maximum C_p. Fluctuations in the rotor speed of the generator varies the TSR of the system. Other common control methods for performing maximum power point tracking (MPPT) for WTSs are power signal feedback (PSF) control and hill climbing searching (HCS) control [6, 7].

There are limitations for every energy conversion technology. For heat engines, there is Carnot efficiency; band gap for photovoltaic cells; and Gibbs free energy for fuel cells. For WTSs, there is a limit for the conversion of kinetic energy into mechanical power, which is known as the *Betz limit*. With the Betz limit, $C_{p\text{-}max} = 16/27$ is the maximum theoretically possible turbine power coefficient. In practice, it is about 40–45% [6].

17.2.2 Generators in WTS

For fixed-speed WTS, squirrel cage induction generators (SCIGs) were mostly used. As the system moved towards capturing power at variable wind speeds, doubly fed induction generators (DFIGs), wound rotor induction generators (WSIG), and synchronous generators (SGs) began to be preferred [8, 9]. Presently, SGs are widely used in wind energy conversion systems as their outputs range from a few kilowatts to several megawatts, as well as owing to their relatively smaller size. The rotor flux of wound rotor SGs (WRSGs) is generated by field windings, and permanent magnets produce rotor flux in permanent magnet SGs (PMSGs).

17.2.3 Configurations of WTSs

Wind turbine generation configuration can be categorized into four types for the different combinations of generator and power converter topologies [6]. These topologies mainly consider ac power collection and transmission. For MVDC or HVDC transmission systems, the configurations would be void of inverters. The types of WTS configurations are as follows:

- *Type A*: Fixed speed.
- *Type B*: Limited variable speed or semi-variable speed.
- *Type C*: Variable speed with partial-scale frequency converter.
- *Type D*: Variable speed with full-scale frequency converter.

17.2.4   Type A: Fixed-speed WTS

In the fixed-speed WTS, the SCIG is connected to the grid through a soft starter [1], as shown in Fig. 17.1(a). A step-up transformer follows the soft starter, and the system is free of any power converter interface. For this type of configuration, a gearbox is essential, which compensates for the rotational speed difference between the generator and the turbine. The objective of the fixed-speed WTS is to maintain a constant frequency, allowing it to inject power directly to the ac grid. After generator start-up, the soft starter is bypassed using a switch, and the system works without any power converter. SCIG draws reactive power from the grid, and thus three-phase capacitor banks are pertinent. These WTSs require stall control, pitch control, or active stall control [6]. The main feature of this type of configuration is simplicity, low costs, and reliability. The key downsides include lower efficiency; variations in wind speed directly affecting the grid; and grid faults causing tremendous stress on the wind turbine. In order to comply with the grid codes, *flexible alternating current transmission system* (FACTS) devices are deployed [10].

17.2.5   Type B: Semi-variable-speed WTS

Variable-speed operation increases the energy conversion efficiency by increasing the power-capturing ability at higher wind speeds. It also reduces the stress caused by sudden wind gusts on the mechanical drivetrain, thereby increasing the longevity of the system. As shown in Fig. 17.1(b), the semi-variable-speed WTS comprises a WRIG, which provides a 0–10% variable-speed operation. The control of rotor resistance using slip rings is called *Optislip control* [11], which allows optimum slip to be maintained for a specific wind speed. As with Type A, this configuration too requires a gearbox, a soft starter, and reactive power compensation devices. The system can capture more power, but it has higher energy losses owing to rotor resistance. Restricted speed range and low energy conversion efficiency are the main drawbacks of such a system.

17.2.6   Type C: Variable-speed WTS with partial-scale power conversion

For Type C configuration, DFIGs are used. Power from the generator is fed through both stator and rotor windings. The size of the converters is approximately 30% of the nominal generator power [12]. For steady-state operations, this configuration (Fig. 17.1(c)) neither requires any soft starter nor reactive power compensation devices. The power converters allow bidirectional power flow in the rotor circuit, increasing the speed range of the generator. The MPPT operation further increases the system efficiency. DFIG-based partial-scale WTS is one of the dominating topologies in today's industry, with about half of the market share [6, 13]. The major drawbacks of these turbines are their limited fault ride through (FRT) capability and the high maintenance cost for offshore applications [14].

17.2.7 Type D: Variable-speed WTS with full-scale conversion

The back-to-back full-scale (100%) power converter topology has gained much popularity for its advantages in grid-connected and standalone ac–ac systems [15]. In such systems, the ac–dc rectifier is followed by a dc-link. The dc-link feeds an inverter, which connects to an ac power collection or any ac load. A full-scale power converter enables the generator to be fully decoupled from the grid and its disturbances. The transients and aberrant outputs of the generator are not reflected to the grid due to the decoupling of the rectifier and inverter through the dc-link block [16] Type D (Fig. 17.1(d)) WTSs have some advantages over the other types, especially in aspects of higher efficiency, FRT capability, and the ability to accommodate higher power. Back-to-back topologies are also suitable for low-wind-speed operations with direct-drive system topologies.

17.3 Multilevel Converters in Wind Power Conversion

The two-level voltage source converters are popular in back-to-back topology for full-scale power conversion owing to their simplicity and high efficiency. Full-scale power conversion can be realized by the use of PMSG, WRSG [15, 17], and SCIG generators. This topology can also be used for Type C WTS with DFIG [12]. For the converter topologies, switching losses are dependent on the switching frequency, and it is maintained within the 1–3 kHz range to achieve lower losses [17]. In order to meet the grid codes requirements, especially to ensure the FRT ability of the system, grid side filters are used. However, VSC are suitable for low voltage and low power levels. In higher-power applications, two-level VSC can be used for MV applications by using them in parallel networks or by connecting the switching devices in series. Although parallel operation has the advantages of increased power handling capability and modularity, it adds complexities such as circulating current, power imbalance, and bigger nacelle for converter accommodation [18]. For medium-voltage and high-voltage applications, the limit of switch voltage rating is curtailed by use of IGBTs in series. However, the physical mismatch between the IGBTs make this solution challenging and partly unreliable.

These challenges have compelled the industry to bring in alternate converter topologies for higher power. The three-level diode-clamped converter and neutral-point-clamped (NPC) converter are some of the converter topologies that have been widely studied in literature [19–21] as plausible solutions. One NPC unit has six sets of dual series-connected IGBTs whose midpoints are clamped with dual series-connected diodes. The midpoint of the clamped diodes meet at a common point that is generally connected to the midpoint of two dc-link capacitors. The output of the NPC inverter has three levels; thus, it offers the advantage of reduced filter size for grid-connected operations [10, 21–23]. Using the NPC as a generator-side converter reduces the step-up transformer size or, in some cases, even eliminates its use. With this high

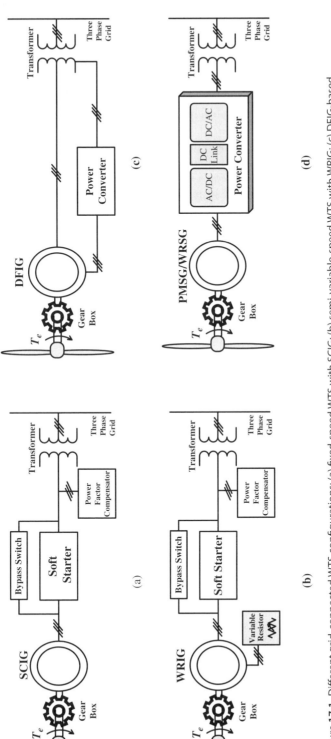

Figure 17.1 Different grid-connected WTS configurations: (a) fixed-speed WTS with SCIG; (b) semi-variable-speed WTS with WRIG; (c) DFIG-based variable-speed WTS with partial power conversion; and (d) variable-speed WTS with full-scale power conversion.

number of switching action capacitor voltages, balancing becomes a challenge [14]. Imbalanced capacitor voltages impose higher stress on switches, affecting operation and lifetime. Literature provides solution to this problem with use of external hardware. In the nacelle of WTS, the addition of any other hardware translates to a larger size and much higher cost of production and installation. It is also possible to alleviate this issue by the use of zero-voltage injection and the selection of redundant switching states. [24, 25]. Thus, back-to-back, full-scale power conversion can be very well supported by NPC converter units.

Issues such as uneven power loss and problems related to variable frequency in NPC can be eliminated by replacing the clamping diodes with IGBTs. This would give the converter the freedom to sustain with equal switching frequency. This topology is termed *active neutral point clamped* (ANPC) converter [26]. This is a recent topology in terms of use in WTS.

Replacing the clamped diodes in the NPC with capacitors gives the multilevel converter topology called *flying capacitor converter* [26, 27]. Although this topology brings in simplicity in comparison to ANPC, it requires the pre-charging of several clamped capacitors. This topology has not yet been realized in WTS practical applications. Other multilevel topologies include five-level and seven-level NPC converters. The higher the level of NPC, the lower the size of output filters, and the better the current quality for grid compliance. However, for a generator-side converter, such a complex device is superfluous.

17.4 Grid-Connected Back-to-Back Three-Phase NPC Converter

Hydropower and wind power are presently the largest contributors of renewable energy. Large wind farms and high-power wind turbines form the backbone of modern renewable energy–based grids. Solar power is also getting high attention. With power production of such high scale, utilities are always considering instantaneous power delivery to the grid. For this reason, grids are imposing stringent regulations for ensuring power quality.

17.4.1 Grid Code Requirements

Renewable energy sources are naturally stochastic. The power quality requirements of the grid vary from country to country [28]. This high penetration of ever-varying wind power has made grid regulators impose various standardized specifications in order to retain grid power quality. Such specifications are termed as *grid codes*. These codes are often under scrutiny and regularly updated [29, 30]. For example, IEEE Std. 1547-2003 and IEEE Std. 929-2000 are required standards that must be strictly complied to for ensuring the power quality of the renewable grid-connected system. Grid codes consider factors such as low-voltage ride-through (LVRT) capability, power regulation, frequency/voltage/power factor control, and power system protection [29]. The grid experiences several kinds of disturbances, such as short-circuit faults, sudden extreme voltage drops, load-inducing harmonics, etc. Such events might require wind power generation units to shut down or disconnect for protection purposes. Again, sudden disconnection of generation units may cause utility network instability and prolonged delay for restart. LVRT is a major grid code that concerns the manufacturers of power converters for WTS. According to LVRT requirements, the WTS should remain connected to the grid at the time of grid fault, regardless of any generator or converter configuration being used. It also applies to the generation unit to supply reactive power in order to help the grid to be able to pull through adverse scenarios [29, 31].

17.4.2 Integration of ESS with Grid-Connected Three-Phase Back-to-Back NPC Converter for WTS

Energy storage systems (ESSs) have been one of the most discussed technologies, with its potential in almost every application associated with power electronics. With increasing renewable energy penetration into the main grid, ESS is becoming a dominant factor of research in the power industry [31]. Large-scale ESS can be a viable solution for stochastic renewable sources. A bidirectional dc–dc converter is designed to commute power in and out of a connected ESS. The stored energy can give backup at times of grid faults or wind speed variations.

Integration of ESS with WTS, can also reduce power fluctuations and enhance system flexibility, by charging up the ESS during periods of low demand from user load and providing stored power during times of higher demand or reduced-wind-speed scenarios. This is one of the reasons for considering ESS. Various topologies such as the electric double-layer capacitor (EDLC) or super-capacitor have been analyzed for integration with WTS [32, 33]. High energy density with fast dynamic response along with long life cycle of super-capacitors have made them suitable for WTSs. However, the super-capacitor cannot be deployed independently owing to the issues of controllability and limited voltage range.

17.4.3 Circuit Configuration

In this chapter's analysis part, a WTS is analyzed that is integrated with bidirectional ESS. The dynamics of grid-connected WTS are observed at times of grid fault and wind speed variations. Although the implementation of a conventional PI current controller is simple, it suffers from multiple frame transformation, and thus high computational complexity. For this work, a proportional resonant (PR) controller is applied for the control of NPC converters at both generator and grid ends. A three-level NPC rectifier is deployed to convert the output ac-power of the PMSG to dc-power, charging up the dc-link. Another three-level grid-tied NPC inverter is considered to convert the dc-power to ac-power and inject to the grid, as shown in Fig. 17.2. Both rectifier and inverter circuits consist of two pairs of IGBT switches (S_{Ra1} and S_{Ra2} for the rectifier and S_{Ia1} and S_{Ia2} for the inverter), along with their complementary switches. In the circuit, other annotations would follow the same pattern of nomenclature as above. The rectifier controls PMSG by controlling its output currents in accordance with the MPPT controller. The NPC rectifiers are connected to two series of capacitors (C_{Rec1} and C_{Rec2}), as are the NPC inverters (C_{Inv1} and C_{Inv2}). The NPC operates with three different switching states to synthesize the three output pole voltage stepped levels at $-V_{DCLINK}/2$, 0, and $V_{DCLINK}/2$ with respect to the midpoint m of dc-link.

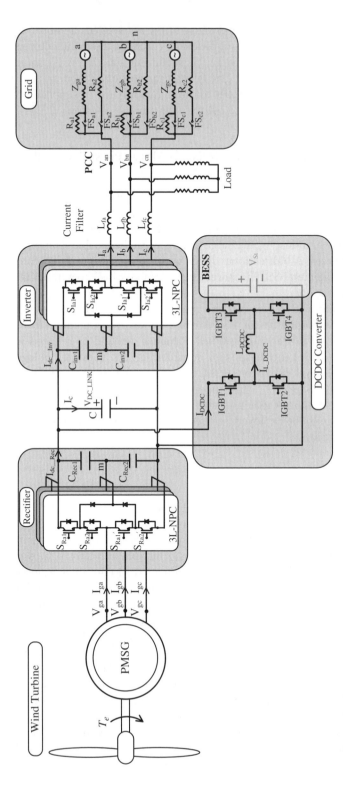

Figure 17.2 A PMSG-connected WTS with back-to-back three-level NPC converter with integrated ESS connected via bidirectional dc–dc converter.

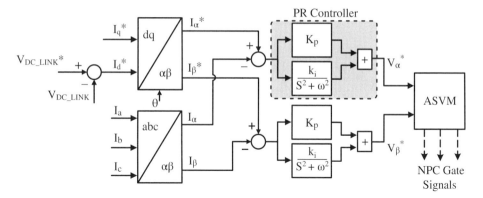

Figure 17.3 Proportional resonant controller used for NPC converter for full-scale WTS.

17.4.4 Control of the NPC Converters

Rectifier control is performed based on the traditional perturb and observe method of MPPT. As shown in Fig. 17.3, the control of grid-connected inverter- and generator-side rectifier is performed with a phase-locked loop (PLL), voltage-oriented controller (VOC), and switching based on adaptive space vector modulation (ASVM).

The measured dc-link voltage (V_{DCLINK}) is compared to the pre-defined reference value, and the difference is input to a PI controller. The output of the PI controller gives the reference d-axis current (I_d^*). The q-axis reference current (I_q^*) is set to zero for unity power factor operation. The PR controller is used for controlling the converter currents in stationary coordinate (αβ-coordinate) [22]. The measured three-phase currents of the converter (I_{ga}, I_{gb}, and I_{gc} for rectifier, and I_a, I_b, and I_c^* for grid-tied inverter) are converted to stationary frame through the Clark transformation. The transfer function (GPR) of the PR controller is given below:

$$G_{PR} = K_P + \frac{K_i}{(S^2 + \omega^2)} \tag{17.4}$$

Here, K_P is the proportional gain term; K_i is the integral gain term; and ω is the resonant frequency.

The reference voltages (V_α^* and V_β^*) are obtained from the PR controller and can be directly used as input to the ASVM without any requirement for inverse Park or Clark transformation. The voltage balancing of the capacitors are found better in the ASVM than the classical SVM. Thus, the controlling can be performed with less computational complexity. This technique can balance the dc-link capacitor voltages by selecting the appropriate switching state [22, 34].

The space vector representation for the NPC inverter or rectifier is shown in Fig. 17.4. It can be observed that there are different vector pairs in each internal voltage state that generate the same output voltage level. The classical SVM uses the pair of vectors during every switching cycle period, which increases the switching loss due to the high switching frequency operation. The ASVM technique is applied in combination with the PR to improve the efficiency of the specific converters [22]. The overall switching loss is decreased by reducing the switching frequency since one of the vector pairs is utilized in each switching cycle.

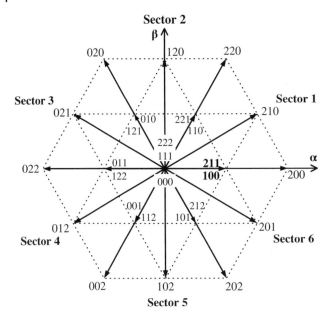

Figure 17.4 Space vector representation of the NPC inverter.

For inverters, even though both vectors generate the same output voltage level, there are different effects on the dc-link capacitor voltages. For instance, when the state vector is (100), the dc-link capacitors C_{Inv1} and C_{Inv2} are in charging and discharging states, respectively, according to the voltage potentials.

The ASVM voltage balancing method is independent of the main controller. The capacitor voltages can be balanced in all operating situations, including unbalance grid faults. On the other hand, the ASVM technique can directly use the stationary frame reference voltages to generate the switching signals, which results in less computational complexity and higher accuracy of the PR controller. The balancing of capacitors is depicted in Fig. 17.5, corresponding to the following equations:

$$(100) \Rightarrow \begin{cases} V_{CInv1} \leq V_{CInv2}, & and \ I_a \geq 0 \\ V_{CInv1} > V_{CInv2}, & and \ I_a < 0 \end{cases} \tag{17.5}$$

$$(211) \Rightarrow \begin{cases} V_{CInv1} \geq V_{CInv2}, & and \ I_a \geq 0 \\ V_{CInv1} < V_{CInv2}, & and \ I_a < 0 \end{cases} \tag{17.6}$$

17.4.5 Bidirectional dc–dc converter connecting ESS to WTS

The bidirectional dc–dc converter consists of four IGBTs and an inductor (L_{DCDC}), as shown in Fig. 17.2. Connecting the ESS to WTS, the dc–dc converter must be able to operate in both buck mode and boost mode. Buck–boost capability can be achieved by controlling switches in different combinations. Based on the voltage of the dc-link, the converter controller performs the buck or boost operation, either charging or discharging the ESS. During normal condition of the grid, the dc–dc converter is inactive. Under

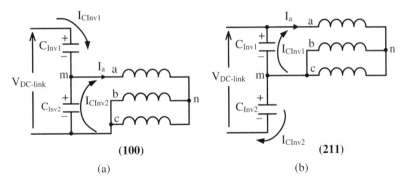

Figure 17.5 Circuit diagram and current path for vectors of one vector pair.

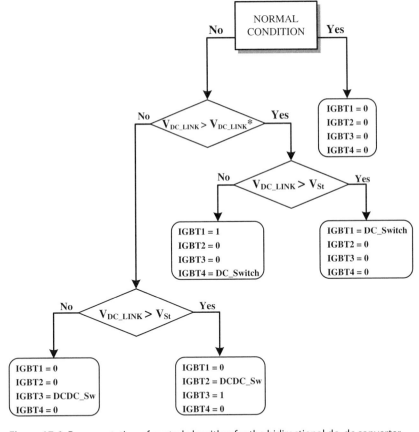

Figure 17.6 Representation of control algorithm for the bidirectional dc–dc converter.

the grid fault or wind speed dynamics, there are four operating conditions that can be visualized through the control algorithm, as shown in Fig. 17.6.

When dc-link voltage is greater than the reference dc-link voltage ($V_{DCLINK} > V_{DCLINK}^*$), we know that the extracted power from the PMSG (P_{O_PMSG}) is larger than the injected power to the grid (P_{O_Inv}). Therefore, the difference between these two values has to be injected to the ESS – i.e. the dc–dc converter will operate in ESS charging mode. The direction of power in the dc–dc converter will be from the dc-link to the ESS, with IGBT4 remaining off. Based on V_{DCLINK} and the voltage of ESS (V_{St}), the converter should perform boost or buck conversion, and the switches will be controlled accordingly. Hence, four different operation modes can be implemented as follows:

- If $V_{DCLINK} > V_{DCLINK}^*$ and $V_{DCLINK} > V_{St}$, then IGBT1 is controlled, and all other IGBTs are set to 0.
- If $V_{DCLINK} > V_{DCLINK}^*$ and $V_{DCLINK} < V_{St}$, then IGBT1 is set to 1, IGBT2 and IGBT3 are set to 0, and IGBT4 is controlled (*dc switch*).
- If $V_{DCLINK} < V_{DCLINK}^*$ and $V_{DCLINK} > V_{St}$, then IGBT4 and IGBT1 are set to 0, IGBT3=1, and IGBT2 is controlled.
- If $V_{DCLINK} < V_{DCLINK}^*$ and $V_{DCLINK} < V_{St}$, then only IGBT3 is controlled, and all others are set to 0.

Instantaneous dc-link voltage is compared to its nominal or reference voltage and the error signal is generated. This error is controlled through a PI controller to obtain the reference capacitor current (I_C^*). It can be seen from Fig. 17.2 that the dc–dc inductor reference current (I_{L_DCDC}) can be calculated as follows:

$$I_{L\_DCDC} = I_{DC\_Rec} - I_{DC\_Inv} - I_{C^*} \tag{17.7}$$

Here, I_{DC_Inv} is the instantaneous inverter input current, and I_{DC_Rec} is the instantaneous rectifier output current.

The current of the dc–dc inductor is controlled through the PI controller, which gives the reference inductor voltage (V_{LDCDC}^*). Subsequently, PWM generates the switching signal from the obtained I_{LDCDC}^* for the IGBT control.

17.4.6 Analysis of Wind and Grid Dynamics for WTS with Connected ESS

A PMSG is connected to a back-to-back (ac–dc–dc) full-scale converter. The performance of the system in three different conditions are analyzed – Case A being the normal operating condition; Case B the performance of the system under grid fault; and Case C the system response for sudden drop in wind speed. In order to observe the behavior of the described system, the grid voltage is considered to be 400 V/50 Hz; the line inductors are considered to be of 5 mH; and the inductor of the dc–dc converter is considered to be 1 mH. For a dc-link voltage of 770 V, capacitors rated at 2.2 µF were chosen. The switching frequencies for both rectifier and inverter are set to 10 kHz.

Case A

The steady state performance of the system is shown in Fig. 17.7. The output phase currents of the PMSG are shown in Fig. 17.7(a), which is sinusoidal with less harmonic contents, owing to the operation of the three-level NPC rectifier. The inverter output currents are depicted in Fig. 17.7(b). It is observed that there is no phase shift between Phase A current (I_a) and its voltage (V_{an}), which proves the unity power factor operation of the proposed controller. Capacitor voltage balancing performance of the ASVM is shown in Fig. 17.7(c). Additionally, it can be observed that the dc-link voltage is at a steady state value, owing to the controlled operation of the NPC inverter in adjusting

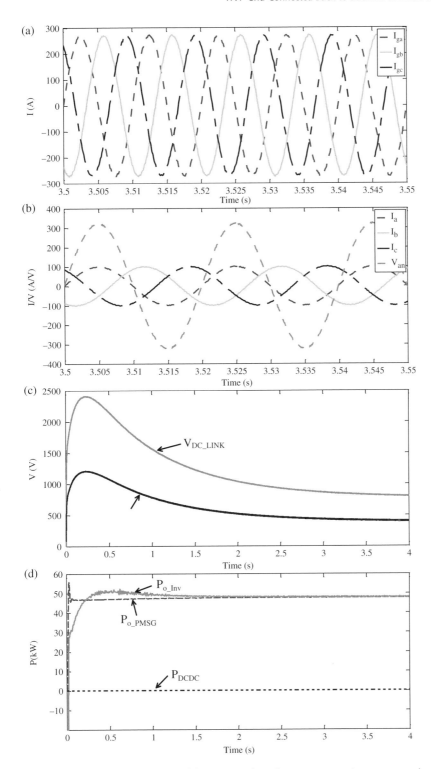

Figure 17.7 Case A: Performance of the proposed wind energy conversion system under steady state: (a) PMSG three-phase currents; (b) NPC inverter three-phase currents and phase A voltage; (c) dc-link voltage; and (d) extracted power from PMSG injected to the grid and storage system.

the P_{O_inv} with the P_{O_PMSG}, as depicted in Fig. 17.7. During steady-state operation, all power from the PMSG is transferred to the grid, and the dc-link voltage is constant.

Case B

In order to observe the response of the system for grid dynamics, a fault is generated in the grid that results in a voltage sag in the *point of common coupling* (PCC) at time $t = 4.0$ s. The response of the system without the dc–dc converter is depicted in Fig. 17.8(a). The inverter injects reactive power to the grid, and active power is consequently reduced, even though the PMSG has been drawing maximum active power from wind. Due to the difference of P_{O_PMSG} and P_{O_inv}, the voltage at the dc-link capacitor exceeds its nominal value, as shown in Fig. 17.8(a). This situation would trigger the dc-link capacitor protection devices and disconnect the WTS.

With the dc–dc converter operating and bringing in the support of ESS, certain advantages are observed. During the steady state, there is no phase shift between the voltages and current, because the inverter only injects active power and the injected reactive

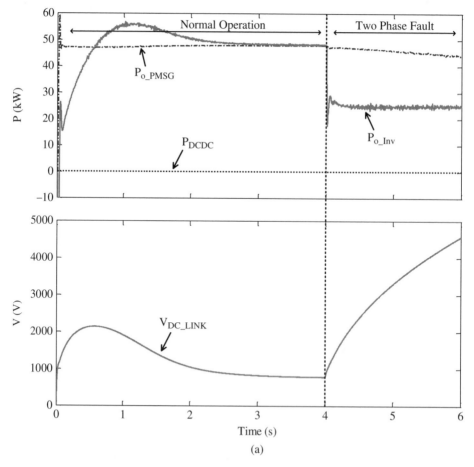

Figure 17.8 Case B: Performance of the proposed WTS under two phase faults: (a) without dc–dc converter operation; (b) with dc–dc converter operation.

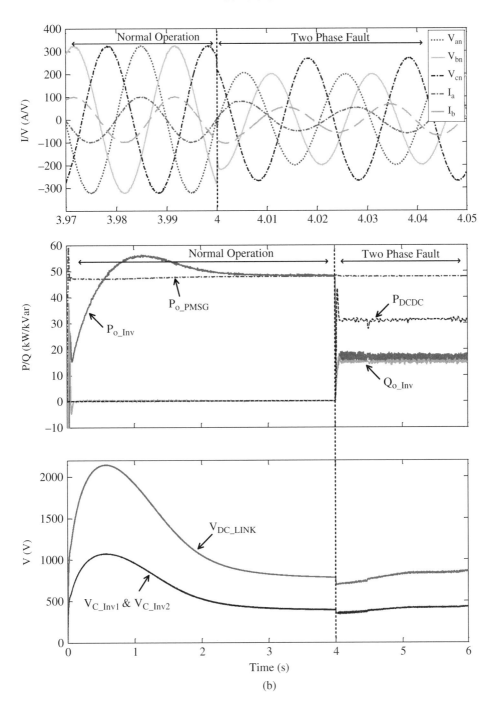

Figure 17.8 (*Continued*)

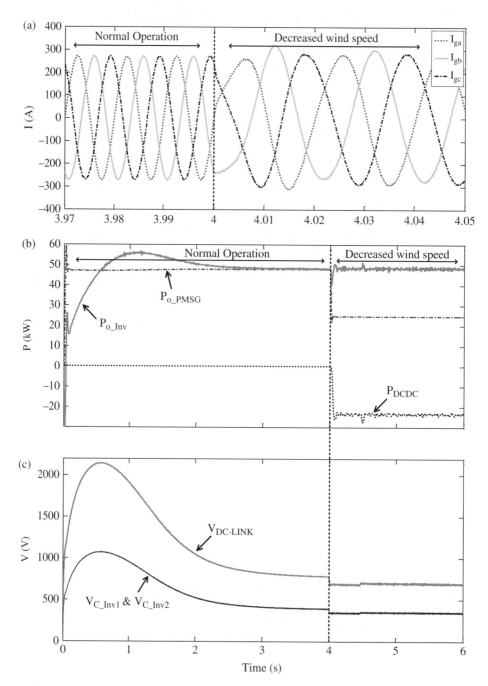

Figure 17.9 Case C: Performance of the proposed WTS during decreased wind speed: (a) PMSG output currents; (b) extracted power from PMSG, injected power to the grid and the storage system; and (c) dc-link capacitor voltage.

power is zero, as shown in Fig. 17.8(b). At $t = 4.0$ s, the inverter starts to inject reactive power to the grid, and its active power is consequently reduced. The power difference is injected to the ESS through the dc–dc converter, which is around 30 kW. The dc-link capacitor voltage is fairly constant, and the inverter capacitor voltages are also balanced due to the performance of the ASVM. As a result, the WTS does not pose any threat to the grid, and thus this operation provides the system FRT capability.

Case C

The dynamics of the system is observed with a sudden drop in wind speed. Wind speed drops at $t = 4.0$ s. Reduced wind speed causes reduced mechanical power, and thus reduced power from PMSG output. Fig. 17.9 shows the reduced frequency of the voltage and current due to the reduced wind speed at $t = 4.0$ s. The aim of any grid-connected renewable source is to achieve FRT capability and to operate undisrupted. For the reduced power produced, the difference of the PMSG power and optimal active power is drawn from ESS. As P_{O_PMSG} reduces, the dc–dc converter is activated and provides the inverter with the required power for continuous constant active power supply to the grid. As shown in Fig. 17.9, the difference in the PMSG power is compensated by the storage system. The inverter output power P_{O_inv} has some initial transient glitches due to the reduction of input power from the generator and variation of the dc-link voltage. However, the compensation of P_{DCDC} ensures that the output of the inverter is fairly constant by maintaining a constant dc-link voltage. During this operation, the dc-link voltage remains near its nominal value, although it decreases minimally. The NPC inverter and the dc–dc converter operate successfully.

17.5   Summary

Gigantic wind turbine generators are expected to play a big role in the future of renewable power generation. The technological development of power converters will be a major contributor to this advancement. The integration of energy storage systems would make renewable sources more stable, reliable, and sturdy. This chapter primarily focused on the application of multilevel converters in large WTSs. This chapter also discussed the control of a bidirectional dc–dc converter connected to the energy storage system to support the dynamics of wind variations and grid faults. As the results suggest, the dc-link voltage remained at its nominal value due to performance of the dc–dc converter. Moreover, at times of reduced wind speed, the output power of the grid-connected inverter remained constant due to the compensation of power from the ESS. In closure, we can say that, for large grid-connected WTSs, multilevel inverters can pave the way for future applications.

References

1 V. Yaramasu, B. Wu, P. C. Sen, S. Kouro, and M. Narimani. "High-power wind energy conversion systems: State-of-the-art and emerging technologies," in *Proceedings of the IEEE*, vol. 103, no. 5, pp. 740–788, May 2015.

2 F. Blaabjerg and K. Ma. "Future on power electronics for wind turbine systems," in *IEEE Journal of Emerging and Selected Topics in Power Electronics*, vol. 1, no. 3, pp. 139–152, Sep 2013.

3 M. S. U. Khan, A. I. Maswood, H. D. Tafti, M. M. Roomi, and M. Tariq. "Control of bidirec- tional dc/dc converter for back to back NPC-based wind turbine system under grid faults," in *2016 4th International Conference on the Development in the in Renewable Energy Technology (ICDRET)*, pp. 1–6, Jan 2016.

4 A. Venkataraman, A. I. Maswood, N. Sarangan, and O. H. P. Gabriel. "An efficient upf rectifier for a stand-alone wind energy conversion system," in *IEEE Transactions on Industry Applications*, vol. 50, no. 2, pp. 1421–1431, Mar 2014.

5 C. E. A. Silva, D. S. Oliveira, L. H. S. C. Barreto, and R. P. T. Bascopê. "A novel three- phase rectifier with high power factor for wind energy conversion systems," in *2009 Brazilian Power Electronics Conference*, pp. 985–992, Sep 2009.

6 H. Li and Z. Chen. "Overview of different wind generator systems and their compar- isons," *IET Renewable Power Generation*, vol. 2, no. 2, pp. 123–138, Jun 2008.

7 C. Huang, F. Li, and Z. Jin. "Maximum power point tracking strategy for large-scale wind generation systems considering wind turbine dynamics," *IEEE Transactions on Industrial Electronics*, vol. 62, no. 4, pp. 2530–2539, Apr 2015.

8 S. Kouro, J. Rodriguez, B. Wu, S. Bernet, and M. Perez. "Powering the future of industry: High-power adjustable speed drive topologies," *IEEE Industry Applications Magazine*, vol. 18, no. 4, pp. 26–39, Jul 2012.

9 S. Kouro, M. Malinowski, K. Gopakumar, J. Pou, L. G. Franquelo, B. Wu, J. Rodriguez, M. A. Perez, and J. I. Leon. "Recent advances and industrial applica- tions of multilevel converters," *IEEE Transactions on Industrial Electronics*, vol. 57, no. 8, pp. 2553–2580, Aug 2010.

10 M. Molinas, J. A. Suul, and T. Undeland. "Extending the life of gear box in wind generators by smoothing transient torque with statcom," *IEEE Transactions on Industrial Electronics*, vol. 57, no. 2, pp. 476–484, Feb 2010.

11 M. R. Khadraoui and M. Elleuch. "Comparison between optislip and fixed speed wind energy conversion systems," in *2008 5th International Multi-Conference on Systems, Signals and Devices*, pp. 1–6, Jul 2008.

12 R. Pena, J. C. Clare, and G. M. Asher. "Doubly fed induction generator using back-to-back pwm converters and its application to variable-speed wind-energy gen- eration," *IEE Proceedings – Electric Power Applications*, vol. 143. no. 3, pp. 231–241, May 1996.

13 J. A. Baroudi, V. Dinavahi, and A. M. Knight. "A review of power converter topolo- gies for wind generators," in *IEEE International Conference on Electric Machines and Drives*, pp. 458–465, May 2005.

14 A. Calle-Prado, S. Alepuz, J. Bordonau, J. Nicolas-Apruzzese, P. CortÃ©s, and J. Rodriguez. "Model predictive current control of grid-connected neutral-point-clamped converters to meet low-voltage ride-through requirements," *IEEE Transactions on Industrial Electronics*, vol. 62, no. 3, pp. 1503–1514, Mar 2015.

15 W. Xin, C. Mingfeng, Q. Li, C. Lulu, and Q. Bin. "Control of direct-drive permanent-magnet wind power system grid-connected using back-to-back pwm converter," in *2013 Third International Conference on Intelligent System Design and Engineering Applications*, pp. 478–481, Jan 2013.

16 H. Geng, D. Xu, B. Wu, and G. Yang. "Active damping for PMSG-based wecs with dc-link current estimation," *IEEE Transactions on Industrial Electronics*, vol. 58, no. 4, pp. 1110–1119, Apr 2011.

17 M. Chinchilla, S. Arnaltes, and J. C. Burgos. "Control of permanent-magnet generators applied to variable-speed wind-energy systems connected to the grid," *IEEE Transactions on Energy Conversion*, vol. 21, no. 1, pp. 130–135, Mar 2006.

18 M. S. U. Khan, A. I. Maswood, and M. Tariq. "Operation of parallel unity power factor passive front end rectifier for PMSG-based offshore wind turbine system," in *2016 IEEE International Conference on Power Electronics, Drives and Energy Systems (PEDES)*, pp. 1–5, Dec 2016.

19 A. Yazdani and R. Iravani. "A neutral-point-clamped converter system for direct-drive variable- speed wind power unit," *IEEE Transactions on Energy Conversion*, vol. 21, no. 2, pp. 596–607, Jun 2006.

20 S. Alepuz, A. Calle, S. Busquets-Monge, S. Kouro, and B. Wu. "Use of stored energy in PMSG rotor inertia for low-voltage ride-through in back-to-back NPC converter-based wind power systems," *IEEE Transactions on Industrial Electronics*, vol. 60, no. 5, pp. 1787–1796, May 2013.

21 V. Yaramasu and B. Wu. "Predictive control of a three-level boost converter and an NPC inverter for high-power PMSG-based medium voltage wind energy conversion systems," *IEEE Transactions on Power Electronics*, vol. 29, no. 10, pp. 5308–5322, Oct 2014.

22 H. Dehghani Tafti, A. I. Maswood, A. Ukil, O. H. P. Gabriel, and L. Ziyou. "NPC photovoltaic grid-connected inverter using proportional-resonant controller," in *2014 IEEE PES Asia-Pacific Power and Energy Engineering Conference (APPEEC)*, pp. 1–6, Dec 2014.

23 J. Rodriguez, Jih-Sheng Lai, and Fang Zheng Peng. "Multilevel inverters: a survey of topologies, controls, and applications," *IEEE Transactions on Industrial Electronics*, vol. 49, no. 4, pp. 724–738, Aug 2002.

24 N. Celanovic and D. Boroyevich. "A fast space-vector modulation algorithm for multilevel three-phase converters," *IEEE Transactions on Industry Applications*, vol. 37, no. 2, pp. 637–641, Mar 2001.

25 A. Mohamed A. S., A. Gopinath, and M. R. Baiju. "A simple space vector pwm generation scheme for any general n-level inverter," *IEEE Transactions on Industrial Electronics*, vol. 56, no. 5, pp. 1649–1656, May 2009.

26 O. S. Senturk, L. Helle, S. Munk-Nielsen, P. Rodriguez, and R. Teodorescu. "Power capability investigation based on electrothermal models of press-pack IGBT three-level NPC and ANPC VSCS for multimegawatt wind turbines," *IEEE Transactions on Power Electronics*, vol. 27, no. 7, pp. 3195–3206, Jul 2012.

27 K. Antoniewicz, M. Jasinski, and S. Stynski. "Flying capacitor converter as a wind turbine interface – modulation and MPPT issues," in *2012 IEEE International Symposium on Industrial Electronics*, pp. 1985–1990, May 2012.

28 M. Tsili and S. Papathanassiou. "A review of grid code technical requirements for wind farms," *IET Renewable Power Generation*, vol. 3, no. 3, pp. 308–332, Sep 2009.

29 F. Blaabjerg, R. Teodorescu, M. Liserre, and A. V. Timbus. "Overview of control and grid synchronization for distributed power generation systems," *IEEE Transactions on Industrial Electronics*, vol. 53, no. 5, pp. 1398–1409, Oct 2006.

30 Mansour Mohseni and Syed M. Islam. "Review of international grid codes for wind power integration: Diversity, technology and a case for global standard," *Renewable and Sustainable Energy Reviews*, vol. 16, no. 6, pp. 3876–3890, 2012.

31 J. M. Carrasco, L. G. Franquelo, J. T. Bialasiewicz, E. Galvan, R. C. PortilloGuisado, M. A. M. Prats, J. I. Leon, and N. Moreno-Alfonso. "Power-electronic systems for the grid integration of renewable energy sources: A survey," *IEEE Transactions on Industrial Electronics*, vol. 53, no. 4, pp. 1002–1016, Jun 2006.

32 A. L. Allegre, A. Bouscayrol, and R. Trigui. "Influence of control strategies on battery/supercapacitor hybrid energy storage systems for traction applications," in *2009 IEEE Vehicle Power and Propulsion Conference*, pp. 213–220, Sep 2009.

33 C. Abbey and G. Joos. "Supercapacitor energy storage for wind energy applications," *IEEE Transactions on Industry Applications*, vol. 43, no. 3, pp. 769–776, May 2007.

34 H. Dehghani Tafti, A. I. Maswood, Z. Lim, G. H. P. Ooi, and P. H. Raj. "Proportional-resonant controlled NPC converter for more-electric-aircraft starter-generator," in *2015 IEEE 11th International Conference on Power Electronics and Drive Systems*, pp. 41–46, Jun 2015.

18

Z-Source Inverter–Based Fuel Cell Power Generation
Muhammad M. Roomi

18.1 Introduction

An alternative enhancement to the traditional methods of electric power generation is provided by distributed energy resources (DERs) systems. The power generation from these sources is directly connected to medium-voltage (MV) and low-voltage (LV) distribution systems rather than bulk transmission systems [1]. DER systems include generation units such as fuel cells, micro-turbines, photovoltaics, etc. In addition to the aforementioned components, DER systems typically consist of other devices, including storage units such as batteries, flywheels, superconducting magnetic energy storage, etc. Figure 18.1 provides a schematic representation of the technologies that support DER systems.

DER systems use various renewable sources such as solar, wind, tidal farms, bio-fuels, solar thermal collectors, hydropower, etc. DER plays a vital role in the electric power generation and distribution system. Inherently, these sources of energy are non-polluting and abundantly available. Furthermore, the outputs from these sources are electrical. Therefore, the transmission and distribution of energy over long distances can be achieved in an efficient way with relative ease. Currently, the most commonly used renewable energy sources are solar and wind. These renewable sources are available abundantly, but provide limited efficiency. Therefore, current electrical grids that generally provide electricity generated by conventional sources are preferred due to their reliability and decades of operational experience. In order to compare conventional and renewable sources of energy, their levelized cost of electricity (LCOE) needs to be compared. To ensure grid parity, the power generated by DER systems must match the LCOE of conventional power generation. With the implementation of semiconductor devices, the issue of efficiency in DER systems can be addressed. This would further enhance the applicability of DER systems in current power generation grids. A technology that is increasingly adopted (regardless of the low power generation) is the fuel cell technology, which is the primary focus of this chapter.

Fuel cells are electrochemical converters whose performance is governed by their anode and cathode reactions, and the heat and energy transfer due to these reactions. The result of these chemical reactions is the production of electricity in the external

Advanced Multilevel Converters and Applications in Grid Integration, First Edition.
Edited by Ali I. Maswood and Hossein Dehghani Tafti.
© 2019 John Wiley & Sons Ltd. Published 2019 by John Wiley & Sons Ltd.

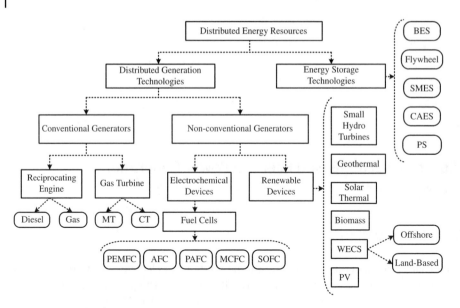

Figure 18.1 Technologies supporting DER systems.

circuit connected to the cell. Fuel cells are classified into many types – based on the fuel used, electrodes, and construction. Some of the commonly used fuel cells are the alkaline fuel cell (AFC), phosphoric acid fuel cell (PAFC), molten carbonate fuel cell (MCFC), solid oxide fuel cell (SOFC), and proton exchange membrane fuel cell (PEMFC) [2]. Table 18.1 tabulates the different types of fuel cell technologies and their characteristics.

Power generation attributes by fuel cells are characterized by chemical reactions. AFC generates a power-utilizing redox reaction, wherein hydrogen is oxidized at the anode and oxygen is reduced at the cathode. The redox reaction produces water molecules from oxygen and hydrogen molecules. PAFC uses phosphoric acid as an electrolyte with a silicon carbide matrix, and the platinum-coated carbon acts as an electrode. MCFC comprises a molten carbonate salt electrolyte in an inert lithium aluminum oxide matrix. A dense layer of ceramic forms the electrolyte in SOFC, wherein the oxidation of fuel takes place in the anode (usually a ceramic–metal composite), and oxygen reduction occurs at the cathode. PEMFC uses hydrogen as a fuel to generate power, by hydrogen oxidation at the anode, and water is produced by the reduction of oxygen at the cathode. Heat and electricity are the by-products of fuel cells. Of these types, PEMFC is chosen as the focus in this chapter, since these fuel cells can operate at relatively lower temperatures while possessing higher power density. The most important feature of PEMFC is that the output of these fuel cells can be changed quickly to meet the shifts in load demand. The output range of a single PEM unit varies from several watts to several kilowatts, thereby allowing this fuel cell to be used in large-scale applications. In recent years, PEM systems have been available to provide primary and backup power to forklifts, many material-handling vehicles, and portable consumer electronics [3].

Table 18.1 Comparison chart of fuel cell technologies.

Type	Operating Temperature (°C)	Reactions	Electrolyte	Power Range (kW)	Electrical Efficiency (%)	Applications
Alkaline fuel cell (AFC)	90–100	Oxidation: hydrogen Reduction: oxygen	Aqueous potassium hydroxide	10–100	60	Military, space
Phosphoric acid fuel cell (PAFC)	150–200	Oxidation: hydrogen Reduction: oxygen	Phosphoric acid in a matrix	200–1000	>40	DG
Molten carbonate fuel cell (MCFC)	600–700	Oxidation: hydrogen, carbon monoxide Reduction: oxygen	Molten carbon salt	250–2000	45–47	Electric utility, DG
Solid oxide fuel cell (SOFC)	600–1000	Oxidation: hydrogen, carbon monoxide and methane Reduction: oxygen	Oxide ion conducting ceramic	225–2000	35–60	Auxiliary power, DG, electric utility
Proton exchange membrane fuel cell (PEMFC)	50–100	Oxidation: Hydrogen Reduction: Oxygen	Proton conducting polymer	3–250	Transportation: 53–58 Stationary: 22–25	Backup power, transportation, DG

18.2 Fuel Cell Power Conversion Principles

When compared to other distributed generation (DG) sources such as solar and wind energy sources, fuel cells do not circumscribe to any geographic limitations. Therefore, they can be transferred and installed at any location. PEMFC demonstrates superior performance when used as a DG source. One of the applications where the PEM fuel cell plays a vital role is the electric vehicle. Increasing research in these vehicles and the reduction in carbon emissions make PEM fuel cells attractive for vehicular applications. The main feature of the PEM fuel cell is that it aids in quick starts, making it highly suitable for electric vehicles. These fuel cells are considered to be good sources of energy that provide reliable power at a steady state. Nevertheless, these are not effective in responding to the load transients promptly. This is attributed to the slow internal processes. Therefore, to make a fuel cell operate efficiently, its transient properties should be analyzed for different applications. Some research studies have been reported on the steady-state [4–10] and dynamic [4, 5, 11–15] fuel cell modelling. As these works usually focus on the electrochemical reactions, various researches have been conducted to analyze the dynamic characteristics of the fuel cell, focusing mainly on their electrical characteristics [16–19].

Similar to other electrochemical cells, PEM fuel cells consist of an anode, a cathode, and an electrolyte medium for the charge carriers to travel. Figure 18.2 provides a simplified representation of a PEMFC along with the corresponding internal voltage drops. Equations (18.1), (18.2), and (18.3) represent the anode, cathode, and the total reactions occurring within the fuel cell:

$$H_2 \rightarrow 2H^+ + 2e^- \tag{18.1}$$

$$2H^+ + 2e^- + {}^1\!/2O_2 \rightarrow H_2O \tag{18.2}$$

$$H_2 + {}^1\!/2O_2 \rightarrow H_2O \tag{18.3}$$

The fuel cell output voltage is a combination of reversible and irreversible voltages. The irreversible voltage represents the losses in the fuel cell. The main losses in the output voltage of the fuel cell are activation losses, ohmic losses, and concentration losses. The output of the fuel cell is given by equation (18.4) [2]:

$$V = E_{rev} + E_{loss} \tag{18.4}$$

$$E_{loss} = V_{act} + V_{ohmic} + V_{conc} \tag{18.5}$$

$$V_{act} = A.\ln\left(\frac{i + i_n}{i_0}\right) \tag{18.6}$$

$$V_{ohm} = i.r \tag{18.7}$$

$$V_{conc} = m.\exp(ni) \tag{18.8}$$

Here, E_{rev} stands for reversible cell potential; $E_{loss} = V_{act} + V_{ohmic} + V_{conc}$ represents irreversible cell potential or losses in the cell (V); A is the Tafel equation constant; i is the current density (mA/cm^2); i_0 is the exchange current density; i_n is the internal current density; r is the area-specific resistance (kΩcm^2); and m, n represent the constants of the concentration loss parameters.

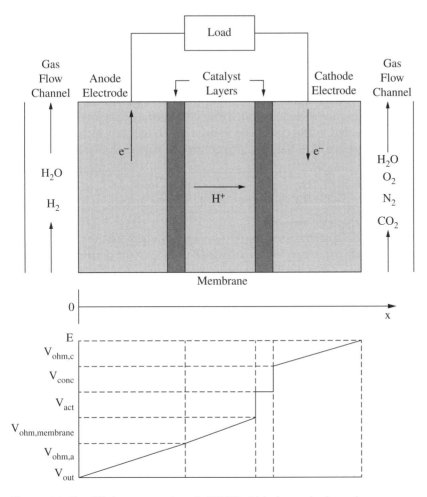

Figure 18.2 Simplified representation of a PEMFC with its internal voltage drops.

18.3 Modelling of the PEMFC

A simple model for the PEMFC based on [20, 21] is used and is illustrated in Figure 18.3. The model consists of a fuel cell stack represented by a voltage source and a resistance connected in series with the fuel cell stack. E and E_{OC} represent the effective and open circuit voltages of the fuel cell, respectively. The output voltage of the fuel cell is:

$$V_{fc} = E - R.i_{fc} \tag{18.9}$$

In general, the controlled voltage of the fuel cell is given by the following equation:

$$E = E_{OC} - N.A. \ln \frac{i_{fc}}{i_0} \tag{18.10}$$

Here, V_{fc} is the fuel cell voltage (V); R is the internal resistance (Ω); i_{fc} is the fuel cell current (A); E_{OC} is the open circuit voltage (V); N is the number of cells; A is the Tafel slope (V); and i_0 is the exchange current (A).

The controlled voltage is obtained from [20], and the electrical model of the losses, primarily the activation loss, is included by connecting an RC branch in parallel. The output of the fuel cell voltage will observe a delay, which is approximated as twice the time constant ($\tau = RC$), when there is a sudden change in the fuel cell current. The effect of the delay is represented in the controlled voltage of the fuel cell with a first-order transfer function. Therefore, the changes in the controlled voltage of the fuel cell are:

$$E = E_{OC} - N.A.\ln\frac{i_{fc}}{i_0} \cdot \frac{1}{s.\frac{T_d}{3}+1} \tag{18.11}$$

Here, $T_d - R$ is the response time (sec).

The fuel cell stack is designed considering the temperature and pressure of the fuel cell. In addition, the flow rate of the fuel and air are varied to analyze the performance of the inverter. These variations affect the Tafel slope, exchange current, and the open circuit voltage. A typical depiction of the system is presented in Figure 18.3. The Nernst voltage and exchange current are represented by E_n and i_0, respectively, which are functions of the utilization of hydrogen (H_2) and oxygen (O_2) as shown in equation (18.12).

$$E_n, i_0 = f(U_{H_2}, U_{O_2}) \tag{18.12}$$

The utilization of fuel and air are given by the following equations:

$$U_{H_2} = \frac{60000.R.T.i_{fc}}{z.F.P_{fuel} \cdot V_{fuel}.x\%} \tag{18.13}$$

$$U_{O_2} = \frac{60000.R.T.i_{fc}}{2.z.F.P_{air} \cdot V_{air}.y\%} \tag{18.14}$$

Here, P_{fuel} is the absolute supply pressure of fuel (atm); P_{air} is the absolute supply pressure of air (atm); V_{fuel} is the fuel flow rate (l/min); V_{air} is the airflow rate (l/min); x

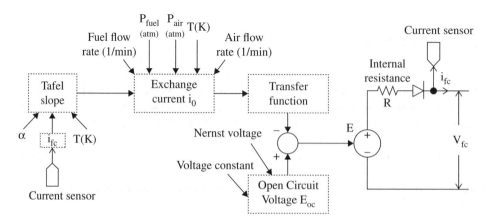

Figure 18.3 Schematic representation of fuel cell stack model

is the percentage of hydrogen in the fuel (%); and y is the percentage of oxygen in the oxidant (%).

Hence, the open circuit voltage, exchange current, and Tafel slope are given by the following updated series of equations (as in [22], [23]):

$$E_{OC} = c \cdot E_n \tag{18.15}$$

$$i_0 = \frac{z.F.k\,(P_{H_2} + P_{O_2})}{R.h} \cdot \exp\left(\frac{-\Delta G}{RT}\right) \tag{18.16}$$

$$A = RT/z\alpha F \tag{18.17}$$

Here, E_n is the Nernst voltage (V); c is the voltage constant at nominal condition of operation; P_{H_2} is the partial pressure of hydrogen inside the stack (atm); P_{O_2} is the partial pressure of oxygen inside the stack (atm); z is the number of moving electrons ($z = 2$); k is Boltzmann's constant (1.38×10^{-23} J/K); F is 8.3145 J/(mol-K); α is the charge transfer coefficient; h is Planck's constant (6.626×10^{-34} J-s); T is the operating temperature (K); and ΔG is the activation energy barrier (J).

Based on the preceding equations, the fuel cell is modelled and the diode is included in the circuit, as illustrated in Figure 18.3, to prevent any negative current flowing into the stack. Furthermore, a flow rate regulator is incorporated to regulate the fuel flow rate based on the output current of the fuel cell, while the airflow is maintained at a constant rate of 401 liters per minute (lpm). Under this operating condition, the fuel can be utilized at more than 99% efficiency [21].

18.4 Circuit Configuration

This section deals with the integration of the fuel cell stack modelled in Figure 18.4 with the Z-source inverter (ZSI). The control strategy for inserting shoot-through (ST) states for the ZSI involves using the maximum boost control (MBC) method, which will be described in the following section. The block diagram of the system is illustrated in Figure 18.4. The system consists of a fuel cell stack, a fuel regulator, a three-phase ZSI controlled by MBC pulse width modulation (PWM), and a corresponding RL load. The three-phase inverter can be a two-level configuration, as depicted in Figure 18.5, or a three-level configuration, as in Figure 18.6. In multilevel inverters, the formation and balancing of the neutral point is of higher importance. Therefore, any of the multilevel Z-source configurations explained in Chapter 15 can be implemented. For easy reference, Figure 18.6 depicts the integration of a single ZSI to the fuel cell stack. The neutral

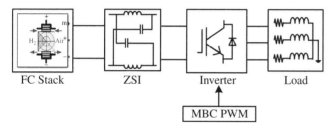

Figure 18.4 Block diagram of the fuel cell stack model with ZSI

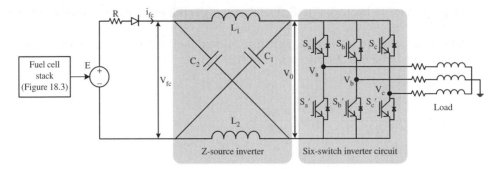

Figure 18.5 Schematic representation of the fuel cell integrated with the two-level ZSI

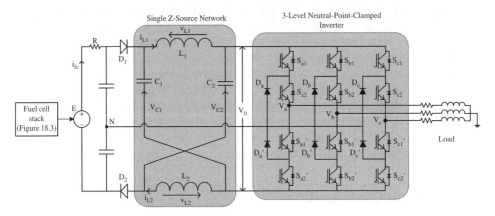

Figure 18.6 Schematic representation of the fuel cell integrated with the three-level ZSI

point for the inverter is formed in between the split capacitors connected to the terminal side of the fuel cell stack. The focus of the study is to analyze the effect of the ZSI on the fuel cell portrayed in Figure 18.3. Therefore, the two-level configuration of ZSI is integrated with the fuel cell stack and simulated to gain better knowledge about the system.

18.5 Control Strategy

As mentioned in section 18.4, the control strategy utilized in the circuit is the MBC PWM method. PWM techniques for the traditional inverter include six active states during the loaded condition and two zero states when the load is shorted through the upper or lower switches of any leg. The gate pulses for the switches in the traditional inverter are achieved by comparing the sine wave with the carrier wave. The traditional PWM technique is generally used in cases where the dc voltage is sufficiently high to generate the desired ac voltage. However, when the ac voltage is large, additional modification is required to boost the dc voltage to equalize the ac output voltage. The modification includes the insertion of an ST state in the PWM control, as described in section 15.4. Inserting this new state does not cause any difference in the operation of the traditional inverter. The switches of each phase are gated "on" and "off" as per switching cycle. The

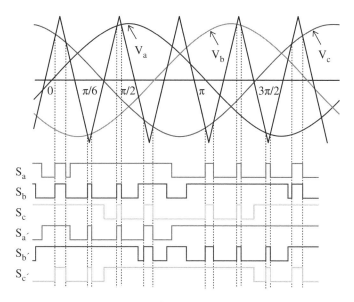

Figure 18.7 Switching pattern for the MBC method

active states of the traditional inverter remain unchanged. The ST state is evenly inserted into the zero states at each phase without actually changing the time interval. These inserted states provide the boosting capability to the inverter. The ST state is the unique property of the ZSI. A few methods have been proposed for the insertion of ST states in the inverter. For the analysis of fuel cell operation, the MBC method of inserting the additional state is preferred due to its capability of utilizing the entire zero state as an ST state. This leads to an increase in the ST duty ratio, thereby reducing the voltage stress on the switches [24]. The ST state insertion for the MBC method is depicted in Figure 18.7. As observed from Figure 18.7, the circuit enters the ST state whenever the triangular carrier wave is greater or lesser than the maximum or minimum value of the reference curves (V_a, V_b, and V_c), respectively.

The MBC method has the ST time interval, which repeats every $\pi/3$. If the switching frequency is higher than the modulation frequency, the ST duty ratio over a switching cycle for the interval ($\pi/6$, $\pi/2$) is given by:

$$D_0 = \frac{2 - \left(M\sin\theta - M\sin\left(\theta - \frac{2\pi}{3}\right)\right)}{2} \tag{18.18}$$

In order to obtain the voltage gain, the average value of the ST duty ratio can be expressed by simplifying equation (18.18) within the limits ($\pi/6$, $\pi/2$) and with respect to θ. Upon simplification, the average value of ST duty ratio is derived as in equation (18.19):

$$D_0(\theta) = 1 - \frac{\sqrt{3}}{2}M\cos\left(\theta - \frac{\pi}{3}\right) \tag{18.19}$$

From which the boost factor is obtained as:

$$B = \frac{1}{1 - 2D_0} = \frac{\pi}{3\sqrt{3}M - \pi} \tag{18.20}$$

With the MBC method, the voltage gain can be obtained by using the modulation index M.

$$G = \frac{\widehat{V_{ac}}}{V_i/2} = MB = \frac{\pi M}{3\sqrt{3}M - \pi} \qquad (18.21)$$

From the voltage gain in equation (18.21), the maximum modulation index for a given voltage gain G is:

$$M = \frac{\pi G}{3\sqrt{3}G - \pi} \qquad (18.22)$$

From equation (18.22), the output voltage grows to infinity when the modulation index reaches $\pi/3\sqrt{3}$. The voltage stress (V_s) across the switches for MBC is given by:

$$V_s = B.V_i = \frac{\pi V_i}{3\sqrt{3}M - \pi} \qquad (18.23)$$

18.6 Validation

The PEMFC stack model (described in section 18.4) and the ZSI implementing the MBC method (explained in section 18.5) are considered for validation. The performance of the fuel cell with the ZSI implementing the MBC method is studied, and the corresponding changes in the fuel cell outputs are analyzed. The changes in fuel flow rate, fuel utilization, stack consumption, and the stack efficiency are studied. The output voltage and output current of the fuel cell, the dc-link voltage, and the three-phase output current and voltage are evaluated using MATLAB/Simulink. The validation is performed for the following paradigms:

- Constant fuel flow rate with balanced load for ZSI
- Constant fuel flow rate with unbalanced load for ZSI
- Variable fuel flow rate with ZSI

In this study, PEMFC has a nominal operating power of 6 kW. Table 18.2 tabulates the PEMFC stack and the operating parameters used for validation.

18.6.1 Constant Fuel Flow Rate

For the first analysis, the fuel cell stack is operated with a constant fuel rate of 5 lpm. The hydrogen flow is maintained at a constant rate throughout the operation. Under this condition, the operation of the fuel cell stack is studied with the ZSI. Furthermore, with a constant fuel flow rate, the fuel cell is operated for two different load conditions, and the performances analyzed. The simulation results are presented in the following text for both balanced and unbalanced load conditions.

Operation of Fuel Cell Stack with ZSI under Balanced Load
A block diagram of the fuel cell integrated with the ZSI is depicted in Figure 18.4. The output voltage of the fuel cell stack is represented as E, as shown in Figures 18.5 and 18.6. The parameters that control the output voltage of the fuel cell are depicted in Figure 18.3.

Table 18.2 Operating parameters of the system

Parameters	Value
Nominal power (W)	6000
Nominal voltage (V)	45
Nominal current (A)	133.3
Maximal power (W)	8325
Nernst voltage of a single cell (V)	1.1288
Exchange current, i_o (A)	0.0160
Number of cells, N	65
Internal resistance of the PEMFC stack, R_{fc} (Ω)	0.0757
Nominal operating temperature (K)	338
Absolute supply pressure of fuel (atm)	1.5
Absolute supply pressure of air (atm)	1
Amount of hydrogen (%)	99.95
Amount of oxygen (%)	21
Z-source inductor (mH)	2
Z-source capacitor (μF)	1000
Switching frequency (kHz)	1
Load inductor (mH)	15
Load resistance (Ω)	2
Fundamental frequency (Hz)	50

Figure 18.8 presents the simulation results of the fuel cell stack when operated with the ZSI by implementing the MBC method.

The fuel flow rate is maintained constant during the experiment. Even though the fuel flow is maintained constant, a fuel flow regulator is used to regulate the flow of hydrogen gas into the PEMFC. The corresponding oxygen and hydrogen utilization rates are presented in the figure. The purpose of this analysis is to compare the performances of ZSI with the traditional dc–dc converter [25]. After the commencement of the experiment, a drop in the utilization rate occurred when the fuel cell was integrated with the traditional dc–dc converter. This initial change might be due to the transient state of the voltage regulator. After that instant, the utilization reverted to normal operation and to the typical consumption rates of the hydrogen and oxygen fuels. However, the utilization rate of oxygen is maintained constant by the ZSI. The fuel consumed by the ZSI is found to be less than the fuel consumption of the boost converter. The consumption is reduced, and the results are observed in Figures 18.8 (b), (c), (d), and (f). The efficiency of the stack is determined as 20%. When compared to the efficiency of the fuel cell with boost converter, the efficiency of the fuel cell with Z-source decreases. Although there is a drop in efficiency, the output voltage of the fuel cell remains constant. In addition, the ZSI performs the boost operation utilizing the ST states, and the dc-link voltage is deduced to 95 V. However, with less amount of fuel utilized, the dc-link voltage is boosted to the expected value of 100 V.

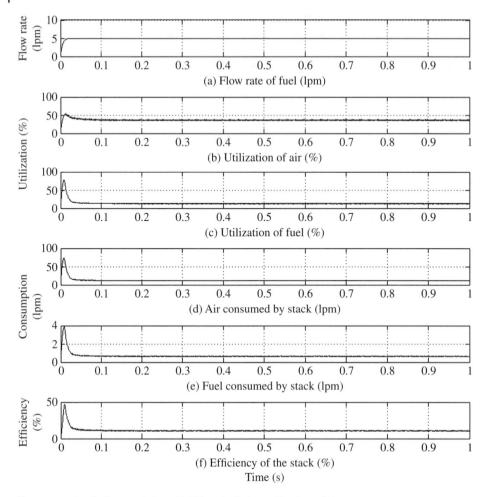

Figure 18.8 Stack characteristics with ZSI under balanced load condition

The fuel cell output voltage, output current, and the dc-link output voltage are depicted in Figure 18.9. The dc-link voltage is maintained at a desirable value to meet the load demand. The output voltage of the traditional dc–dc converter experienced a 30% spike, when the system faced a drop in utilization [25]. However, from the dc-link voltage in Figure 18.9, no peak is observed at the beginning of the experiment. This proves that the ZSI eliminates the effect of the voltage regulator. The dc-link voltage can be further boosted by controlling the ST states in the inverter. The three-phase ac output voltage and the current of the fuel cell system with ZSI are illustrated in Figures 18.10 and 18.11, respectively. From Figure 18.11, it is surmised that the output current is almost sinusoidal and the three-phase currents are identical. This serves as confirmation that the system operates in a balanced load condition.

Figure 18.12 illustrates the current Fast Fourier Transform (FFT) of the ZSI with a balanced load condition and a constant fuel rate. For the balanced load condition that is shown in Figure 18.12, it can be observed that the fundamental component at frequency

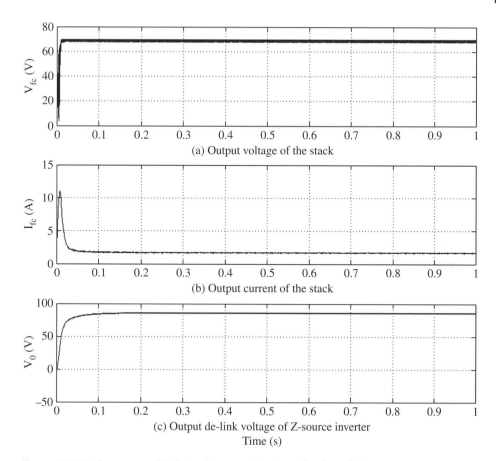

Figure 18.9 Stack output and dc-link voltage under balanced load condition

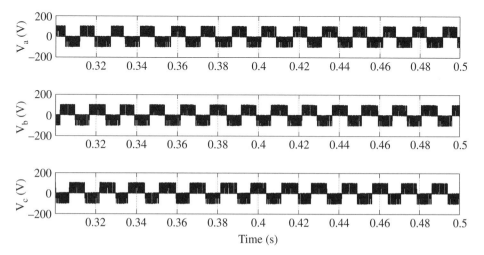

Figure 18.10 Output phase voltages of the ZSI under balanced load condition

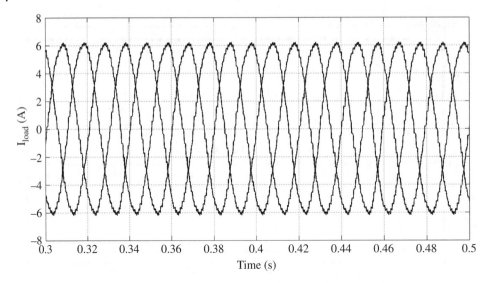

Figure 18.11 Three-phase output currents of the ZSI under balanced load condition

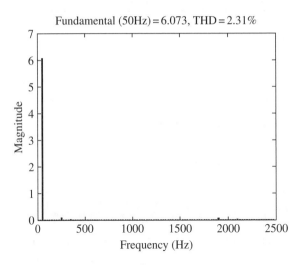

Figure 18.12 Current FFT for phase "a" of the ZSI under balanced load condition

50 Hz has a magnitude of 6.073 A, and a total harmonic distortion of 2.31%. When compared to the current FFT waveform of the traditional dc–dc converter (3.36%) in [25], the total harmonic distortion of the ZSI is low. This ensures that the operation of the ZSI is improved when compared to the traditional dc–dc converter.

Operation of Fuel Cell Stack with ZSI under Unbalanced Load

The fuel cell stack is still maintained at a constant fuel flow rate, and the performance with the ZSI under unbalanced load conditions is investigated in this section. The simulation results demonstrating the change in utilization, consumption, and efficiency of the stack are presented. Figure 18.13 illustrates the characteristics of the fuel cell stack. Similar to the balanced load condition, a drop in the utilization of the stack when implemented with boost converter was noted in [25]. The utilization rate of oxygen is

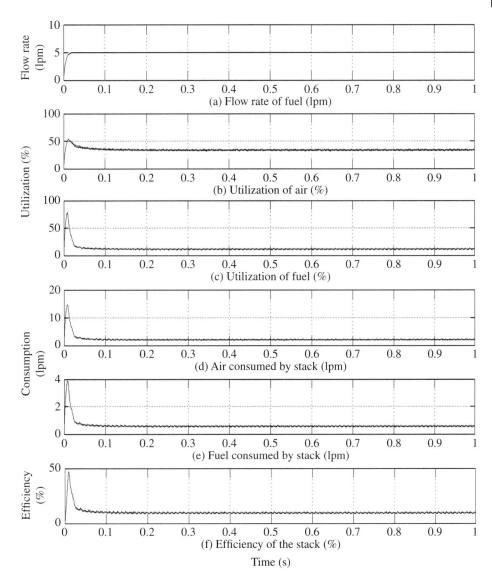

Figure 18.13 Stack characteristics with ZSI under unbalanced load condition

maintained at identical magnitudes for the balanced and unbalanced conditions, which is observed from Figures 18.9 (b) and 18.13 (b), respectively. Notably, the utilization rate does not decline to zero at any instance. There is a change in the percentage of hydrogen utilized, as observed from Figure 18.13 (c), which still remains less than the traditional dc–dc converter. The consumption of air and fuel solely depends on the load requirements, as depicted in Figure 18.13 (d) and (e), respectively. The efficiency of the fuel cell stack during the unbalanced load is 20%. This is less than the traditional dc–dc converter during the unbalanced load condition. However, the ZSI managed to deliver around 95 V even though the efficiency underwent reduction.

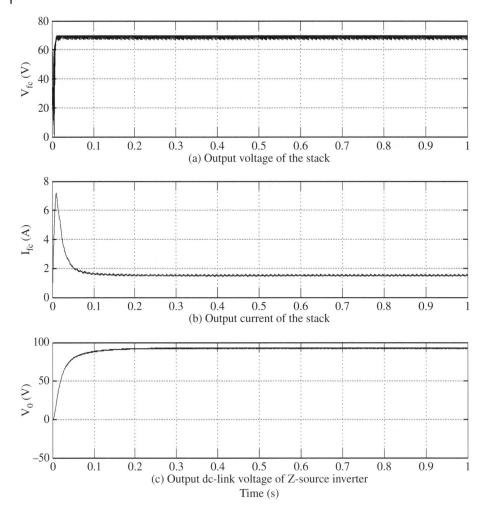

Figure 18.14 Stack output and dc-link voltage under balanced load condition

The fuel cell output voltage, fuel cell output current, and dc-link voltage are shown at the top, middle, and bottom of Figure 18.14, respectively. As explained earlier, the dc-link voltage can be further increased by adjusting the ST duty ratio of the ZSI. ZSI again eliminates the peak at the beginning of the dc-link voltage during unbalanced condition. The dc-link voltage is boosted to the desired value despite less usage of fuel. Figures 18.15 and 18.16, respectively, portray the three-phase output voltage and current waveforms. From Figure 18.16, it can be observed that the magnitudes of the three-phase currents are dissimilar, thereby demonstrating the performance of the fuel cell stack for the unbalanced load condition.

Figure 18.17 portrays the current FFT of ZSI that is integrated with the fuel cell stack for the unbalanced load condition. The fundamental component has a magnitude of 6.957 A and a total harmonic distortion of 2.26%. When compared with the results for the traditional converter (3.8%), the ZSI provides less harmonic distortion than the

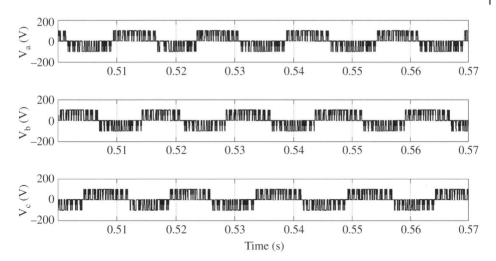

Figure 18.15 Output phase voltages of the ZSI under unbalanced load condition

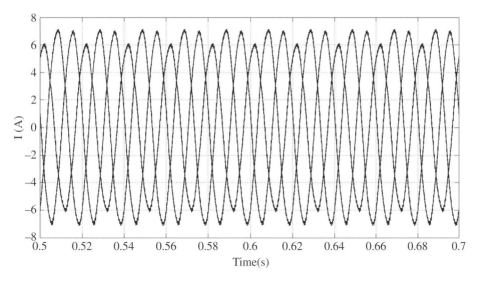

Figure 18.16 Three-phase output currents of the ZSI under unbalanced load condition

traditional converter even under unbalanced load condition. With the results for the performance of the ZSI under constant fuel flow rate discussed, a similar analysis under variable fuel flow rate is presented in the following section.

18.6.2 Variable Fuel Flow Rate

The depiction in Figure 18.4 for ZSI remains the same for this analysis. This case is analyzed by introducing a small change in the input – that is, fuel flow rate. In this case, the hydrogen flow rate to the fuel cell stack is varied after a particular interval of time.

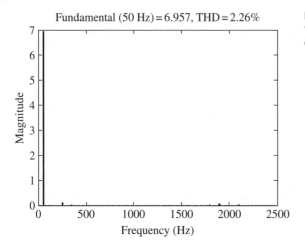

Fundamental (50 Hz) = 6.957, THD = 2.26%

Figure 18.17 Current FFT for phase "a" of the ZSI under unbalanced load condition

With the varying flow rates, the performance of the fuel cell stack with the ZSI is investigated. Furthermore, the changes in the output of the fuel cell and dc-link voltage are quantified. The corresponding waveforms are presented in the following section.

Operation of Fuel Cell Stack with ZSI

In this section, the performance of the system with variable fuel rate is analyzed. A fuel flow regulator is used to maintain the fuel (hydrogen) utilization at a value of 99.56% for the first 1 sec. By bypassing the fuel flow rate after 1 sec, the rate of fuel utilization is allowed to vary, and is shown in Figure 18.18 (a). Now, the fuel flow rate is increased to 10 lpm. Subsequently, the changes in the voltage of the stack are observed. The changes in the utilization of fuel would definitely impact the amounts of fuel and oxygen consumed, and the efficiency of the fuel cell stack. With increase in the fuel flow rate, hydrogen utilization is commensurately reduced. This causes an increase in the Nernst voltage, decreasing the output current of the fuel cell [21]. Hence, the fuel consumption by the fuel cell stack and the efficiency of the fuel cell stack decrease. The changes in utilization, consumption, and efficiency of the fuel cell stack for different flow rates are shown in Figure 18.18 (b) and(c); (d) and (e); and (f), respectively. The change in the fuel flow rate impacts the fuel cell stack output. The output voltage, output current of the fuel cell, and the dc-link voltage are illustrated in Figure 18.19. From Figure 18.19, it can be observed that the dc-link voltage is maintained at around 95 V by the ZSI. This ensures the deliverance of an uninterrupted voltage to the load. As ZSI maintains the dc-link voltage, the inverter output remains the same as the constant fuel flow rate. If the load demand increases, then the change in the fuel flow rate will have a considerable impact on the characteristics of the fuel cell stack.

The performance results of the PEMFC stack with the ZSI under variable fuel flow condition are depicted in Figures 18.18 and 18.19. With a fuel flow rate of 5–10 lpm as observed from Figure 18.18(a), the percentage of oxygen utilized, the utilization of hydrogen, and the efficiency are plotted. The utilization rate of hydrogen in Figure 18.18 (c) decreases when the fuel flow rate is increased. The efficiency of the fuel cell stack is calculated based on the utilization of the fuels for the chemical reaction that generates power. An important conclusion of this study is that the hydrogen consumption and utilization rate affect the efficiency of the fuel cell stack for the conditions of constant

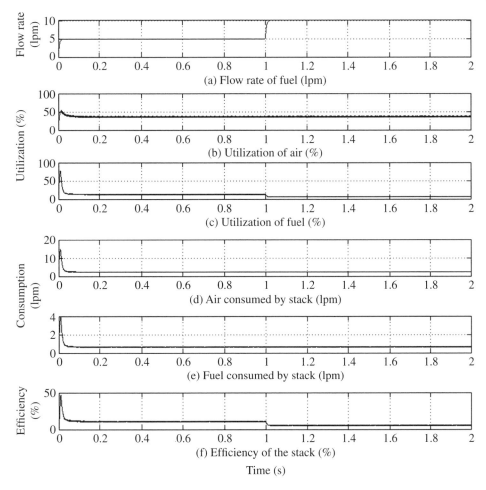

Figure 18.18 Stack characteristics under variable fuel flow rate condition

and variable fuel flow rates. However, when compared to the traditional converter, the drop in utilization and the spike in dc-link voltage are eliminated using ZSI, even during unbalanced load condition [26]. From Figure 18.19, it can be established that the output voltage of the fuel cell and the dc-link voltage remain unaffected. The three-phase output voltages and current waveforms are not presented in this section as the ZSI maintains the dc-link voltage regardless of change in the fuel flow rate.

18.7 Summary

The fuel cell is an important energy source, and is a part of the promising and emerging technology for DER applications. Given the importance of this technology, the fuel cell is considered as a dc source. The initial part of this chapter dealt with the modelling of the fuel cell. The modelled fuel cell was then integrated with the ZSI, and the performances were evaluated. The utilization and consumption of fuel and air were monitored to track

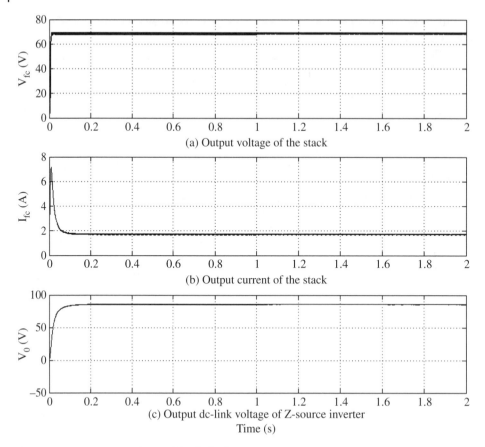

Figure 18.19 Output waveforms of the fuel cell and dc-link under variable fuel flow rate condition

the changes in fuel flow and efficiency. The fuel flow regulator performs the important task of maintaining the fuel flow at a sufficient rate to evaluate the operation of the fuel cell. Therefore, the fuel flow regulator plays a primary role in altering the fuel rate based on demand. The results of the balanced and unbalanced load conditions prove that the transient effect of the regulator was more prominent (dc-link peak voltage of 125 V_0) in the results of the traditional converter, and has been eliminated by the ZSI. The ZSI also provides a smooth transition due to the capacitor network. Another notable point from the validated results is that the ZSI uses less fuel as compared to the traditional converter while generating similar magnitudes of output. The ZSI delivers the desired voltage irrespective of the change in the fuel flow rate, providing an impetus for its usage. The major concern about the traditional boost converters is the static output voltage. However, in the ZSI, the output can be boosted by adjusting the ST duty cycle. Therefore, the ZSI is a viable and improved alternative to the traditional boost converter in DER applications.

References

1 M. F. Akorede, H. Hizam, and E. Pouresmaeil, "Distributed energy resources and benefits to the environment," *Renewable and Sustainable Energy Reviews*, 2010.

2 R. P. O'hayre, Suk-won Cha, Whitney G. Colella, and Fritz B. Prinz, *Fuel Cell Fundamentals*. John Wiley & Sons, Inc. USA, 2009.

3 C. Spiegel, *PEM Fuel Cell Modelling and Simulation using MATLAB*, Elsevier Publications, 2008.

4 J. C. Amphlett, R. M. Baumert, R. F. Mann, B. A. Peppley, P. R. Roberge, and T. J. Harris, "Performance Modeling of the Ballard Mark IV Solid Polymer Electrolyte Fuel cell, I. Mechanistic model development," *ECS – The Electrochemical Society*, vol. 142, pp. 1–8, Jan. 1995.

5 "Performance modeling of the Ballard Mark IV solid polymer electrolyte fuel cell, II. Empirical model development," *Journal of the Electrochemical Society*, vol. 142, no. 1, pp. 9–15, Jan. 1995.

6 R. Cownden, M. Nahon, and M. A. Rosen, "Modeling and analysis of a solid polymer fuel cell system for transportation applications," *International Journal of Hydrogen Energy*, vol. 26, no. 6, pp. 615–623, Jun. 2001.

7 C. E. Chamberlin, "Modeling of proton exchange membrane fuel cell performance with an empirical equation," *Journal of the Electrochemical Society*, vol. 142, no. 8, pp. 2670–2674, Aug. 1995.

8 A. Rowe and X. Li, "Mathematical modeling of proton exchange membrane fuel cells," *Journal of Power Sources*, vol. 102, no. 1–2, pp. 82–96, Dec. 2001.

9 D. Bevers and M. Wöhr, "Simulation of a polymer electrolyte fuel cell electrode," *Journal of Applied Electrochemistry*, vol. 27, no. 11, pp. 1254–1264, Nov. 1997.

10 G. Maggio, V. Recupero, and L. Pino, "Modeling polymer electrolyte fuel cells: an innovative approach," *Journal of Power Sources*, vol. 101, no. 2, pp. 275–286, Oct. 2001.

11 J. C. Amphlett, R. F. Mann, B. A. Peppley, P. R. Roberge, and A. Rodrigues, "A model predicting transient responses of proton exchange membrane fuel cells," *Journal of Power Sources*, vol. 61, no. 1–2, pp. 183–188, Jul./Aug. 1996.

12 J. Hamelin, K. Agbossou, A. Laperrière, F. Laurencelle, and T. K. Bose, "Dynamic behavior of a PEM fuel cell stack for stationary applications," *International Journal of Hydrogen Energy*, vol. 26, no. 6, pp. 625–629, Jun. 2001.

13 M. Wöhr, K. Bolwin, W. Schnurnberger, M. Fischer, W. Neubrand, and G. Eigenberger, "Dynamic modeling and simulation of a polymer membrane fuel cell including mass transport limitation," *International Journal of Hydrogen Energy*, vol. 23, no. 3, pp. 213–218, Mar. 1998.

14 H. P. L. H. van Bussel, F. G. H. Koene, and R. K. A. M. Mallant, "Dynamic model of solid polymer fuel cell water management," *Journal of Power Sources*, vol. 71, no. 1–2, pp. 218–222, Mar. 1998.

15 M. Wang and M. H. Nehrir, "Fuel cell modeling and fuzzy logic-based voltage control," *International Journal of Renewable Energy Engineering*, vol. 3, no. 2, August 2001.

16 M. D. Lukas, K. Y. Lee, and H. Ghezel-Ayagh, "Performance implications of rapid load changes in carbonate fuel cell systems," in *Proc. IEEE Power Engineering Society Winter Meeting*, vol. 3, 2001, pp. 979–984.

17 R. Lasseter, "Dynamic models for micro-turbines and fuel cells," in *Proc. IEEE Power Engineering Society Summer Meeting*, vol. 2, 2001, pp. 761–766.

18 P. Srinivasan, A. Feliachi, and J. E. Sneckenberger, "Proton exchange membrane fuel cell dynamic model for distributed generation control purposes," in *Proc. 34th North American Power Symposium*, Tempe, AZ, Oct. 2002, pp. 393–398.

19 P. Famouri and R. S. Gemmen, "Electrochemical circuit model of a PEM fuel cell," in *Proc. IEEE Power Engineering Society Summer Meeting*, Toronto, ON, Canada, Jul. 2003.

20 S. N. Motapon, O. Tremblay, and L. -A. Dessaint, "Development of a generic fuel cell model: application to a fuel cell vehicle simulation," *International Journal of Power Electronics*, vol. 4, no. 6, 2012, pp. 505–522.

21 Njoya, S. M., O. Tremblay, and L. -A. Dessaint, "A generic fuel cell model for the simulation of fuel cell vehicles," *Vehicle Power and Propulsion Conference, 2009, VPPC '09*, IEEE®. Sept. 7–10, 2009, pp. 1722–1729.

22 J. C. Amphlett, R. M. Baumert, R. F. Mann, B. A. Peppley, P. R. Roberge, and T. J. Harris, "Performance modeling of the Ballard Mark IV solid polymer electrolyte fuel cell, I. Mechanistic model development," *ECS – The Electrochemical Society*, vol. 142, pp. 1–8 Jan. 1995.

23 J. Larminie, A. Dicks, "Fuel cell systems explained," 2nd ed. John Wiley and Sons Ltd.

24 F. Z. Peng, S. Miaosen, and Q. Zhaoming, "Maximum boost control of the Z-Source inverter," *IEEE Transactions in Power Electronics*, vol. 20, pp. 833–838, 2005.

25 M. M. Roomi, A. I. Maswood, and H. Dehghani Tafti, "Performance evaluation of boost and Z-source converters for fuel cell application," *IEEE Innovative Smart Grid Technologies – Asia (ISGT ASIA)*, 2015, pp. 1–7.

26 M. M. Roomi, A. I. Maswood, M. S. U. Khan, and M. Tariq, "Study of traditional and Z-source converter under variable fuel-flow-rate of PEMFC," *International Conference on the Development in the Renewable Energy Technology (ICDRET)*, 2016, pp. 1–6.

19

Multilevel Converter-Based Flexible Alternating Current Transmission System

Muhammad M. Roomi and Harikrishna R. Pinkymol

19.1 Introduction

Power converters are classified, based on the power flow directions, as *voltage source converters* (VSCs) and *current source converters* (CSCs). In most of the flexible ac transmission systems (FACTSs), VSCs are preferred over CSCs due to their low overall cost and good performance. The STATic synchronous COMpensator (STATCOM) is a shunt-connected advanced FACTS controller used for voltage control, VAR compensation, and for stabilizing voltage in electric power systems [1–3]. The VSC-based STATCOM generates a controllable synchronous three-phase output voltage from an energy storage capacitor that forces reactive power exchange with the ac system required for compensation. The transformer used in conventional two-level VSC-based STATCOM for voltage matching between converter ac output voltage and utility grid contributes to major power losses and increases the system size and cost. Multilevel inverters generate high voltages with less harmonics and EMI emissions by using devices of smaller voltage ratings, which makes them suitable in STATCOM application [4, 5]. Hence, multilevel inverter-based STATCOM eliminates the need for a transformer and can be directly connected to the utility grids for improved efficiency. In [6–8], multilevel inverter-based commercial STATCOM systems are discussed. Hence, the maintenance of a constant voltage across the dc-link capacitors during all operating conditions is necessary. The voltage balancing circuit presented in [9] consists of two bidirectional buck/boost choppers connected to the dc-link of the converter. The midpoint control employing the sixth harmonic zero-sequence voltage injection technique provides well-balanced capacitor voltages for a 200V, 10kVA STATCOM based on a five-level neutral-point-clamped (NPC) converter. However, the need for this additional hardware increases the system size and cost. The voltage balancing issues in classical diode-clamped multilevel inverters (DCMIs) and some voltage equalization methods are addressed in [6, 10–12]. The space vector modulation (SVM)–based voltage balancing strategy in [13] utilizes the redundant voltage vector property of five-level NPC inverters to maintain a constant voltage across the dc-link capacitors in all operating conditions. Though this method eliminates the auxiliary hardware, it requires the sector information in which the reference space vector is located. This involves an increased number of calculations at each sampling interval, which increases system complexity.

Advanced Multilevel Converters and Applications in Grid Integration, First Edition.
Edited by Ali I. Maswood and Hossein Dehghani Tafti.

Therefore, the performance of five-level multiple-pole multilevel diode-clamped inverter (M²DCI) using the new SVM method to balance dc-link capacitor voltages is investigated in this chapter for STATCOM application. This new methodology reduces the number of calculations required at each sampling time and eliminates the need for voltage balancing circuits. This new topology, along with the SVM technique, allows better efficiency and enhances the performance of STATCOM with less number of power diodes as compared to the classical five-level DCI.

19.2 A Space Vector Modulated Five-Level Multiple-pole Multilevel Diode-Clamped STATCOM

The circuit configuration of the five-level M²DCI explained in Chapter 10 can be used for the STATCOM converter. The configuration of the five-level M²DCI connected to a three-phase load illustrated in Figure 19.1 was originally depicted in section 10.3.2 as Figure 10.4.

19.2.1 Circuit Configuration

Using Figure 19.1, the five-level M²DCI can be configured as a STATCOM converter, as demonstrated in Figure 19.2.

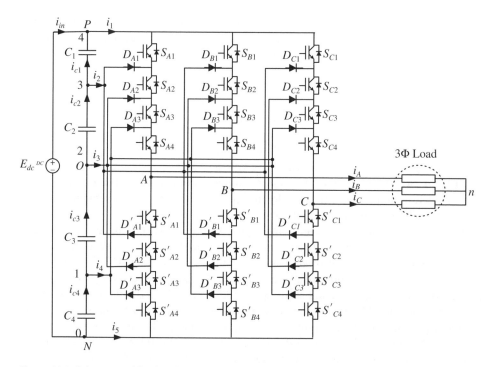

Figure 19.1 Schematic of five-level diode-clamped inverter (DCI) with three-phase load.

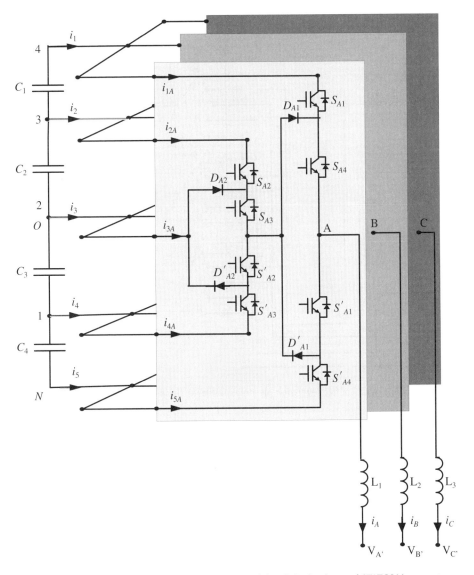

Figure 19.2 Schematic of five-level multiple-pole multilevel diode-clamped STATCOM converter.

19.2.2 Operation Principles

Figure 19.3 shows the current model of the STATCOM converter. From Figure 19.3, the dc-link capacitor currents can be expressed by the following equations:

$$i_{C2} = i_{C1} + i_2$$
$$i_{C3} = i_{C2} + i_3$$
$$i_{C4} = i_{C3} + i_4 \tag{19.1}$$

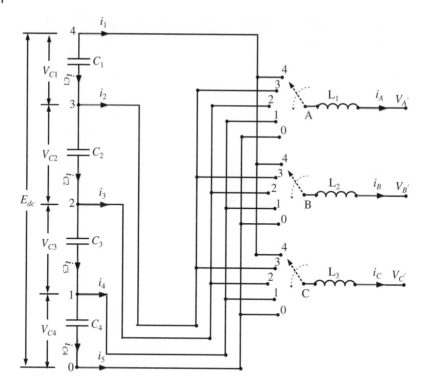

Figure 19.3 Current model of five-level multiple-pole multilevel diode-clamped STATCOM converter.

In general, the sums of the capacitor voltages are equal to the dc-link reference voltage of magnitude E_{dc}. Hence, by applying KVL at the dc-link and derivatives:

$$\frac{1}{C_1}i_{C1} + \frac{1}{C_2}i_{C2} + \frac{1}{C_3}i_{C3} + \frac{1}{C_4}i_{C4} = 0 \tag{19.2}$$

If all the capacitors are assumed to have equal magnitudes, equation (19.2) becomes:

$$i_{C1} + i_{C2} + i_{C3} + i_{C4} = 0 \tag{19.3}$$

Therefore, from equations (19.1) and (19.3), the current through the capacitors is calculated from midpoint currents as:

$$i_{Cj} = \frac{1}{4}\sum_{x=1}^{3} x\, i_{x+1} - \sum_{x=j+1}^{4} i_x \quad for \quad j = 1, 2, 3, 4. \tag{19.4}$$

In addition, the midpoint currents are given by:

$$i_2 = (S_{A2} - S_{A1})i_A + (S_{B2} - S_{B1})i_B + (S_{C2} - S_{C1})i_C$$
$$i_3 = (S_{A3} - S_{A2})i_A + (S_{B3} - S_{B2})i_B + (S_{C3} - S_{C2})i_C$$
$$i_4 = (S_{A4} - S_{A3})i_A + (S_{B4} - S_{B3})i_B + (S_{C4} - S_{C3})i_C \tag{19.5}$$

The detailed modellings of the converter and SVM-based voltage balancing control have been presented in Chapter 10. The cost function used for implementing voltage

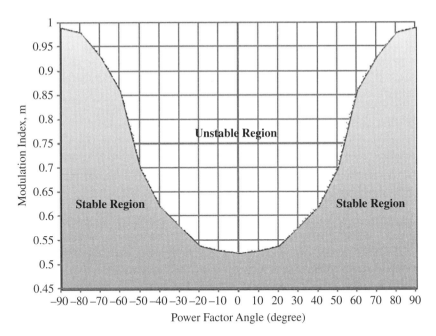

Figure 19.4 Voltage balance limits for the SVM-based capacitor voltage balancing strategy of five-level M²DCI.

balancing is given by:

$$\sum_{j=1}^{3} \Delta V_{Cj}(k) \left(\sum_{x=j+1}^{4} \bar{i}_x \right) \geq 0 \tag{19.6}$$

The SVM-based voltage-balancing algorithm that is presented in Chapter 10 will select the switching vector combination during each sampling interval. The converter switches are thereby controlled in order to balance the voltage across the four dc-link capacitors at a reference value of $(E_{dc}/4)$.

From the stability plot (Figure 19.4) obtained for the five-level converter, it can be noted that the SVM-based balancing control eliminates the voltage drift phenomena for a range of linear modulation indices in low power factor operations. Hence, the SVM-based balancing control can be used for controlling the reactive power in transmission/distribution lines.

19.2.3 STATCOM Modelling

STATCOM is a shunt-connected reactive power compensator capable of generating or absorbing the desired amount of reactive power at its output terminals when an energy storage device is connected as an input. It consists of a VSC, which generates the desired three-phase voltages and currents from an energy-storage capacitor to improve power system performance. In this section, a five-level M²DCI is employed instead of the conventional two-level VSC. This ensures a reduction in harmonic pollution due to the addition of STATCOM with the power system. In addition, the STATCOM output is coupled with the ac system voltage through a small reactance, as shown in Figure 19.5.

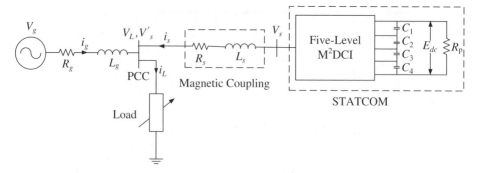

Figure 19.5 Five-level M²DCI–based STATCOM connected to the ac system.

The first-order differential equation for the ac-side of the system in Figure 19.5 is expressed as:

$$pi_{sA} = \frac{1}{L_s}(-R_s i_{sA} + V_{sA} - V_{LA})$$

$$pi_{sB} = \frac{1}{L_s}(-R_s i_{sB} + V_{sB} - V_{LB})$$

$$pi_{sC} = \frac{1}{L_s}(-R_s i_{sC} + V_{sC} - V_{LC}) \tag{19.7}$$

Here, p is a (d/dt) operator.
STATCOM terminal voltages are defined as:

$$\begin{bmatrix} V_{sA} \\ V_{sB} \\ V_{sC} \end{bmatrix} = \frac{mE_{dc}}{\sqrt{3}} \begin{bmatrix} sin(\omega t + \theta_1) \\ sin\left(\omega t + \theta_1 - \frac{2\pi}{3}\right) \\ sin\left(\omega t + \theta_1 + \frac{2\pi}{3}\right) \end{bmatrix} \tag{19.8}$$

The dc-side circuit equation is expressed as:

$$pE_{dc} = \frac{-1}{C_{equ}}\left(I_{dc} + \frac{E_{dc}}{R_p}\right) \tag{19.9}$$

Here, R_p represents the capacitor losses and switching losses, and $C_{equ} = \frac{1}{\frac{1}{C_1} + \frac{1}{C_2} + \frac{1}{C_3} + \frac{1}{C_4}}$.

The ac-side equations are converted on a synchronous rotating reference frame using this transformation:

$$[f_{dq0}] = T \cdot [f_{ABC}] \tag{19.10}$$

Here,

$$T = \sqrt{2/3} \begin{bmatrix} sin(\theta_{pcc}) & sin\left(\theta_{pcc} - \frac{2\pi}{3}\right) & sin\left(\theta_{pcc} + \frac{2\pi}{3}\right) \\ cos(\theta_{pcc}) & cos\left(\theta_{pcc} - \frac{2\pi}{3}\right) & cos\left(\theta_{pcc} + \frac{2\pi}{3}\right) \\ 1/\sqrt{2} & 1/\sqrt{2} & 1/\sqrt{2} \end{bmatrix}; [f_{dq0}] = [f_d \ f_q \ f_0]^t.$$

$[f_{ABC}] = [f_A \ f_B \ f_C]^t$ and $\theta_{pcc} = \omega t + \theta_L$

Here, θ_L is the initial phase angle of PCC voltage V_L.
STATCOM terminal voltages and load terminal voltages in the $d–q$ axis are given by:

$$\begin{bmatrix} V_{sd} \\ V_{sq} \end{bmatrix} = \frac{mE_{dc}}{\sqrt{2}} \begin{bmatrix} cos(\theta_1 - \theta_L) \\ sin(\theta_1 - \theta_L) \end{bmatrix} \tag{19.11}$$

$$\begin{bmatrix} V_{Ld} \\ V_{Lq} \end{bmatrix} = \sqrt{\frac{3}{2}} \begin{bmatrix} V_{Lpeak} \\ 0 \end{bmatrix} \tag{19.12}$$

V_{Lpeak} is the amplitude of PCC voltage V_L. Equation (19.7) is converted into the $d–q$ reference frame by:

$$\begin{bmatrix} pi_{sd} \\ pi_{sq} \end{bmatrix} = \begin{bmatrix} -R_s/L_s & \omega \\ -\omega & -R_s/L_s \end{bmatrix} \begin{bmatrix} i_{sd} \\ i_{sq} \end{bmatrix} + \frac{1}{L_s} \begin{bmatrix} V_{sd} - V_{Ld} \\ V_{sq} - V_{Lq} \end{bmatrix} \tag{19.13}$$

The power balance equation of the converter is given by:

$$E_{dc}I_{dc} = (V_{sd} \times i_{sd} + V_{sq} \times i_{sq}) \tag{19.14}$$

Substituting equation (19.11) in equation (19.14) gives:

$$I_{dc} = \frac{m}{\sqrt{2}} [cos(\theta_1 - \theta_L)\, i_{sd} + sin(\theta_1 - \theta_L)i_{sq}] \tag{19.15}$$

The dc-side circuit equation expressed in equation (19.9) modifies into:

$$pE_{dc} = \frac{-1}{C_{equ}} \left(\frac{n}{\sqrt{2}} [cos(\theta_1 - \theta_L)i_{sd} + sin(\theta_1 - \theta_L)i_{sq}] + \frac{E_{dc}}{R_p} \right) \tag{19.16}$$

19.2.4 Control Strategy

i_{sd} and i_{sq} are coupled with each other through an inductive reactance, which is presented in the state-space model of STATCOM, and given by equation (19.13).

To provide a decoupled control, the STATCOM terminal voltage vector is controlled by:

$$\begin{aligned} V_{sd} &= -\omega L_s i_{sq} + V_{Ld} + L_s V_{sd}^* \\ V_{sq} &= \omega L_s i_{sd} + V_{Lq} + L_s V_{sq}^* \end{aligned} \tag{19.17}$$

Substituting equations (19.12) and (19.17) in equation (19.13) leads to:

$$\begin{bmatrix} pi_{sd} \\ pi_{sq} \end{bmatrix} = \begin{bmatrix} -R_s/L_s & 0 \\ 0 & -R_s/L_s \end{bmatrix} \begin{bmatrix} i_{sd} \\ i_{sq} \end{bmatrix} + \begin{bmatrix} V_{sd}^* \\ V_{sq}^* \end{bmatrix} \tag{19.18}$$

The transfer function of equation (19.18) is expressed as:

$$G_s(s) = \frac{I_{sd}(s)}{V_{sd}^*(s)} = \frac{I_{sq}(s)}{V_{sq}^*(s)} = \frac{1}{s + (R_s + L_s)} \tag{19.19}$$

Two independent PI controllers can be used to produce the control signals V_{sd}^* and V_{sq}^*. STATCOM control defined in equation (19.17) is implemented, and the block schematic is depicted in Figure 19.6.

$$V_{sd}^* = K_{pd}(i_{sd}^* - i_{sd}) + K_{id} \int (i_{sd}^* - i_{sd})$$

$$V_{sq}^* = K_{pq}(i_{sq}^* - i_{sq}) + K_{iq} \int (i_{sq}^* - i_{sq}) \qquad (19.20)$$

The reactive current reference i_{sq}^* is obtained from the quadrature component of the load current. The dc-link voltage control loop provides the direct axis current reference i_{sd}^* to regulate the dc-link voltage of the STATCOM to its reference magnitude.

19.2.5 Validation

The reactive power compensation of the ac system in transient and steady-state conditions is achieved by controlling the five-level M²DCI as illustrated in Figure 19.6. The performance of the new system is evaluated using Matlab/Simulink® and PSIM for various operating conditions. The circuit and controller parameters are listed in Table 19.1.

Performance of STATCOM with Inductive Load

The system in Figure 19.5 is initially at no-load condition, and the STATCOM connected to the grid supplies no reactive power. At 0.1s, a balanced three-phase load of value $R_L/phase = 1.4\Omega$ and $L_L/phase = 22$mH (load PF = 0.47) is connected, and the system response to the reactive power change is shown in Figure 19.7. Figure 19.7 (a) depicts the individual capacitor voltages when 400V is given as the dc-link voltage reference. The control algorithm provides a balanced voltage of magnitude 100V across the dc-link capacitors. Additionally, M²DCI line-to-line voltage, which is illustrated in Figure 19.7 (b), has nine distinct levels due to the increase in modulation index. The grid voltage and

Table 19.1 Control Parameters of the System.

UTILITY	
Source voltage, V_{gAB}	190.52 V
Resistance, R_g	0.5 Ω
Reactance, X_{Lg}	0.031 Ω
Supply frequency, f	50 Hz
STATCOM	
dc-link voltage, E_{dc}	400 V
dc-link capacitor, C_j, j=1,2,3,4.	8800 uF
Inductor reactance, X_{Ls}	1.57 Ω
Inductor resistance, R_s	0.9 Ω
Sampling Frequency, f_s	2.5 kHz
PI Controllers	
dc-link voltage controller	$K_p=1; K_i=30$
d-axis current loop controller	$K_{pd}=2.5; K_{id}=25$
q-axis current loop controller	$K_{pq}=10; K_{iq}=200$

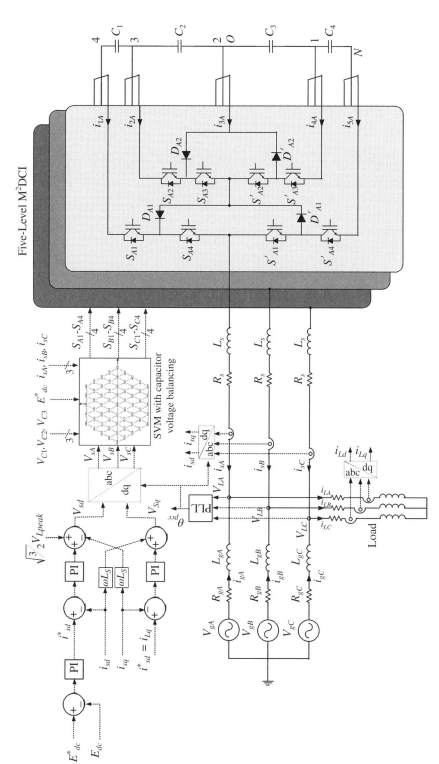

Figure 19.6 Control scheme for five-level M²DCI–based STATCOM connected to the ac system.

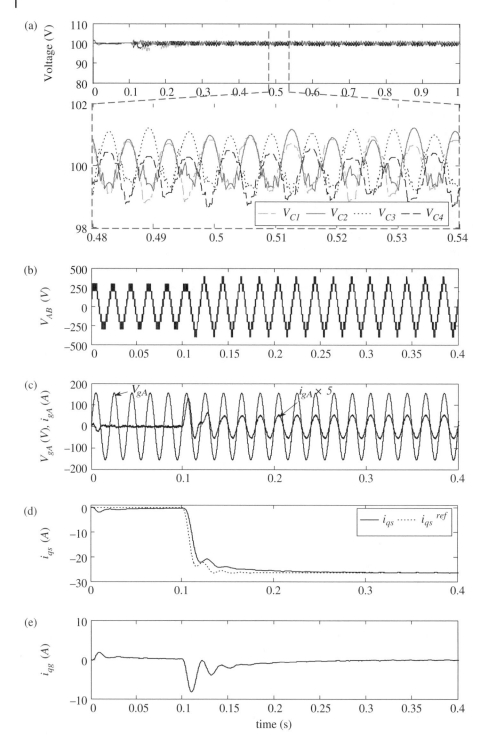

Figure 19.7 System response for an addition of inductive load at 0.1s. (a) dc-link capacitor voltages V_{Cj} for j=1, 2, 3, 4; (b) line voltage of M²DCI, V_{AB}; (c) grid voltage V_{gA} and grid current I_{gA} in phase "A"; (d) STATCOM reactive current component, i_{qs}; (e) grid reactive current component, i_{qg}.

current waveforms are in-phase with each other due to the reactive current provided by the STATCOM, and are shown in Figure 19.7 (c). The reference reactive current component and the STATCOM reactive current component are displayed in Figure 19.7 (d). The STATCOM delivers the reactive current demand of the load, and hence the source reactive current component is zero, as demonstrated in Figure 19.7 (e).

Step change from Lagging to Leading Load

Figure 19.8 depicts the system performance with a change in load from lagging to leading at 0.4s. When the load is changed at 0.4s, the compensating current tracks the reference, and the SVM-based balancing control method balances voltage across the dc-link capacitors. Figure 19.8 (a) and (b) display the individual dc-link capacitor voltages, while Figure 19.8 (c) demonstrates grid voltage and current in phase "a." The STATCOM controller generates the desired reactive current component for maintaining unity power factor at the grid after a load change. The reactive current component of the STATCOM and grid are illustrated by Figure 19.8 (d) and (e), respectively.

Unbalanced Source Condition

To verify the effectiveness of the new voltage balancing control strategy during unbalanced source condition, three-phase source voltage of amplitudes $V_{gA} = 147.78$V, $V_{gB} = 155.55$V, and $V_{gC} = 163.32$V (Figure 19.9 (c)) are used. The dc-link capacitor voltages V_{C1}, V_{C2}, V_{C3}, and V_{C4} are shown in Figure 19.9 (a). The dc-link capacitor voltages converge towards the steady-state value of 100V, which ensures that the balancing algorithm eliminates the voltage drift phenomena in M^2DCI during unbalanced source conditions. M^2DCI line-to-line voltage is shown in Figure 19.9 (b). The grid voltage and current waveforms are in-phase with each other and are demonstrated in Figure 19.9 (d). The reactive current component of STATCOM and source are demonstrated in Figures 19.9 (e) and (f), respectively. The ripples in the reactive current component are due to the imbalance in source voltages.

Change in Linear Load from Balanced to Unbalanced

To verify the effectiveness of the new voltage balancing control strategy during unbalanced load currents, the load at the PCC changes from balanced to unbalanced condition at 0.4s. Figure 19.10 illustrates the corresponding system response. Although there is a change in load at 0.4s, the capacitor voltages remain stable and balanced at around 100V, and the M^2DCI line-to-line voltage has nine distinct levels. The low-frequency oscillations found in the reactive current components (Figure 19.10 (e) and (f)) after 0.4s is due to unbalanced load currents.

Change in Balanced Load from Linear to Non-linear

The operating condition of the system in Figure 19.5 is described in section 5.4.1. At 0.4s, the three-phase non-linear load is also connected to the system, which generates harmonics in load currents (Figure 19.11 (c)). The system response due to the change in load is depicted in Figure 19.11. The individual dc-link capacitor voltages displayed in Figure 19.11 (a) remain stable and balanced. The M^2DCI line-to-line voltage shown in Figure 19.11 (b) has nine distinct levels. The STATCOM delivers the reactive power demand of the load, and provides unity power factor at the source, as illustrated in Figure 19.11 (d)–(f).

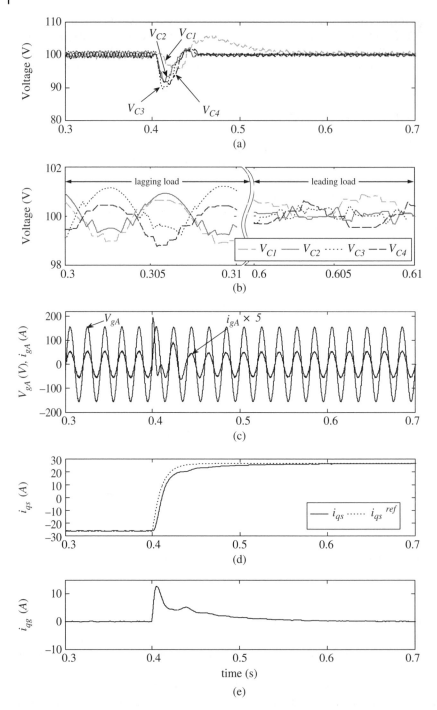

Figure 19.8 System response for a step change in load from lagging to leading at 0.4s. (a) dc-link capacitor voltages V_{Cj} for j=1, 2, 3, 4; (b) zoomed waveforms of dc-link capacitor voltages; (c) grid voltage V_{gA} and grid current I_{gA} in phase "A"; (d) STATCOM reactive current component, i_{qs}; (e) grid reactive current component, i_{qg}.

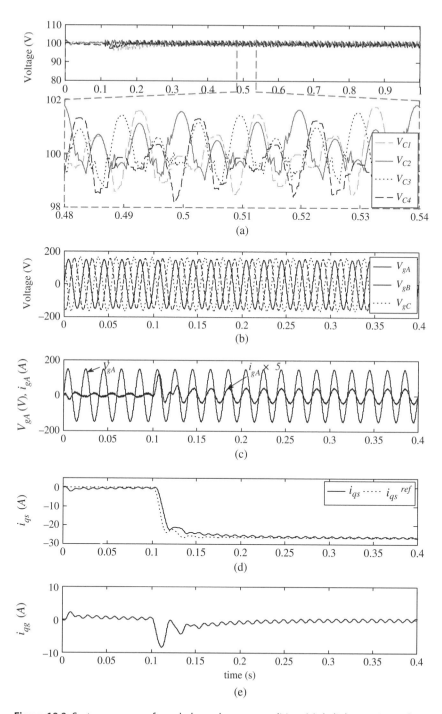

Figure 19.9 System response for unbalanced source condition. (a) dc-link capacitor voltages V_{Cj} for j=1, 2, 3, 4; (b) line voltage of M²DCI, V_{AB}; (c) grid voltage waveforms, V_{gA}, V_{gB}, and V_{gC}; (d) grid voltage V_{gA} and grid current I_{gA} in phase "A"; (e) STATCOM reactive current component, i_{qs}.

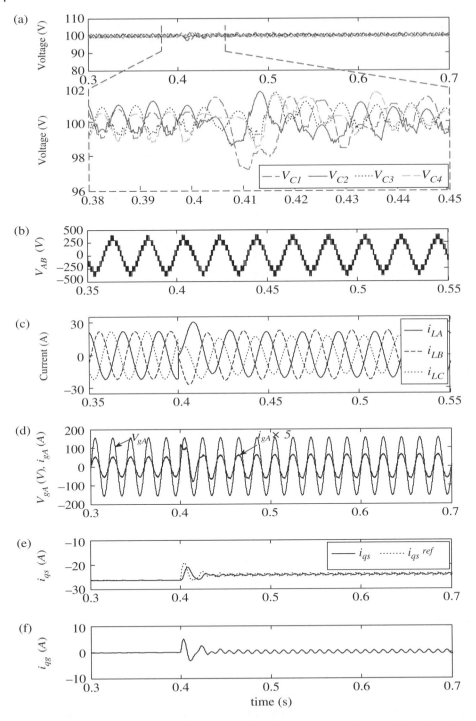

Figure 19.10 System response for a change in load from balanced to unbalanced at 0.4s. (a) dc-link capacitor voltages V_{Cj} for j=1, 2, 3, 4; (b) line voltage of M²DCI, V_{AB}; (c) load current waveforms i_{LA}, i_{LB}, and i_{LC}; (d) grid voltage V_{gA} and grid current I_{gA} in phase "A"; (e) STATCOM reactive current component, i_{qs}; (f) grid reactive current component, i_{qg}.

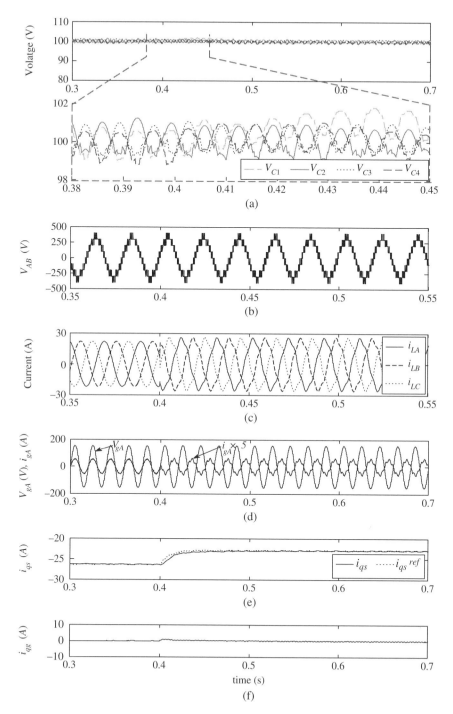

Figure 19.11 System response for a step change in balanced load from linear to non-linear at 0.4s. (a) dc-link capacitor voltages V_{Cj} for j=1, 2, 3, 4; (b) line voltage of M²DCI, V_{AB}; (c) load current waveforms i_{LA}, i_{LB}, and i_{LC}; (d) grid voltage V_{gA} and grid current I_{gA} in phase "A"; (e) STATCOM reactive current component, i_{qs}; (f) grid reactive current component, i_{qg}.

19.3 Summary

Multilevel inverter-based STATCOM configuration is preferred over two-level VSC due to its improved power quality. In addition, it provides the flexibility to connect high-efficiency-utilizing devices of smaller power ratings directly to the grid in high-power conversions. The number of power diodes in the new five-level M^2DCI–based STATCOM is fewer than the classical five-level DCI. Additionally, the SVM-based voltage balancing strategy presented in this chapter reduces the required number of calculations at each sampling period by predicting the dc-link capacitor current. The performance attributes of the five-level M^2DCI–based STATCOM in transient and steady-state conditions, such as balanced, unbalanced, and non-linear operating conditions, were evaluated using the Matlab/Simulink® and PSIM environments. Results ensure that the voltage balancing control maintains balanced voltage across the dc-link capacitors under all operating conditions. This new methodology provides better efficiency and enhances the performance of STATCOM.

References

1 J. Dixon, L. Moran, J. Rodriguez, and R. Domke, "Reactive power compensation technologies: state-of-the-art review," *Proceedings of the IEEE*, vol. 93, no. 12, pp. 2144–2164, 2005.

2 N. G. Hingorani, L. Gyugyi, and M. El-Hawary, *Understanding FACTS: concepts and Technology of Flexible AC Transmission Systems*. Wiley Online Library, 2000.

3 B. Singh, R. Saha, A. Chandra, and K. Al-Haddad, "Static synchronous compensators (STATCOM): a review," *IET Power Electronics*, vol. 2, no. 4, pp. 297–324, 2009.

4 J. Rodriguez, J.-S. Lai, and F. Z. Peng, "Multilevel inverters: a survey of topologies, controls, and applications," *IEEE Transactions on Industrial Electronics*, vol. 49, no. 4, pp. 724–738, 2002.

5 S. Kouro *et al.*, "Recent advances and industrial applications of multilevel converters," *IEEE Transactions on Industrial Electronics*, vol. 57, no. 8, pp. 2553–2580, 2010.

6 Y. Cheng, C. Qian, M. L. Crow, S. Pekarek, and S. Atcitty, "A comparison of diode-clamped and cascaded multilevel converters for a STATCOM with energy storage," *IEEE Transactions on Industrial Electronics*, vol. 53, no. 5, pp. 1512–1521, 2006.

7 ABB. Available: www.abb.com

8 SIEMENS. Available: www.siemens.com

9 H. Akagi, H. Fujita, S. Yonetani, and Y. Kondo, "A 6.6-kV transformerless STATCOM based on a five-level diode-clamped PWM converter: system design and experimentation of a 200-V 10-kVA laboratory model," *IEEE Transactions on Industry Applications*, vol. 44, no. 2, pp. 672–680, 2008.

10 C. Newton, M. Sumner, and T. Alexander, "The investigation and development of a multi-level voltage source inverter," *Proceedings of Sixth International Conference on Power Electronics and Variable Speed Drives*, pp. 317–321, Sep 1996.

11 K. Sano and H. Fujita, "Voltage-balancing circuit based on a resonant switched-capacitor converter for multilevel inverters," *IEEE Transactions on Industry Applications*, vol. 44, no. 6, pp. 1768–1776, 2008.

12 Z. Shu, N. Ding, J. Chen, H. Zhu, and X. He, "Multilevel SVPWM with DC-link capacitor voltage balancing control for diode-clamped multilevel converter based STATCOM," *IEEE Transactions on Industrial Electronics*, vol. 60, no. 5, pp. 1884–1896, 2013.

13 M. Saeedifard, R. Iravani, and J. Pou, "Control and DC-capacitor voltage balancing of a space vector-modulated five-level STATCOM," *IET Power Electronics*, vol. 2, no. 3, pp. 203–215, 2009.

Index